THE KINETICS OF SIMPLE MODELS IN THE THEORY OF OSCILLATIONS

KINETIKA PROSTYKH MODELEI TEORII KOLEBANII

КИНЕТИКА ПРОСТЫХ МОДЕЛЕЙ ТЕОРИИ КОЛЕБАНИЙ

The Lebedev Physics Institute Series

Editors: Academicians D.V. Skobel'tsyn and N.G. Basov

P. N. Lebedev Physics Institute, Academy of Sciences of the USSR

Recent Volumes in this Series

Proceedings (Trudy) of the P. N. Lebedev Physics Institute

Volume 90

The Kinetics of Simple Models in the Theory of Oscillations

Edited by

N. G. Basov

P.N. Lebedev Physics Institute
Academy of Sciences of the USSR
Moscow, USSR

Translated from Russian by
Donald H. McNeill

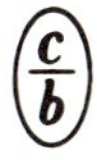

CONSULTANTS BUREAU
NEW YORK AND LONDON

Library of Congress Cataloging in Publication Data

Akademiia nauk SSSR. Fizicheskiĭ institut.
The kinetics of simple models in the theory of oscillations.

(Proceedings (Trudy) of the P. N. Lebedev Physics Institute; v. 90)
"A special research report."
Translation of Kinetika prostykh modeleĭ teoriĭ kolebaniĭ.
Includes bibliographies and index.
1. Laser plasmas – Addresses, essays, lectures. 2. Matter, Kinetic theory of – Addresses, essays, lectures. I. Basov, Nikolaĭ Gennadievich, 1922- II. Title. III. Series: Akademiia nauk SSSR. Fizicheskiĭ institut. Proceedings; v. 90.
QC1.A4114 vol. 90 [QC718.5.L3] 530′.08s [531′.322] 78-5936
ISBN 0-306-10948-4

Aus dem Bestand der
Universitätsbibliothek
Siegen ausgeschieden

The original Russian text was published by Nauka Press in Moscow in 1976 for the Academy of Sciences of the USSR as Volume 90 of the Proceedings of the P. N. Lebedev Physics Institute. This translation in published under an agreement with the Copyright Agency of the USSR (VAAP).

A Division of Plenum Publishing Corporation
227 West 17th Street, New York, N.Y. 10011

Printed in the United States of America

PREFACE

This volume deals with models in the theory of oscillations and the prospects for using the associated analytic techniques in a wide variety of scientific problems. It is primarily intended to be pedagogical. Thus, in discussing specific problems which illustrate and explain the general concepts of the theory of oscillations, we shall consider the significance of both the simple models of phenomena introduced at the beginning and the evolution of these models during the course of research. We do not attempt to hide, but, rather, emphasize the incompleteness of the results obtained here and dwell in each case on the difficulties which arise along the way, noting the short-term goals of the work while not overlooking the prospects for the research as a whole.

Both the pedagogical aspects of these problems and the specific results derived here are, in our opinion, of definite value and may hence already be of interest to specialists in the appropriate areas of science. Each chapter has been written in such a way that it may be read separately from the others. Every one of the nine chapters is therefore a small, self-contained review of the work done by its authors over the past few years. We have not tried to link the chapters with subjective and logical "bridges" as we felt such an effort at this stage would be of little use and have thus limited ourselves solely to the "Introduction" written by L. I. Gudzenko. Chapter I was written by L. I. Gudzenko together with V. V. Evstigneev and S. I. Yakovlenko; Chapter II, by L. I. Gudzenko, Yu. I. Syts'ko, and S. I. Yakovlenko; Chapter III, by L. I. Gudzenko with I. S. Lakoba and S. I. Yakovlenko; Chapter IV, by L. I. Gudzenko and V. S. Marchenko; Chapter V, by L. I. Gudzenko and S. I. Yakovlenko; Chapters VI, VIII, and IX, by L. I. Gudzenko and V. E. Chertoprud; and Chapter VII, by L. I. Gudzenko and A. E. Sorkina.

The book consists of two parts. The Introduction discusses the place of the theory of oscillations in modern science and describes the contents of this volume. A principle for isolating dynamic couplings is formulated and then applied in developing a correlation method for analyzing the inverse problems of the theory.

In Part I several related problems in applied physics are used as an example for discussing the simplest variant of the traditional (direct) problems of the theory of oscillations. Then the motion of a dynamic system near a stable stationary point is examined.

The purpose of this part is to seek active media for powerful coherent light sources (plasma lasers at various wavelengths) and to analyze the prospects for building power reactor-lasers. The first chapter discusses the conditions which lead to an inversion in the atomic level populations in a dense supercooled (relative to the free electrons) plasma. In Chapter II a theoretical analysis is made of several methods for producing a supercooled plasma. Chapter III deals with the recombination relaxation of plasmas made up of inert gases and their combinations with other chemical elements. The requirements on the chemical composition and parameters of a plasma which have to be satisfied in order to produce an active medium that will amplify the resonant emission of atoms are discussed in Chapter IV. Finally, in Chapter V the theoretical possibility of obtaining a substantial portion of the energy from power plants (such as nuclear reactors) in the form of laser radiation is discussed.

Part II is more unusual from a methodological standpoint and deals with some rather diverse specific problems. Some chapters are included there which illustrate the usefulness of the correlation technique for solving inverse problems in the theory of oscillations. In this sense, Chapter IX, which does not use this method directly, is an exception. It is essentially a supplement to Chapter VIII (on the cyclic activity of the sun) which arose during the analysis but has an independent significance. Chapter VI deals with a scheme for measuring the probabilities of nonradiative atomic (molecular, ionic) transitions in a low-temperature plasma using the correlation functions of readouts. The elements of a correlation method for medical diagnostics using the pulsations of large blood vessels are discussed in Chapter VII.

CONTENTS

INTRODUCTION

Statement of the Problem

The level of development of every natural science is determined by the depth to which the subject has been penetrated by measurement techniques and by the detail with which the relationships among the resulting quantitative expressions have been studied. This is just what is meant when it is said, for example, that genetics, psychology, or linguistics have become competent exact sciences. The center of gravity of these sciences has been moving ever more noticeably from general considerations, allowing a substantial nonuniqueness in the conclusions drawn, toward the discovery of controlling quantitative relationships. Of course, increasing the accuracy of the agreement between theory and experiment requires improvements not only in the measurement techniques used but, especially, in the theory. The aim of a theory is taken to be the analysis of relationships among the observed quantities, starting ultimately with some fundamental irrefutable statements, i.e., laws of nature. In physics (mathematics must be given special consideration from this standpoint), the most developed natural science, it is customary to first examine the simplest models, idealized schemes each of which only involves one or two of these laws. In real experiments an increase in accuracy is always associated with a departure from the initial idealization; however, in physics this has usually led to a gradual improvement in the approximation as the model of the phenomena becomes more complex, rather than to a rejection of the previous description.

The situation changes when from the very beginning one has to take into account many diverse factors which contribute equally to the phenomenon being investigated. The difficulties then arise even at the stage of choosing the simplest model and, in any case, when attempting to solve the resulting equations. This is especially typical of such sciences as astrophysics, biology, or sociology; however, now even in physics and chemistry it more often becomes necessary to deal with processes so far removed from simple systems that a theoretical analysis is possible only as progress is made in high-speed computing techniques and the appropriate computational methods are developed. Digital computers acquire an ever-greater role in the processing of the immense numerical bulk that results. A new research psychology arises which leads, in particular, to computer testing of the possible physical models, that is, to the so-called mathematical experiment. Numerical methods have their disadvantages not only in that a numerical answer is always specific to the parameters of the problem as opposed to analytic treatments (in which a wide class of problems of a single type are examined at once). Even in analyses of the simplest cosmological problems, the now obvious inadequacy of approximations to particular solutions of the equations of celestial mechanics has been noted. Thus, in the three-body problem, numerical integration does not provide an explanation of whether the distance between the bodies remains bounded. Here it is necessary to analyze the motion "as a whole." Similar questions arise in many problems with a comparatively small number of variables and parameters, for example, in studies of the containment of "collisionless" charged particles in accelerator storage units and thermonuclear fusion devices. When there are a large number of variables the results of a computer calculation are practically incomprehensible in that the relationships among the model characteristics are not explicit.

In the first step of a theoretical investigation, the choice of the simplest possible satisfactory model, it is necessary to be able to evaluate the general trends in the behavior of different models under various conditions in order to compare the motion given by the model with the observed changes of state of the object. When the model has been chosen and its equations written down, it is necessary to analyze the solutions in sufficient detail. Often the statement of and methods for treating problems have analogies not only in other branches of physics but in other natural sciences, for example, chemistry, economics, astronomy, etc. Stability analysis of equilibrium states, searching for periodic motions, replacement of a dynamic description of many factors of a single type by stochastic characteristics, evaluation of correlations, interpretation of asymptotic solutions, calculation of the bifurcation parameters at which the entire character of a motion changes, studies of the dependence of the number of significant degrees of freedom on the observation conditions, and many similar problems become impractical if they must be solved anew for every specific scientific problem. These questions lie within the scope of the theory of oscillations.

The description of this science is best begun with a discussion of its approach to the study of phenomena. The theory of oscillations deals chiefly with the qualitative characteristics and laws of motion of an object. For example, in damped oscillations the most important characteristic is not the variable phase and amplitude of the oscillations but the frequency and damping constant characteristic of the motion as a whole (or, at any rate, of the motion over long periods of time) [1]. It may be said that the theory of oscillations analyzes the set of motions of an entire ensemble of identical pendulums under all possible initial conditions rather than the behavior of a given pendulum. It is thus clear how important the methods of the qualitative theory of differential equations were for the theory of oscillations from the very outset and how close the apparatus of the theory of random processes is to it.

Motions which repeat themselves to some degree or other, in particular, periodic ones, are taken to be oscillatory. This property is no longer the basic defining feature of the theory of oscillations. From this standpoint the name "theory of oscillations" does not even reflect the significance of this science although in its time the fruitfulness of a unified approach to the study of oscillatory processes of different kinds was the reason for its isolation as an independent branch of physics. From its very foundation it was clear that the generality of the terminology and methods of the theory of oscillations makes it possible to transfer results from mechanics to electrodynamics and acoustics or conclusions from electronics to mechanics and optics. Over the past half century the range of topics in the theory of oscillations has increased greatly. Its principal aim has become the isolation and study of macroscopic models of the same type belonging to widely different areas of knowledge.

From the above it is already clear that the theory of oscillations must not be regarded as a mathematical discipline although it is often treated that way in textbooks and monographs. The characteristic problems of the theory of oscillations are not analyzed using one logic alone and proceeding from a chosen system of postulates. Experiment plays an important, if not defining, role† in these problems, at least at the model building stage. Besides selecting known and developing new mathematical methods the theory of oscillations is characterized by a distinct realization that the actual analysis is always associated with an idealization, a radical simplification of the phenomena. An analysis of the relationship between models and phenomena and a discussion of the dependence of the number of significant degrees of freedom of an object on the experimental conditions characterize the theory of oscillations to no less a degree than do the mathematical methods used in it [2].

† This is an indication of the arbitrariness in the first part of the term "theory of oscillations."

It is difficult at present to evaluate fully the role of the ideas and concrete results of the theory of oscillations in modern science, a role which appears literally in all areas. It is sufficient to simply list some fields which have very recently sprung up from the theory of oscillations, such as radioastronomy and radiospectroscopy, quantum electronics, and nonlinear optics, to make the importance of its role obvious. There is no doubt about the progress over the last few decades in the individual fields and in the theory of oscillations itself, for example, in the theory of dynamic Hamiltonian systems and in certain aspects of the statistical theory of oscillations. However, if we judge from the scientific literature, interest in the general problems of the theory of oscillations has fallen noticeably over this period. This is unjustified since their significance to science is far from exhausted but continues to grow.

In traditional scientific publications some initial discussion of the timeliness of the topic and the shortcomings of previous work is followed by the principal content which deals with a measurement (or computational) technique and the results obtained from it. It is nevertheless important from time to time to discuss in sufficient detail the research strategy, including, in particular, the conditions which stimulated new directions in the research and the reasons for choosing one or another model of the processes being analyzed. There are, it is true, very few such books and they are written by mathematicians or about mathematics [3-5]. The role of this type of writing is no less important in other widely different fields of science. The need for an analysis of only recently accumulated experience with a reasonable choice of models is perhaps indisputable in the case of "young" exact sciences which have only recently entered (or are still entering) the path of "extensive mathematization," such as physiology psychology, or economics. But even in long-established exact sciences such as physics, chemistry, or astrophysics, progress in important areas is frequently slowed down precisely due to the absence of appropriate models at a given time. Equally objectionable is the irrational expenditure of time and energy in an attempt to solve problems which have not yet made corresponding progress in neighboring fields. It is clear that the choice of a model is determined not only by the intended degree of detail in the investigation but also by the purpose and level of knowledge.

A rather detailed discussion of the reasons for choosing a given model, and perhaps even a valid discussion of why it was necessary to avoid choosing other descriptive schemes (which at first seemed natural), cannot, of course, serve as a recipe for the investigation of new problems, and soon there must be some assistance from that secret feature of the researcher himself which we may refer to as "model intuition" (or simply intuition?). Such books should, it seems, occasionally be written by the greatest scientists in, shall we say, the form of "scientific-methodological memoirs." But this does not happen. Thus, we have tried here to describe our limited experience and the ways which gradually led us to a choice of model which "worked" in several fields. This applies to the traditional direct problems. Primarily we are speaking here of relaxation of extremely nonequilibrium (supercooled) dense plasmas as possible media for efficient amplification of light (including x rays). This also refers to the question now plaguing makind of the rational use of available energy sources. In fact, even of itself the behavior of a dense decaying plasma is complex in the simplest imaginable conditions. It is determined by a large number of interrelated parameters. Solving the corresponding equations directly makes it possible to analyze the simplest situations only with the aid of a high-speed computer. If we try to study the prospects for effective conversion (in nuclear reactor-lasers) of the energy from fissions of nuclei into directed light extracted from the apparatus, the difficulties become practically insuperable, at least in the near future. Besides nuclear processes, we have here to include recombination relaxation of the plasma formed by the fission fragments, the rapid chemical reactions taking place in such a medium, heat transfer to the walls of a vessel filled with gaseous fissionable material, etc.

One direct problem discussed here very briefly is an exception. In the final chapter we discuss the dynamic character of feedback in the phase of oscillations from the well-known relaxation oscillator with a gaseous discharge lamp. The statistical limitation in the growth of the phase-shift dispersion in this device was first observed in analyses of the cyclic activity of the sun. This effect served for a time as the basis of an erroneous concept of the mechanism for this activity and even later, after the discovery of the meaning of this effect, gave substantial support to the development of models for the processes which control cyclic activity. In this book the effect is discussed in the framework of a simple experiment which could be done in the laboratory.

The ideas and methods of the theory of oscillations have clearly come from the study of processes which are nearly periodic. Between "oscillatory" and "nonoscillatory" motions there is no significant fundamental boundary starting at which we would have to place only motions with components which regularly change sign in the domain of the theory of oscillations. It is unreasonable to separate oscillation and limiting relaxation problems, setting, for example, a critical value of damping as the boundary for the theory of oscillations. Thus, the oscillatory problems would include only dynamic systems lying in the regions of their phase spaces adjacent to limiting cycles or to focus or center stationary points. It would be no less artificial to require the existence of oscillatory motions among the trajectories since that would exclude consideration of some parameter intervals of the system. But even the simplest variants of nonoscillatory motions are often associated with striking effects which can be analyzed naturally by the methods of the theory of oscillations. Of the wide variety of such problems, in the first section we have chosen to study the intense recombination of a supercooled dense plasma in order to construct lasers with high energy densities. A shortage of reliable data on the probabilities of a number of elementary events controlling the population kinetics has limited further progress in this important field. In one chapter of the second part we discuss an inverse problem for the theory of oscillations which arises from this situation – an experimental technique for determining transition probabilities.

A discussion of inverse problems in the theory of oscillations is the main subject of two other chapters in this part of the book. In one chapter a general scheme for these inverse problems is applied to the mechanism for the cyclic activity of the sun, and in the other chapter the possibility of medical diagnosis by means of statistical analysis of the external pulsations of large blood vessels is discussed.

Direct and Inverse Problems

The approaches to research in the natural sciences are grouped under direct or indirect schemes. Both in mathematics and in other exact sciences the discussion of the direct problems was begun much earlier; hence, they are more familiar than the inverse problems. Briefly speaking, the direct problems of mathematics consist of seeking the solutions of a given equation, while trying to find the equation itself from known solutions is the object of the inverse problems. An analogous division into direct and indirect schemes holds for other branches of science. Essentially all schemes that have been worked out in detail belong to the traditional (direct) problems. The examination of the inverse schemes has begun very recently, the terminology has not yet been settled, and the statement of the main problems is not always clear. This by no means indicates that the inverse problems are of secondary significance. Precisely because of the existence of direct methods for analyzing them new problems in science are often studied using an excessive number of models. It is clear that this method is not always effective. A gradual réalization of this has led in the last ten or fifteen years to the appearance of a number of publications (including review articles and monographs) on the statement and analysis of inverse problems in mathematical physics, cybernetics, and the theory of oscillations.

We must note that the concept of inverse problems often used now in applied mathematics does not completely coincide with that noted in the previous paragraph and used in this book.

Thus, we shall discuss some definitions, starting with some examples from "pure" mathematics in which we limit ourselves solely to some illustrative terminology.

The solution $\mathbf{x} \equiv (x_1, x_2, \ldots, x_p)$ of the equation $\mathbf{f_A(x)} = \mathbf{y}$, where $\mathbf{A} \equiv (A_1, A_2, \ldots, A_q)$ is a parameter and $\mathbf{y} \equiv (y_1, y_2, \ldots, y_r)$ is the right-hand side, will be written in the form $\mathbf{x} = \mathbf{f}^{-1}_{\mathbf{A}}(\mathbf{y})$. For various values of $\mathbf{a}$ the operation $\mathbf{f_a(x)}$ forms a family, $\{\mathbf{f_a(x)}\}$, of transformations of the solution $\mathbf{x}$ into the right side $\mathbf{y}$. We shall write this in the form

$$x, A \xrightarrow{\{f_a\}} y. \tag{1}$$

Finding the behavior of the inverse transformation

$$y, A \xrightarrow{\{f_a^{-1}\}} x \tag{2}$$

is equivalent to finding the solution of the equation. The inverse problem is now another transformation of Eq. (1) which involves finding the parameter $\mathbf{A}$ from known values of the image $\mathbf{y}$ and object $\mathbf{x}$:

$$x, y \xrightarrow{\{f_a\}} A; \qquad y, x \xrightarrow{\{f_a^{-1}\}} A. \tag{3}$$

If for given values of $\mathbf{x}$ and $\mathbf{y}$ it is not possible to find the parameter (due to nonuniqueness or inconsistency of the corresponding equations) then the problem (3) is said to be incorrectly stated.

The meaning of $\mathbf{A}$ and the right-hand side, $\mathbf{y}$, in scientific applications of this scheme varies. Staying within a mathematical framework we now illustrate this with a more specific example. In the theory of ordinary differential equations the direct problem is the Cauchy problem. The equation

$$\frac{d^q x}{dt^q}(t) + D_A\left[x(t), \frac{dx}{dt}(t), \ldots, \frac{d^{q-1}}{dt^{q-1}}x(t)\right] = y(t)$$

with the initial condition

$$\frac{d^k x}{dt^k}(t_0) = x_0^{(k)}, \qquad k = 0, 1, \ldots, q-1 \quad \text{or} \quad x(t_0) = x_0$$

maps x(t) into the function y(t):

$$x(t), A, (x_0) \xrightarrow{\{D_a\}} y(t). \tag{4}$$

The reverse of this transformation,

$$y(t), A \xrightarrow{\{D_a^{-1}\}} x(t), \tag{5}$$

is a solution of the Cauchy problem. The inverse problem is to find the parameter $\mathbf{a} \pm \mathbf{A}$ which distinguishes that differential operator within a given family, $\{D_a\}$, of operators which transforms x(t) into another, also known, function y(t). The two variants of the inverse problem are written symbolically in the form

$$\begin{aligned} &x(t), y(t) \xrightarrow{\{D_a\}} A, \\ &y(t), x(t) \xrightarrow{\{D_a^{-1}\}} A. \end{aligned} \tag{3'}$$

Problem (3') may also not have solutions if the chosen family of operators $\{D_a\}$ does not work for the given functions x(t) and y(t) or, we say, is too narrow. If the family $\{D_a\}$ is too wide, it impedes the search for the desired value of the parameter and may make the solution to (3') nonunique.

The cybernetic "black box" analysis technique is akin to the two preceding examples. For a set of known effect y_n fed into the box and the reactions x_n received at the output of the box it is necessary to find an operator D_A from among a class of operators $\{D_a\}$ which transforms the responses x_n of the object into values y' which are sufficiently close to the signals, y_n, at the input to the box. The operator D_A, which transforms x into y', serves in effect as a description of the object being studied, which remains enclosed in the black box, showing itself to the researcher in no other way than through its responses x. For a reasonable choice of classes $\{D_a\}$ which sort the functions y_n it is possible in principle by constantly improving the accuracy of the operators $\{D_A\}$ relative to the responses x_n to construct a detailed mathematical model of the object which will permit prediction of its behavior under new conditions with sufficient accuracy. The ultimate goal of cybernetic techniques is just this kind of description of an object, independently of the specific (physical, chemical, biological) processes actually taking place within it.

Completing our illustration of the difference between the inverse problem and the inverse transformation, we conclude that where the goal of the inverse problem is the elucidation of the characteristics of the object itself, the significance of the inverse transformation is to find an external effect on the object. Inverse transformations correspond to the problem of reducing observational data when it is necessary to reproduce the actual signal entering the receiver and the distorting of the receiving and recording apparatus are known. This problem is usually complicated by noise, inaccurate knowledge of the characteristics of the apparatus, and so on. What we have said here is probably sufficient to clarify the difference in the concepts of direct and inverse problems as used in the theory of oscillations and in mathematical physics. In the latter, the direct problem refers to solving a given equation, or more generally, to finding the consequences of a known cause. In mathematical physics, the inverse problem means not only looking for the operator (which describes the object) in the left-hand side of the equation, but also determining (for a known form of operator) the right-hand side which describes an external effect. Both these problems have in common a general principle, finding the cause of a known consequence, as well as a number of similar difficulties in their solution. This, of course, is sufficient for the two problems to be considered jointly. However, the difference between them is so great that they merit separate names. Inverse problems are more characteristic of research problems while inverse transformations are more characteristic of direct processing of measurement results. In this book the term "inverse problem" refers everywhere to a search for a description of a phenomenon or (speaking up to now only of the mathematical side) to an analysis of the structure of an object in terms of its signal. In cybernetics such problems are referred to as "object identification problems." We must now discuss the difference in the approach to inverse problems in the theory of oscillations and cybernetics.

The traditional method of studying new phenomena may be divided into several stages:

1. Isolation of the most important processes under conditions similar to those in the object, taking the interactions of these processes into account, and deriving equations which describe the chosen arrangement of phenomena.
2. Analyzing the solution of the equations for various supplementary conditions (boundary, initial, etc., conditions) leading to uniqueness in the mathematical problem. If the model has been chosen correctly these solutions will yield an acceptable picture of the phenomenon. However, the appropriateness of the idealization chosen is not known in advance. Hence the following stage is necessary.
3. Comparing the resulting solutions with the observed behavior of the object.

If there is qualitative disagreement between the theoretical and observed motions the model is scrapped. Then it is generally necessary to begin everything from the first stage. This research approach in fact involves an excess of different descriptions. On the other hand, experience with successful models of similar phenomena usually allows us to reduce the number of variants attempted.

This approach, which is typical for direct problems in the theory of oscillations, is inappropriate when too little is known about the processes producing the observed properties of the object. Then it is natural to change the order of doing things. Without specifying the model too much at first it is possible to try to find the equations directly from the signal by analyzing the signal numerically in order to then interpret these equations in terms of specific processes. We note the stages of research in the inverse problems of the theory of oscillations in greater detail as follows:

1. Mobilization of a priori information about the object in order to narrow as much as possible the class of operators (equations) $\{D_a\}$ used to describe it.
2. Analysis of the signal from the object in order to isolate from the class $\{D_a\}$ the operator D_A which agrees best with the observations, i.e., construction of a mathematical model of the phenomenon.
3. Construction of a concrete model by interpreting the mathematical description in terms of the specific processes which produce this phenomenon.

The last stage is evidently the least amenable to any kind of regular recipe and thereby the most difficult, but in the theory of oscillations it is just this stage which is the actual goal. And, perhaps, only this stage is capable of evoking the enthusiasm necessary for laborious problem (meaning, in general, nonapplied problems, that is, those which promise no early practical yield) solving in order to surmount the first two less creative stages. This third stage is not included in the cybernetic "black box" scheme.

It would be incorrect to think that only the presence of the third stage differentiates the inverse-problem method for the theory of oscillations from the cybernetic-object-identification problem. Often the first and, almost in every case, the second stages in the theory of oscillations are substantially different from the cybernetic treatment, since isolation of the operator D_A is determined not only by the signal from the object but also by the class of operators and, most importantly for the question at hand, by the criterion for optimizing the operator. The choice of criterion is dictated by the purpose of the discussion and these purposes, finding the most exact mathematical description of the signal (in cybernetics) and constructing a concrete process model of the phenomenon (in the theory of oscillations), do not immediately coincide.

Isn't something incompatible here? Isn't the purpose of any theory to find the most exact description of the observed motion and then to predict the behavior of the object under new conditions? Isn't the approach of cybernetics then the most reasonable? This point of view ("the direct way is always the shortest") simplifies the actual situation. It has been repeatedly noted that the generality of the methods and the "universality of the language" of the theory of oscillations permits the use of analogies. It is appropriate to begin a brief listing of the degrees of analogy from our standpoint by recalling the analogy between a phenomenon and the models which approximate it. A generalization of the principle widely used in the theory of similarity allows us to transfer the results of a specific study not only to phenomena at different scales but also to different scientific fields. The agreement among the dimensionless equations corresponds both to the mathematical isomorphism and to the similarity of the concrete models of various phenomena. Similar dimensionless complexes of parameters and variables are thereby defined. Yet richer is the following degree of analogy in which qualitatively similar models are discussed. This refers to dimensionless equations which are not identical but have "phase portraits" of the same type.

From this listing there comes still one more very common degree of analogy. It follows from the similarity of component processes of a single type which make up fairly diverse phenomena. This isolation of similar processes makes it possible to classify them in terms of physical, chemical, biological, social, and other processes. Within physics, for example, this classification distinguishes electromagnetic, nuclear, gravitational and other processes. In more detail, electromagnetic processes may be divided, for example, into those which take place in plasmas, gases, liquids, and solids. A still more detailed classification distinguishes relaxing free plasmas, electric discharges, shock waves in plasmas, etc. The greatest detail makes it possible to do a unique theoretical analysis. Thus, if the parameters of a free plasma are known at $t = t_0$, it is possible to predict their variation for $t > t_0$. The level of detail in classifying processes which makes it possible to analyze their later variation corresponds to the term "specific (or concrete) processes" used in this book.

Of course, a general statement that it is necessary to explain the specific processes in every case would be wrong. Thus, for example, if in biology basic interest is usually in interpreting specific mechanisms, then in medicine it would be an inadmissible luxury to make a complete analysis of the biological processes which are taking place every time, while avoiding use of a prescription which would reliably lead to recovery from a dangerous disease until this analysis was made.

But it is no less obvious that knowledge of the controlling processes in the end always aids in obtaining an exact quantitative description of a phenomenon. At the same time, until identification and analysis of the most important processes, excessive detail and accuracy in the analysis are most often an obstacle to discovering the essence of the observed phenomenon and constructing an appropriate (at first crude) model. Clearly, even the first physical laws (for example, the laws of dynamics) could not be discovered without that understanding of the unity of the specific processes which made it possible for the first time to avoid using both a series of distinctly different factors and the less important factors. It is just this isolation of the specific component processes of phenomena which has led to the present highly accurate agreement between experiment and theory in physics. And where has there been a science able to analyze every phenomenon anew with a mathematically optimal description, that is, by closely following the cybernetic scheme described above? Thus, in medicine, having not yet found other ways to objectively diagnose heart ailments in detail, we most often turn to purely "mechanical" methods in interpreting cardiograms and electrocardiograms. The results would undoubtedly be improved if there were reasonable preliminary preparation of the material to be analyzed. For example, knowing that the heart plays the basic role of a pump in the body to drive blood through two blood circulation loops, it is natural to first use a method for obtaining (for example, from the electrocardiogram traces of the major blood vessels) the individual equations of each person's "own pump." It is certain that with these equations it is easier to set up a computer diagnosis (although with the same means of identifying the features in charts) than to do this directly from unprepared electrokymograph traces, i.e., almost blindly. The principle that a "smart machine should understand everything by itself" is as yet premature.

The poverty of the pragmatic, transient cybernetic approach is especially obvious in those areas of science without practical significance such as astrophysics or cosmology in which the meaning of the mathematical description usually appears in the confirmation or refutation of specific process models of phenomena. These branches of science, which have grown rapidly in the last few decades, are also noteworthy in that the objects of the analysis are not subject to intervention by the scientist. To use the terminology of cybernetics, we may say the black box has no input in this case. In such fields it is appropriate to use the word "observation" in place of "experiment." About ten years ago a statistical method in the framework of an inverse problem in the theory of oscillations was proposed for the study of uncontrollable objects [6]. Its later development and application to several astrophysical objects led to a number of instructive, and at times unexpected, results.

Complex logical structures which begin from precise definitions and postulates are typical of modern theories. Usually the corresponding apparatus comes from mathematics in more or less complete form. In other cases the apparatus initially developed for analyzing a specific problem is gradually abstracted from its scientific specifics and becomes a new mathematical discipline of itself. This entwinement with mathematics is one of the reasons for the traditional deductive exposition of theories in monographs and textbooks. It produces a distorted concept of the significance of various methods and results in science. The impression is created that adequate statements of scientific problems together with rigorous schemes for quantitatively analyzing them always lead to the analysis of some equations, most often to one of the modifications of the Cauchy problem. Meanwhile, the fundamental achievements of natural science are usually associated with inductive rather than deductive approaches. Nature most often presents the researcher with an inverse problem rather than a direct one. In fact, only in the case of the motion of a known mechanism under given external circumstances is the solution of a direct problem applicable. In investigating new phenomena one always deals with a signal (in a rather general sense) about the equations of which very little is known in advance. Thus, rapidly developing areas of the exact sciences which are solving fundamental problems must correspond more fully to the logic of inverse problems, methods of problem solving which, if they were developed as much as the traditional methods, would be of great benefit. It is to be hoped that the growing understanding of the significance of inverse problems will lead to intensive discussion of the questions which arise in connection with them. The notion of inverse problems is gaining a deserved place in the most varied sciences, including the theory of oscillations, which deals more and more with general methods for understanding and analyzing dynamic laws.

A Principle for Isolating Dynamic Couplings

Separating the characteristics of an object into stochastic and dynamic characteristics is often fairly arbitrary at the start.† It would seem possible to reason as follows: Let $x(t)$ be the signal from our object. If its behavior is sufficiently accurately described by the equation $D[x] = F(t)$, where D is a dynamic operator and $F(t)$ is a fluctuation, then D plays the role of a dynamic description of the phenomenon and $F(t)$ is a stochastic perturbation. But without special refinements such a description is nonunique even in the simplest situations. Let the dynamic operator D belong to a class L_q of linear homogeneous differential operators of the q-th order with constant parameters. Then, besides the equation $D[x] = F$, we can always write $D_1[x] = F_1$, where D_1 is any other operator from the same class L_q and $F_1 \equiv F(t) + D_1[x(t)] - D[x(t)]$ is again a fluctuating process. What kind of limitations need to be placed on D and F to ensure uniqueness of these concepts?

The problem of uniquely isolating the stochastic component loses its sharpness when the fluctuations in the signal are sufficiently small and can be neglected when analyzing the phenomenon. If the class of equations for the dynamic model has been determined, then it is possible in principle to find an effective description for any sufficiently complete set of external perturbations of the object in whose responses it is possible to recognize all the degrees of freedom of the observed phenomenon. We now illustrate this with an elementary example. Let the signal $x(t)$ from an autonomous object be described by an ordinary linear differential equation

$$\frac{d^q x}{dt^q}(t) + \sum_{m=0}^{q-1} A_m \frac{d^m x}{dt^m}(t) = 0 \tag{6}$$

† Here and in the following it is understood that the phenomena under consideration are macroscopic.

with a stable stationary point x(t) = 0. We suppose that a perturbation of this steady-state "motion" occurs in an initial deviation in $d^m x/dt^m(t_0) = x_0^{(m)}$ or $\mathbf{x}(t_0) = \mathbf{x}_0$, where $\mathbf{x}_0 \equiv (x_0^{(1)}, x_0^{(2)}, \ldots, x_0^{(q-1)})$. Assuming for simplicity that the characteristic equation

$$\lambda^q + \sum_{m=0}^{q-1} A_m \lambda^m = 0 \tag{7}$$

does not have multiple roots, we write $x(t) = \sum_{l=1}^{q} C_l \exp(\lambda_l t)$ for $t > t_0$. Equation (7) establishes the relationship between the sets $\{\lambda_l\}$ of q characteristic numbers and the sets $\{\mathbf{A}\} = \{(A_1, A_2, \ldots, A_{q-1})\}$ of coefficients of Eq. (6). From the form of the motion x(t) it is possible at first to evaluate the characteristic numbers $\{\lambda_l\}$ and then the set $\{\mathbf{A}\}$, which describes the motion dynamically. But the form of the signal x(t) makes it possible in principle to find those (and only those) λ_l for which $C_l \neq 0$. Thus, every initial deviation in x_0 for which, according to

$$\sum_{l=1}^{q} C_l (\lambda_l)^m = x_0^{(m)}, \qquad m = 0, 1, \ldots, q-1 \quad \text{or} \quad m = \overline{0, q-1}$$

all C_l are nonzero, can be used to find all the λ_l, and therefore **A**. However, such an initial deviation from equilibrium on the part of the object is realized in this simplest example by the entire set of perturbations.

These discussions basically lose their meaning for observations of the dynamically established motion of an uncontrolled object, for which it is no longer appropriate to speak of perturbations in its initial conditions. It should be noted that the domain of application of this type of problem is very wide. The class of uncontrollable objects is vastly greater than might appear at first sight. It includes more than the typical astrophysical objects (stars, galaxies, quasars, pulsars), in which no noticeable change of events can be caused by man. It must be taken to include many medical, industrial, and sociological objects for which deviations from their ordinary operations are undesirable. It is natural also to include the question of analyzing unique recorded observations. Furthermore, at the start of an investigation it is sometimes convenient to regard as uncontrolled an object which, although accessible to the experimenter, is rather complicated and has a number of couplings with the outside world which, while not very important for the phenomenon being studied, may be disrupted by perturbations and severely complicate the analysis. It should be noted that it is not usually clear in advance how complete the set of perturbations used by the experimeter is.

Thus, there are a number of interesting objects the analysis of which by conventional methods is quite difficult. For example, let the steady-state emission of an unknown point source consisting of one or two spectral components be observed in a telescope. As a rule, too many equally unconvincing hypotheses about the nature of this emission will appear. And the less detailed the record of the phenomenon is, the larger the number of different "theories" which may compete on an equal basis. Formal resort to the inverse-problem schemes of the dynamic theory of oscillations only emphasizes the hopelessness of the analysis. In fact, an observed state corresponds to a stationary point of a dynamic system. Due to the unforeseeable arbitrariness of the solution the problem of constructing a mathematical model whose equation has a function with a value determined by an equilibrium state loses its meaning without additional information. The situation is not much easier even in the case of a dynamically steady-state periodic variation in the signal.

To analyze this type of problem reasonably we must return to the unanswered question at the beginning of this section about separating the description into dynamic and stochastic components and turn to the stochastic method of finding the characteristics of the dynamic model. Remaining in the framework of autonomous dynamic models with a finite number of degrees of freedom, we write the equations which describe them in the form

$$\frac{d\mathbf{x}}{dt}(t) = \mathbf{X}_A[x(t)], \qquad \mathbf{x} \equiv (x^1, x^2, \ldots, x^n), \\ \mathbf{X}_A \equiv (X_A^1, X_A^2, \ldots, X_A^n). \tag{8}$$

When an object of this type experiences a frequency-selective interaction with another object, it is possible to conserve the independence (autonomy) of the system by expanding the descriptive system and including both interacting objects in it. A nonselective effect of an external medium on the object is most simply represented as a perturbation of the autonomous dynamic model (8) by steady-state white noise, i.e.,

$$\frac{d\mathbf{x}}{dt}(t) = \mathbf{X}_A[\mathbf{x}(t)] + \mathbf{F}(t), \qquad \mathbf{F} = (F^1, F^2, \ldots, F^{(n)}), \qquad \langle \mathbf{F} \rangle \equiv 0, \\ \langle F^k(t) F^l(t') \rangle = C^{k,l}\delta(t - t'), \qquad k, l = 1, 2, \ldots, n. \tag{9}$$

Let the class $\{X_a\}$ of Eqs. (9) define the a priori information about the dynamic structure of the object. We shall seek the value $\mathbf{a} = \mathbf{A}$ of the parameter which isolates from this class that description for which the observed signal may be represented in the form of a response of the corresponding dynamic system to a stationary fluctuating influence with a short time correlation (i.e., which is broad band in frequency), i.e.,

$$\frac{d\mathbf{x}}{dt} = \mathbf{X}_A[\mathbf{x}] + \mathbf{F}_A[\mathbf{x}, t], \qquad [\mathbf{F}_A(\mathbf{x}, t)] \equiv 0, \\ \langle F_A^k(\mathbf{x}, t) \cdot F_A^l(\mathbf{x}, t) \rangle \cong 0 \quad \text{for } |t - t'| \gg \tau_0, \ \tau_0 \ll \tau_{\min}. \tag{10}$$

Here $\tau_{\min}$ is the minimum characteristic time of the corresponding dynamic system (8). Then, by definition Eq. (8) describes the dynamic model of the system. This is the isolation principle for the dynamic system. When, after long recording of the signal x(t) by a sufficiently sensitive apparatus, a good dynamic description (8) of the signal has been found, the function F_A then [within the more detailed description of Eq. (10)] reflects the effect on the signal of many degrees of freedom which were neglected in Eq. (8) and ultimately reduces to microscopic motions within the object.† It must be emphasized that the principle formulated here makes it possible to find (for favorable observing conditions) the operator $\mathbf{X_A}$ in fairly wide classes $\{\mathbf{X_A}\}$ since the perturbation $\mathbf{F_A}$ is complete [i.e., it excites all degrees of freedom of the dynamic model (8)]. This means that the dynamic characteristics may (in principle) be found from the signal; that is, such a separation of the description into dynamic and stochastic components is unique.

The schemes (8)-(10) once again illustrate the difference in approach to the analysis of phenomena in the theory of oscillations and cybernetics. Following the recipes of the latter it is possible to describe fairly accurately the course of the recorded signal from a regular (nonrandom) time dependence with a large number of parameters which can be evaluated from the signal. This approximation will, as a rule, adequately predict the future behavior of the signal for some time after its being recorded for the analysis. This must not be taken to mean that scheme (10) from the theory of oscillations, as opposed to cybernetics, proposes solely to describe the signal as a fluctuating perturbation of a dynamic model, rather than purely dynamically. The problem of finding the parameter for the dynamic model (8) may be solved by analyzing the response of this model to white noise, that is, by going to model (9) in the cybernetic "black box" scheme. According to principle (10) the theory of oscillations proposes the use of the method of successive approximations for the gradually expanding observational material in order to describe the growing number of significant degrees of freedom in the

† This viewpoint is in accordance with many works which start out with the Langevin description of Brownian motion.

sequence of dynamic models, regarding the motions associated with unimportant (for these conditions) degrees of freedom in each approximation as perturbations of the dynamic model being sought. The correlation time τ_A of each such fluctuation perturbation is small compared to the characteristic times of the dynamic operator for the same approximation. But at no stage do the fluctuations $\mathbf{F}_A(t)$ become white noise with a correlation interval of zero; otherwise, the process of successive approximations would have to stop right there.

Literature Cited

1. L. A. Mandel'shtam, Complete Works [in Russian], Nauka, Moscow (1955), Vol. 4.
2. A. A. Andronov, A. A. Vitt, and S. É. Khaikin, The Theory of Oscillations [in Russian], Fizmatgiz, Moscow (1959).
3. G. Polya, Mathematics and Plausible Reasoning, Princeton University Press (1969).
4. I. Kepler, A New Stereometry of Wine Casks [Russian translation], GTTI, Moscow-Leningrad (1948).
5. R. Courant and H. Robbins, What Is Mathematics? Oxford University Press (1941).
6. L. I. Gudzenko, Izv. Vyssh. Ucheb. Zaved., Radiofizika, 5:572 (1962).

Part I

RECOMBINATION RELAXATION IN DENSE PLASMA

CHAPTER I

PLASMA LASERS USING ATOMIC AND ATOMIC ION TRANSITIONS

Statement of the Problem

The term plasma lasers [1] will be used here to denote oscillators or amplifiers which use a supercooled plasma as an active medium, that is, a plasma in which the temperature of the free electrons is substantially less than the equilibrium (corresponding to the electron and ion densities in the plasma) value. Left to itself, such a plasma will undergo volume recombination. An atomic level population distribution similar to that in a freely decaying plasma is also inherent in a plasma in which a specified level of ionization is maintained with the aid of an auxiliary high-energy source (beam, short-wavelength radiation, etc.). Under such conditions it is customary to speak of a plasma which is supercooled or, equivalently, undergoing recombination relaxation (the term "superionized" is less often used).

In a supercooled plasma the excited levels are filled from the continuum, and during the course of relaxation the "excitation flux" flows toward the ground state of the atoms and molecules. Under these conditions the flow over the more highly excited energy levels is faster. If at the same time it is assured that the lower energy levels can be emptied sufficiently rapidly, then a population inversion will occur over a number of transitions and the medium can effectively amplify the corresponding radiation.

In a supercooled plasma with density $N \gtrsim 10^{20}$ cm^{-3}, often now used as an active medium, the presence of molecules and molecular ions has a very significant effect on the relaxation kinetics and laser characteristics [1]. But a number of plasma laser problems are of interest even in gaseous media at relatively low ("intermediate") pressures, $p \lesssim 0.1$ atm. If the medium is basically atomic (that is, the presence of molecules and molecular ions in the gas may be neglected), then consideration of the processes taking place in it is greatly simplified. In addition to the amplification in a supercooled atomic plasma discussed here, there is interest in the case of multiple ionization. It must be said that [2] even in a fairly dense plasma (chosen in a specific way for its chemical composition) amplification on atomic transitions may be the determining factor but the effect of molecules on the kinetics is great. Here we shall limit ourselves to the case of a low-temperature plasma in which there are practically no molecules.

Due to the theoretical development and practical realization of laser and electron beam techniques for delivering high energy densities in matter, it is becoming more relevant to extend laser principles into the short-wavelength region of the spectrum (by which we shall mean the relatively modest wavelength interval $\lambda \sim 10$–100 Å). In the literature there is a lively discussion of many schemes for obtaining a population inversion over the levels of multiply charged ions and in the corresponding plasma. The variety of these schemes and the imprecision in their initial assumptions makes it impossible to evaluate them in any complete fashion in this small chapter; for this we turn to the reviews by Molchanov [3] and by Bushuev and Kuz'min [4]. Nevertheless, we shall discuss very briefly those few publications whose authors both sought schemes for producing population inversions on short-wavelength transitions and in some way analyzed the prospects for a practical realization of their approaches.

In most papers it is proposed to fill the upper working level by nonequilibrium plasma ionization (i.e., under conditions in which there are relatively few free electrons and they have been heated by nonequilibrium means). These authors do not count on the buildup of populations at some level of an ion due to electronic collisions as here such collisions are mainly deleterious. As a rule the basis of these ariticles is a simple scheme for direct pumping to the working level while the lower level is neglected. Here it is possible in principle only to use narrow-band radiation. This means that one must first have a similar laser, but of shorter wavelength than the one being built, since intense broadband irradiation of a plasma leads, not to population of high ionic states, but to more rapid depopulation with formation of ions of higher charge states. Otherwise, it is now impossible to say that (when there is still no serious experimental and theoretical analysis of radiation transport in such a nonuniform plasma) either the authors of bold predictions or their critics can confidently insist on "their own" solutions.

Another group of articles discusses the possibility of producing a population inversion and efficient amplification during recombination decay of a multiply ionized plasma. Most of these papers envision three-body recombination, i.e.,

$$X^{(z+1)} + e + e \rightarrow X^{(z)*} + e.$$

In some papers [5, 6] it has been proposed that the decay of such a plasma be intensified by adding cold gas atoms which are rapidly mixed in some way with this fairly hot plasma. Some doubts have already been expressed about the feasibility of the proposed methods of mixing. It seems that an ionization wave, which would proceed from the hot plasma and ionize the cold gas atoms, will propagate much more rapidly than the heavy multiply charged ions.

In the following we shall deal only with amplification schemes involving a "superionized" plasma of multiply charged ions and regard recombination as the beginning of the relaxation process. One of the questions which then arises deals with the possibility of sufficiently filling the upper working level in this manner. Here there are substantial difficulties due to the virtual absence (in this wavelength range) of cavities with any significant Q. Under these conditions it is necessary either to count on uniform (in time and space) delivery of a high energy into an active medium of characteristic dimension $L \sim 1$ m, or to find a way of so intensely populating the upper working level that the linear unsaturated gain coefficient will be sufficiently large ($\varkappa \gtrsim 1\ \mathrm{cm}^{-1}$).

Yet another (but still favorable) distinctive feature of amplification of short-wavelength radiation is that the ground state of the "working" ion (on a transition of which lasing occurs) is no longer the final step in the process of recombination relaxation. This state recombines into an ion of the next lower charge at a rate which is only $[z/(z-1)]^3$ times slower than the rate of filling this charge state. Here and in the following z denotes the spectroscopic (charge) number of the working ion.

A Model for the Relaxation of a Plasma with a Low Degree of Ionization

We shall now assume that the chemical composition of the plasma and its "gross" characteristics (primarily the gas density N and its temperature T) have been chosen so that it is possible to neglect the effect of ion collisions with heavy particles on the recombination rate. It is thus in fact assumed that there are almost no molecules, molecular ions, or negative atomic ions in the gas. In such media recombination relaxation begins in a single way. If the plasma is supercooled and not excessively rarefied, it is meaningless to include radiative recombination, $X^+ + e \rightarrow X^* + \hbar\omega$. For a low gas temperature T and low degree of ionization we may also neglect the contribution of dielectronic recombination, $X^+ + e \rightarrow X^{**}$, which leads directly to doubly excited atomic states. Dielectronic recombination is fairly rare and is almost completely compensated by its highly probable inverse process, autoionization. Thus, neutralization of this medium takes place only during three-body collisions of atomic ions with two electrons, $X^+ + e + e \rightarrow X^* + e$. A direct result of these three-body collisions is the formation of highly excited atoms which then create a relaxation flux, on the average moving downward toward the ground state of the atom, among the bound electrons. For a real, strongly supercooled plasma (with a high free-electron density N_e and low electron temperature T_e) this flux is a pump sufficiently intense to produce a high population in any reasonably chosen upper working level. The same flux ejects electrons from levels adjacent to the ground state. Thus, under favorable conditions a population inversion is produced. To do this the emptying of the level chosen to be the lower working level must be sufficiently effective. The analysis of the conditions for rapid recombination depopulation of low-lying levels is one of the basic problems in the theory of plasma lasers. An analysis of the distribution of populations N_n of the atoms in the general case involves extremely tedious calculations and depends on a large amount of data about the rates of elementary processes in plasmas. Thus, the primary problem here is to make the calculation of the level kinetics possible on a modern computer.

An important simplification is the assumption that the plasma is spatially uniform. The problem is set up for an elementary volume of plasma in the absence of collective motions and radiative interchange, both of which are usually taken into account approximately. Evidently the main error introduced by such an approximation is in accounting for the interaction of resonance line emission with atoms.

The system of equations which describes the recombination relaxation of an atomic plasma includes the kinetic equations for the populations of excited atomic levels,

$$\frac{dN_n}{dt} = \sum_{m=1}^{n_1} K_{nm}N_m + D_n \equiv \Gamma_n, \qquad n = 1, 2, \ldots, n_1, \tag{1}$$

the equation for the ion densities,

$$\frac{dN_+}{dt} = -\beta N_e^2 N_+ + sN_eN_1, \tag{2}$$

and the heat balance equations,

$$\begin{aligned} \frac{3}{2}N_e\frac{dT_e}{dt} &= \frac{3}{2}T_e\Big(\sum_{n=1}^{n_1}\Gamma_n\Big) + Q_{\text{inel}} - Q_{\Delta T}, \\ \frac{3}{2}N\frac{dT}{dt} &= Q_{\Delta T}, \end{aligned} \tag{3}$$

as well as the condition $N_e = N_+$ (quasineutrality of the plasma, here one with a homogeneous chemical composition and a low degree of ionization) and a conservation law for the number of heavy particles.

$$N = N_+ + \sum_{m=1}^{n_1} N_m = \text{const}.$$

Here N_n is the population of the n-th level, K_{nm} is the transition-probability matrix (relaxation matrix), D_n is the flux of particles to the n-th level from the continuum, N_+ is the ion density, β and s are the recombination and ionization coefficients, Q_{inel} is the heat released in the electron gas due to inelastic collisions, $Q_{\Delta T}$ is the amount of heat transferred to the heavy particles in the gas (per unit volume per unit time) by free electrons in elastic collisions, and N is the density of heavy particles.

We denote the probabilities of elementary processes by the following: B_n, ionization from the n-th level; B'_n, recombination to this level; V_{nm}, a collisional m → n transition due to free electron impacts; and A_{nm}, radiative m → n (m > n) decay. We write

$$K_{nm} = V_{nm}N_e + A_{nm}, \qquad n \neq m, \qquad K_{nn} = -\Big(\sum_{n\neq m} V_{nm}N_e + \sum_{n<m} A_{nm} + B_nN_e\Big).$$

We note that

$$|K_{nn}| > \sum_{m \neq n} K_{mn}, \quad D_n = B'_n N_e^2 N_+, \quad \beta = \sum_{n=1}^{n_1} B'_n,$$

$$s = \Big(\sum_{n=1}^{n_1} B_n N_e N_n\Big) \Big/ N_e N_1, \; Q_{\text{inel}} = N_e \sum_{n=1}^{n_1} [N_n \sum_{\substack{m=1 \\ m \neq n}}^{n_1} E_{nm} V_{mn} + E_n (B'_n N_e^2 - B_n N_n)],$$

where E_n is the energy of the n-th level reckoned from the continuum downward ($E_n > 0$), $E_{nm} \equiv E_m - E_n$, $Q_{\Delta T} = 3 \frac{m}{M} \nu_e N_e (T_e - T)$, where m is the mass of an electron, M is the mass of a heavy particle, and ν_e is the effective frequency of collisions of electrons with ions and neutral particles.

Equations (1)-(3) form a nonlinear system of ordinary differential equations. This system was solved in a complete form for a hydrogen plasma by Gudzenko et al. [7]. In many important applications, however, the statement of the problem may be simplified.

In particular, in a quasistationary regime of plasma decay (for example, when the plasma contains a "thermostat" of buffer gas particles [1]) such effects as the divergence of the electron temperature T_e from the gas temperature T often do not have to be taken into account. It is possible for regimes (gas pumpthrough) to exist in which all the parameters N, N_+, N_e, T_e, and T may be regarded as constant.

The chief influence on the volume of calculations and their accuracy is the choice of the computational scheme of energy levels included in the system (1). An inherent factor which limits the number of discrete levels of the atom is Debye screening in the plasma, which, if included, leads to a reduction in the ionization potential of the atom (cf., for example, [9]). However, in calculations one usually uses the fact that the populations of highly excited states are close to equilibrium with the continuum, and therefore detailed calculations are not required to determine them. These states are called the quasiequilibrium spectrum and the value of n_1 (or m_1) for the lowest (in energy) of these states in the quasiequilibrium spectrum is its boundary. Still another means for significantly reducing the number of detailed analyses of Eqs. (1) for the population relaxation is to take into account the closeness in energy of the sublevels E_{nl} which differ only in the orbital quantum number l. In a relatively dense plasma with a moderate degree of ionization, levels separated by an energy of not more than the free electron temperature T_e are populated practically in accordance with a Boltzmann law.

In this case it is possible at first to analyze just the population of blocks (n) of states rather than solving Eq. (1) for the populations N_{nl} of every state.

We now discuss briefly a simplified accounting for the interaction of atoms with radiation. In the case of the comparatively dense plasmas with large cross-sectional areas of interest in laser problems, the absorption of resonance emission has a very important effect on the population distribution [10]. Any detailed accounting for reabsorption by solving the equations for the radiation field of atoms together with Eqs. (1) leads to a complicated system of nonlinear integrodifferential equations [11] which for the intense plasma decay conditions of interest in plasma laser theory can hardly be analyzed on available computers. Thus, we use an approximate description of radiative reabsorption according to the Biberman–Holstein scheme [12]. According to this scheme the effect of resonance capture of radiation on the populations of atoms is taken into account by reducing the probability of radiative decay; instead of the probability A_{1n} of spontaneous decay of the n-th level to the ground state of the atom a smaller effective decay probability, $\tilde{A}_{1n}$, is introduced. This quantity is obtained by multiplying the Einstein coefficient by a factor $G(\tau)$, i.e., $\tilde{A}_{1n} = G(\tau) A_{1n}$, where $0 < G(\tau) < 1$ and τ is the optical density of the plasma at the $n \to 1$ transition. In the case of a Stark line profile $G(\tau) \sim 1\sqrt{\tau}$, and in the case of Doppler broadening $G(\tau) \sim 1/\tau \sqrt{\pi \ln \tau}$.

The existence of resonant reabsorption leads to an effective weakening of the strong radiative transitions. Hence as the gas density is increased amplification (lasing) is also reduced on the working transitions in which the lower level is emptied by radiative decay. Thus, in a dense low-temperature plasma it is necessary to include only collisional mechanisms for emptying the lower level, those defined by cold electron impacts, the Penning effect, or loss of an atom in a chemical reaction.

A scheme for Calculating Populations

The characteristic times of various processes in a plasma may differ from one another by orders of magnitude [13]. Already in [14] this was used for the constant-sink approximation, according to which the populations of the excited states are found by solving the system of algebraic equations

$$K'_{nm}N_m + D_n = 0, \qquad n, m = 2, 3, \ldots, n_1, \tag{4}$$

where

$$K'_{nm} = K_{nm} - EK_{1n}, \qquad D'_n = D_n + K_{n1}N_1,$$

in place of Eqs. (1).

This corresponds to an assumption of the instantaneous adjustment of the populations N_n ($n \neq 1$) to values corresponding in a steady state to the plasma parameters N_e, T_e, and N_1. For practical application of this approximation it is important to indicate the conditions for its use, especially since the requirement formulated in [14] for this purpose,

$$N_n \ll N_e, \qquad n \neq 1, \tag{5}$$

itself contains the unknown quantity N_n and, in addition, is only a sufficient condition. Already in it we can see the possibility of using approximations in which derivatives in only some of the equations (1) are set equal to zero. Other criteria, such as

$$\sum_{n \leqslant n_1} N_n \ll N_e \tag{5'}$$

or

$$\beta N_e^2 \ll \sum_{n < n_1} |K_{nn}|, \tag{5''}$$

proposed in [15, 1], require satisfaction of the conditions for a stationary sink for each of the excited states individually and are more strict than condition (5). A rigorous comparison of the characteristic values of the matrices of the system (1) shows that under the conditions discussed here we may use the equation [16]

$$\beta N_e^2 \ll \min |K_{nn}| \tag{6}$$

as a condition for validity of the stationary-sink approximation. Its validity may be easily demonstrated before calculating (for given plasma parameters) the population distributions.

It should be noted that even using the stationary-sink approximation, calculation of the level populations is cumbersome and a computer is needed. Often, however, it is possible to quickly make simple estimates "by hand," using various features of the relaxation matrix K_{nm}.

Among these estimation schemes are the dominant-sink approximation [17], the single-quantum approximation [18], and the open two-level model [19].

We first note that in a supercooled dense plasma the relaxation matrix is close to a right triangle. In the dominant-sink approximation, upward transitions from the levels concerned are neglected; this is equivalent to neglect of the subdiagonal elements of the relaxation matrix K_{nm}. This eases the calculations; however, it limits the domain of applicability of the approximation to low ($T_e \ll \Delta E$, where ΔE is the difference in the energies of the levels) free electron temperatures.

If we neglect all elements in the relaxation matrix except the principal and two neighboring diagonals, we obtain the single-quantum approximation. It is equivalent to including only transitions to neighboring levels. This approximation is best suited to computing the populations in hydrogen and hydrogenlike systems. It cannot be used to evaluate the level populations of elements with more complicated spectra in general.

The open two-level model [1] is frequently used for estimates. In it only the upper (b) and lower (a) working levels are included. This model is suitable only for examining in principle the possibility of a population inversion and for making crude quantitative estimates. In this approximation $N_b = \beta N_a^2 N_+ / |K_{bb}|$, and $N_a = K_{ab} N_b / |K_{aa}|$. The condition for a population inversion among the working levels is then extremely simple:

$$\frac{N_a/g_a}{N_b/g_b} = \frac{K_{ab} g_b}{|K_{aa}|\, g_a} < 1,$$

that is, the rate of emptying the lower working level must be greater than the rate at which it is filled from the upper working level. The magnitude of the gain coefficient ($\varkappa \sim N_b$) is determined by the rate of filling ($\beta N_e^2 N_+$) and the rate of emptying (K_{bb}) the upper working level.

The most appropriate approximation for recombination lasers is the intense-recombination approximation [20]. The idea of a "bottleneck" in the recombination flux (see, for example, [21]) is essential to this approximation. By the "bottleneck" we usually mean a part of the energy spectrum or a quantum number n_γ of a level "below" which "downward" transitions are much more frequent than "upward" transitions. Such a level can exist, it appears, only in a nonequilibrium supercooled plasma.

The location of the "bottleneck" in the spectrum depends on the degree of disequilibrium of the plasma and on the relative roles of spontaneous and collisional transitions in the relaxation. In the case of an optically thin supercooled plasma, when the effect of the ground state on the populations of the excited levels may be neglected, it is possible to evaluate the number n_γ from the behavior of the ratio

$$\eta(n) = \frac{\sum\limits_{m<n} K_{nm}}{\sum\limits_{m>n} K_{nm}}, \qquad n = 2, 3, \ldots,$$

as a function of the number n. If we go from large to smaller n, then in nonequilibrium situations after some $\bar{n}$, η (n) decreases sharply from values near 1 to 0. Thus, $n_\gamma \simeq \bar{n}$.

The populations of levels with $n > n_\gamma$ are close to equilibrium with the continuum (by the definition of n_γ), so the details of these population distributions have little effect on the recombination coefficient [21]. On the other hand, transitions with $n < n_\gamma$ are mostly downward and these levels also have little effect on the recombination rate. Thus, the recombination rate βN_e^2 is determined by the rate of passage through the level $n = n_\gamma$, hence justifying the name given it. The intensive recombination approximation is based on these properties of the population.

First the location of the "bottleneck" in the spectrum of the element being considered is determined. Then, for given values of N_e and T_e in the plasma, we find the number $n_c < n_\gamma$ of the state from which transitions due to electron impacts still predominate over radiative decay, i.e., for which $V_{c-1,c} > A_c$. Its population is found from the expression

$$N_{n_c} = \beta N_e^2 N_+ / |K_{n_c n_c}|. \tag{7}$$

For states with $n \leq n_c$, because they lie below the "bottleneck," the dominant-sink approximation is valid and their populations are found either by using the recurrence relation

$$N_n = \sum_{m>n} K_{nm} N_m / |K_{nn}|, \quad 1 < n < n_c,$$

neglecting upward transitions, or by solving a system of low-order algebraic balance equations.

We note that when single quantum collisional transitions predominate Eq. (7) makes it possible to evaluate the level populations with only two variables, the recombination rate and the decay constant of the levels.

The most exact values of βN_e^2 are apparently to be found in [22]. For $N_e \gtrsim 10^{14}$ cm^{-3} and $0.1 < T_e$[eV] < 0.4, the values of βN_e^2 from [22] are well approximated by

$$\beta N_e^2 \sim 10^{-27} \; z^3 N_e^2 / T_e^{9/2}.$$

A Decaying Alkali Metal Plasma

The use of an alkali metal plasma as an active medium for recombination lasers is of interest, first, because a population inversion can be produced in these atoms in a plasma with a high gas density and free-electron temperature, second, because in them the working transition corresponds to a relatively large portion (up to 40%) of the ionization potential, and, third, because they are chemically active.

Studies of the level populations of this kind of atomic system were begun in [14] using the pseudo-alkaline atom as an example. It was found that the rate of recombination in such systems does not differ strongly from the rate for the hydrogen atom. Since then a number of theoretical and experimental papers (see, for example, [23-25]) have appeared dealing with decaying plasmas of different alkali metals. The principal topic of these papers was the recombination coefficient. The population kinetics were analyzed using various approximations in [26-28]. The relaxation of potassium in a helium plasma was discussed in [28]. For moderate (not too high) potassium concentrations an inversion was found in the 5s–4p and 5p–3d transitions due to radiative decay of the 4p and 3d states. The mechanism for inversions in a cesium plasma is analogous [27]. Reference [26] dealt with solving the stationary-sink equations for the nine lowest energy states of a lithium plasma as an example. There for the first time the possibility was demonstrated of obtaining inversions among excited states in a high-density, supercooled plasma (with the lower state being emptied by collisions with cold plasma electrons). At the same time, the inadequacy of the model used and of the numerical formulas for the transition probabilities leads to a distortion of the numerical results; in particular, the recombination coefficient exceeds that in other papers [29] by more than an order of magnitude. This comment also applies to [27, 28].

From the standpoint of laser applications the following ranges of the plasma parameters N_e and T_e are of interest: $N_e = 10^{13}$–10^{15} cm^{-3} and $T_e = 0.2$–0.3 eV. The electron temperature obviously cannot realistically be set lower than the gas temperature, but the boiling temperature of either the alkali metals or their compounds is fairly high. Thus, values of T_e below

0.15 eV in this sort of plasma are not real, while high values of T_e lead to a disruption of the population inversion. Low ($N_e < 10^{13}$) values of the density are not of interest because of the low gains. For $N_e > 10^{15}$ the role of spontaneous (meaning radiative) transitions in an alkali metal plasma is negligible and thus the populations N_n of the excited states are proportional to N_e^2.

Levels with $7 \leq n \leq 12$ were taken to be in quasiequilibrium and those with $n \leq 6$ were considered in detail. The system of kinetic equations was solved in the stationary-sink approximation. The spontaneous transition probabilities were taken from [30, 31]. The rates for collisional transitions $n'l' \to nl$ ($E_{nl} > E_{n'l'}$) were calculated from the formula [32]

$$V_{nl,\,n'l'} = 10^{-8}\left(\frac{Ry}{\Delta E}\right)^2\left(\frac{E_{nl}}{E_{n'l'}}\right)^{3/2}\frac{e^{-x}G(X)}{2l'+1}, \qquad x = \Delta E/T_e, \tag{8}$$

where

$$G(x) = A\,\frac{\sqrt{x(x+1)}}{x+\chi}\begin{cases}\ln(16+x^{-1}) & \Delta l = \pm 1,\\ 1 & \Delta l \neq \pm 1.\end{cases}$$

The values of the parameters A and χ were found for each transition by interpolating the tabulated data of [32]; several missing data points (for large l) were found by extrapolation. The rates of collisional transitions to levels in the quasiequilibrium spectrum were calculated assuming the levels to be hydrogenlike. The rate of electron collisional ionization, B_{nl}, was calculated lated using the semiempirical formula

$$B_{nl} = 1.74\cdot 10^{-8}\left(\frac{Ry}{\Delta E}\right)^{-3/2}e^{-x}G_1(x) \tag{9}$$

taken from [33] in the Drawin form. Here, as above, $X = \Delta E/T_e$, ΔE is the ionization energy, and $G_1(X)$ is a tabulated function. The characteristics of the transitions of greatest interest are listed in Table 1.

Lithium Plasma

A typical configuration of the excited level populations of lithium is shown in Fig. 1. The 3s–2p and 4s–3p transitions have population inversions regardless of the free-electron density (for $N_e \geq 10^{14}$).† For $N_e > 10^{15}$ cm^{-3} the results are close to curve 2 and are thus not shown on the graph. The calculations showed that the inversion in the 3s–2p transition is determined (along with its dependence on N_e and T_e) by the degree of ionization of the lithium plasma. The threshold free-electron temperature T_e for inversion is higher at higher degrees of ionization of the plasma, α. However, if α is changed by an order of magnitude, T_e changes much more slowly because of the factor exp $(-\Delta E/T_e)$ in the excitation probability.

TABLE 1

Transition	Li		Na		K
	3s — 2p	4s — 3p	4s — 3p	5s — 4p	5s — 4p
ΔE, eV	0.52	0.51	1.09	0.36	0.99
λ, Å	8126	24468	11406	34141	12523
A_{nm}, sec^{-1}	$3.5\cdot10^7$	$7.5\cdot10^6$	$2.51\cdot10^7$	$4\cdot10^6$	$2.35\cdot10^7$

†Here we shall not consider the radiative mechanism for an inversion between these states since it only operates for small densities N_e and N_1.

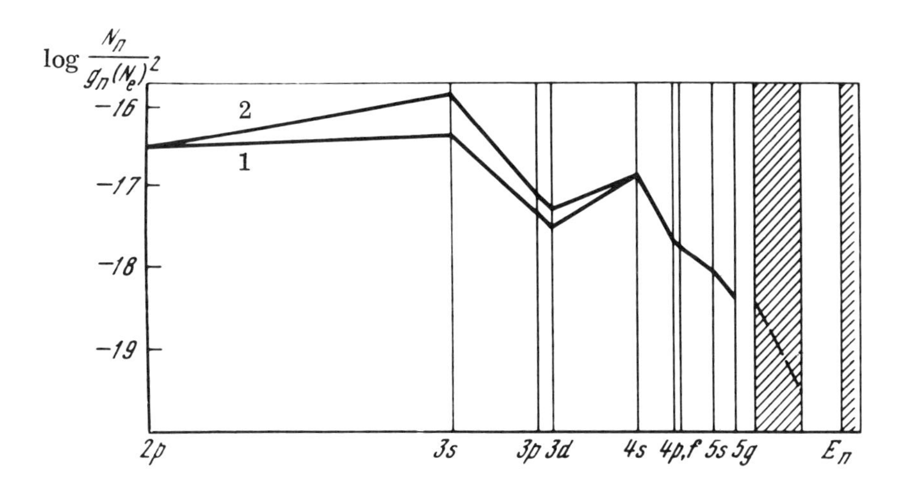

Fig. 1. The level populations of lithium.

The population inversion in the 4s−3p transition is less sensitive to α. The threshold value of T_e for the inversion is about 0.3 eV. Thus, while the inversion on the 3s−2p transition ceases to exist in an optically thick plasma with $\alpha \ll 1$ at low values of N_e, the inversion in the 4s−3p transition remains even for $N_e < 10^{14}$ due to the high rate of the radiative 3d−2p transition. Thus, it is easier to realize lasing on the 4s−3p transition in a recombining lithium plasma than it is on the 3s−2p transition.

By using a rapid chemical reaction to reduce the population of the ground state of the atom it is possible to lighten the demands on N_e and T_e in the plasma for the 3s−2p transition in the recombination regime. The balance condition for the ground-state population N_1 can be rewritten in this case as a condition on the rate γ of the chemical reaction,

$$\gamma \gg \beta N_e^2 N_+ / N_1 = \beta N_e^2 \alpha N_+ / N_e. \tag{10}$$

Sodium Plasma

The populations of the excited states of sodium for one combination of T_e and N_e of the plasma are shown in Fig. 2. It is clear that even for complete reabsorption of resonance emission (curve 2) the inversion does not disappear for all pairs of levels (e.g., the 5s−4p transition). This is because in a supercooled plasma the rate of decay of the 4p state into the 4s

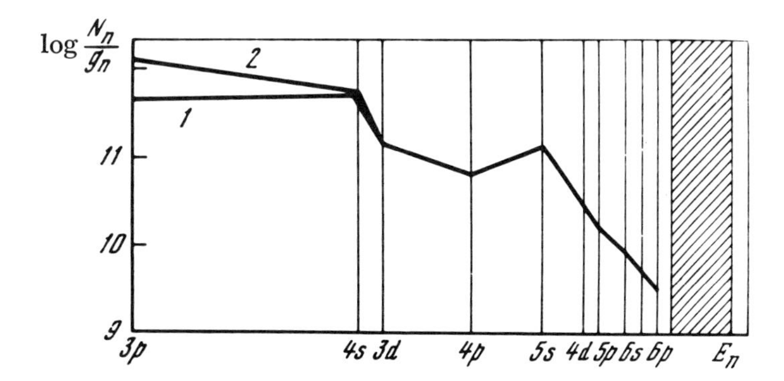

Fig. 2. Sodium level populations for T_e = 0.1 eV and $N_e = 10^{14}$ cm^{-3}. Curves 1 and 2 correspond to optically transparent and optically thick plasmas.

and 3d states due to electron collisions is greater than the rate of decay of the 5s state into the 4p state. The boundary for the inversion in the 5s−4p transition lies in the region of T_e near 0.2 eV.

In a sodium plasma the resonance (3p) state is emptied by free-electron collisions too slowly for an inversion in the 4s−3p transition in an optically dense plasma. We now estimate the rate of a chemical reaction at which radiative depopulation of the 3p state would work. We first note that even in an optically thin sodium plasma with $N_e > 10^{14}$ cm^{-3} there will be no inversion in the 4s − 3p. state. We estimate the required reaction rate from the relation $kR\beta N_e^2 N_+/\gamma \sim 1$, where k is the absorption coefficient in the center of the 3p−3s transition line and R is the characteristic size of the region. In the case of a Doppler broadened line profile we have $\gamma \gtrsim 5.5 \cdot 10^{-13} RN_+\beta N_e^2$.

For example, for $T_e = 0.1$ eV, $N_e = 10^{14}$ cm^{-3}, and R = 1 cm, we have $\gamma \gtrsim 5 \cdot 10^7$ sec^{-1}. In the case of a reaction with the halogens this corresponds to $N \sim 10^{18}$ cm^{-3}.

Potassium Plasma

Calculations showed that in the actually attainable range of plasma parameters no population inversion can occur in an optically thick (optical depth $\tau \gg 1$) pure potassium plasma. In this sense potassium is similar to hydrogen. In an optically thin potassium plasma an inversion in the 5s−4p transition is possible due to radiative depopulation of the 4p state. The possibility noted in [28] of an inversion in the 5p−3d transition was not confirmed by the calculations done here. Estimating the chemical reaction rate needed for an inversion in a potassium plasma with $T_e = 0.1$ eV, $N_e = 10^{14}$ cm^{-3}, and R = 1 cm, we find $\gamma \geq 5 \cdot 10^8$ sec^{-1}, that is, a stricter condition than for sodium. For a reaction with the halogens this value of γ corresponds to $N \sim 3 \cdot 10^{18}$ cm^{-3}.

A population inversion has already been observed experimentally [25, 34] in alkali metal plasmas and in [34] detailed measurements were made of N_e, T_e, and the populations of several excited levels in the afterglow of an ionizing pulse in K and Cs plasmas.

A population inversion was found to exist in transitions of the type (n + 1)s − np in a plasma with a pressure p of about $5 \cdot 10^{-1}$ Torr due to the high rate of radiative decay of the np state. The 6s−5p and 5s−4p transitions in a potassium plasma and the 8s−7p and 7s−6p transitions in a cesium plasma were found to have population inversions. In a low-pressure, low-density plasma ($p \sim 10^{-3}$ Torr, $N_e \sim 10^{12}$ cm^{-3}) population inversions were observed [25-34] in the 5d−7s and 9d−7p transitions of Cs and in the 4d−5p and 4d−4p transitions of K.

In an alkali plasma with a high ($N > 10^{14}$ cm^{-3}) free-electron density, population inversions are possible without impurities in lithium plasmas in the 4s−3p and 3s−2p transitions and in sodium plasmas in the 5s−4p transition. The possibility in principle of building a high-power laser using the 4s−3p transition in a sodium plasma, and the 5s−4p, 6s−5p, and 7s−6p transitions in potassium, rubidium, and cesium plasmas, relies on the use of intense chemical reactions to reduce the population of the lower state.

A Population Inversion over the 3s and 2p States of the Lithiumlike Beryllium Ion

Singly ionized atoms of the alkaline earth elements Be, Mg, Ca, Sr, and Ba have energy-level schemes which satisfy the requirements for effective emptying of the resonance state by free-electron impacts better than the alkali metals. In fact, convincing experiments have been done involving the use of plasma made of these elements (except Be) for plasma lasers in the afterglow of a field pulse [35-37]. In addition, with the aid of some preliminary calculations

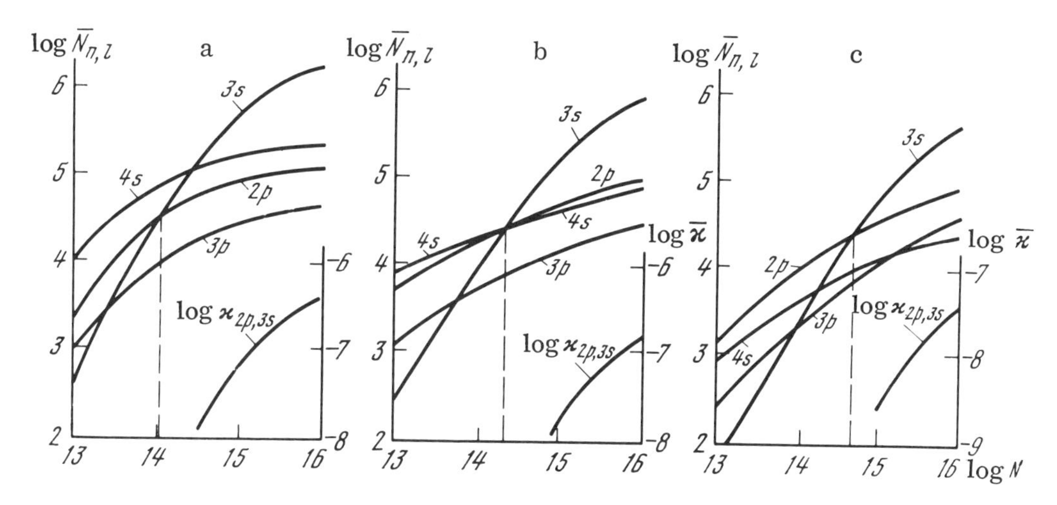

Fig. 3. The populations of the excited levels of the BeII ion. a) $T_e = 0.1$ eV; b) $T_e = 0.2$ eV; c) $T_e = 0.4$ eV.

Latush and Sém [37] showed that the population inversion is due to collisional emptying of the lower working level. Thus a more detailed examination of the possibility of lasing on the 3s–2p transition of the BeII ion (the possibility of such lasing was demonstrated in principle in [38]) is of immediate practical interest.

The intense recombination approximation, beginning with n = 4, was used over the density and free-electron temperature ranges $T_e = 0.1$–0.4 eV and $N_e = 10^{13}$–10^{16} cm^{-3}. The bottleneck in the recombination flux lies substantially above n = 4; thus, of the upward transitions due to collisions with free electrons only those from the 3s level into the 3p and 3d states were included. The states with n = 4 as well as the 3p and 3d states were assumed to be distributed, each within its own n, according to a localized Boltzmann law with temperature T_e.

Therefore, it is possible to write $N_4 = \beta N_e^2 N_{++}/|K_{44}|$. The remaining populations were found by solving a system similar to Eq. (4). Data for evaluating A_{nm} were taken from [9], and for V_{nm}, from [39].

Figure 3 shows plots of the level populations and of $\varkappa_{2p,3s}$ (the linear unsaturated gain coefficient) as functions of N_e. The populations are reduced to the statistical weights of the levels. Both the populations and the gain coefficients are given in units of $\beta N_e^2 N_{++}/N_e$, which is inversely proportional to the decay time of the plasma. A calculation showed that an inversion is achieved only for $N_e \gtrsim 10^{14}$ cm^{-3}, when the collisional depopulation of the 2p state of BeII is at least as effective as its radiative repopulation. We note that an inversion between states with n = 4 and the 3p and 3d states is assured all the way up to $T_e \sim 0.4$ eV and $N_e \sim 10^{15}$ cm^{-3} and that an inversion exists over the 3s–2p transition (at low values of T_e and N_e).

Amplification (lasing) has been achieved experimentally on the 4s–2p transition at this time [40].

The gain coefficients at a wavelength of 1776 Å obtained in plasmas that are currently available are sufficient to produce lasing. Thus, for $N_e = 10^{15}$ cm^{-3}, $T_e = 0.2$ eV, and $N_{++}/N_e = 10^{-2}$ ($\beta N^2 = 10^8$ sec^{-1}) we have $\varkappa_{2p,3s} = 3 \cdot 10^{-2}$ cm^{-1}.

Radiative Depopulation of the Lower Working Level of Multiply Charged Ions

We now proceed to a consideration of the possibility of amplification of short-wavelength radiation in a decaying multiply ionized plasma. The possibility of population inversions among excited levels of multiply charged ions due to the higher rate of spontaneous radiative decay of low-lying states was first pointed out in [41] using the example of the decay of a fully ionized plasma. However, the insufficient accuracy of the statement of the problem in that fairly old work makes it impossible to take any practical prescription directly from it. Thus we shall follow the results of [16].

Assuming the emission from the working transition b → a to be Doppler broadened ($\Delta\omega = 2\pi^{3/2}\lambda^{-1}v_T$, where v_T is the thermal speed of the radiating ions and is taken to be about $3 \cdot 10^6$ cm/sec), we write the gain coefficient in the form

$$\varkappa_{ab} = \frac{\lambda_{ab}^2}{4}\frac{A_{ab}}{\Delta\omega}\left(N_b - \frac{g_b}{g_a}N_a\right) \sim 10^{-8}\lambda_{ab}^3 A_{ab} N_b. \tag{11}$$

From this for $\varkappa_{ab} \gtrsim 1\ \text{cm}^{-1}$ it follows that

$$N_b \gtrsim 10^8\,\lambda_{ab}^{-3} A_{ab}^{-1}. \tag{12}$$

In a supercooled plasma, states b lying below the so-called bottleneck in the recombination flow obey

$$N_b \sim J/|K_{bb}|, \tag{13}$$

where $J = \tau^{-1}N_+$ [cm^{-3} sec^{-1}] is the magnitude of the recombination flux, K_{bb} is the total rate of disintegration of state b, sec^{-1}, τ is the characteristic recombination time, and N_+ is the density of ions in the charge state following that of the working ion, z. Taking $|K_{bb}| \sim A_{bb} \sim A_{ab}$, where A is the spontaneous transition probability, we find, from Eqs. (12) and (13), that

$$J \geqslant J_0 \sim 10^8\,\lambda_{ab}^{-3}\,[\text{cm}] = 10^{32}\,\lambda_{ab}^{-3}\,[\text{Å}]. \tag{14}$$

It is clearly difficult to plan on having $N_+ > 10^{22}$ cm^{-3}. Taking $N_+ = 10^{22}$ cm^{-3} and for now taking $\tau^{-1} \sim 10^7$–10^{10} sec^{-1}, we find that the fundamental wavelength limit for lasing in a quasistationary regime ($\tau^{-1} \ll |K_{bb}|$) lies in the range $\lambda \sim 10$–1 Å.

Since as the charge of the ions is increased in any given isoelectronic sequence the energy spectra of these ions becomes more hydrogenlike, a natural subject for analysis is the hydrogenlike ions.

We shall first clarify the conditions under which the controlling mechanism for decay of a state with principal quantum number n is spontaneous radiative transitions. We shall approximate the values of A_{nn} and of the disintegration rate due to free electron collisions, V_{nn}, by the expressions†

$$A_{nn} \sim 1.6 \cdot 10^{10} z^4 n^{-9/2}\ [\text{sec}^{-1}], \tag{15}$$

and

$$V_{nn} \sim 2.5 \cdot 10^{-7} z^{-3} n^{7/2}\ [\text{cm}^3/\text{sec}]. \tag{16}$$

From these it is clear that for fixed n (for example, n = 3) the value of z at which the level n decays to the radiative ground state is not too large even at extremely large free-electron densities (for example, $N_e = 10^{23}$ cm^{-3}; in that case z = 27). Thus, for a plasma having almost any realizeable density it is possible to find a pair of values n and z such that the prin-

† The data for V_{nm} are for $T_e/z^2 \sim 1$ eV. Data from [42] have been used in the calculation of V_{nm}.

cipal decay mechanism will be spontaneous radiative transitions. For lasers, of course, transitions between states with low n are of greatest interest.

This interest arises primarily because the lower the quantum number n of the lower working level the greater the portion of the energy spent in creating the plasma that can go into the energy of the working transition. It should, however, be noticed that reabsorption by the Lyman series is possible. Strong reabsorption greatly increases the difficulty of depopulating the lower level, but this may be important only for decay of the state with n = 2 since the remaining states (n = 3) may decay into undesirable excited (i.e., less populated) states. Proceeding from this, the optimal transition is $4 \rightarrow 3$.†

The populations of states below the "bottleneck" in the recombination flux are larger for smaller total rates of decay. Equations (15) and (16) make it possible to show (using the condition $A_{nn} \sim V_{nn} N_e$) that the quantum number n* of the most populated level is

$$n^* \sim (6.4 \cdot 10^{16}\ z^7/N_e)^{1/8}. \tag{17}$$

Assuming that n* = 4, we find the relationship between the free-electron density of the plasma and the charge on the working ion at which the level with n = 4 is most efficiently populated:

$$N_e \sim 10^{12} z^7. \tag{18}$$

An estimate of the limiting free-electron densities will now be made with principal reference to a plasma created by "long-wavelength" laser irradiation (at wavelength λ_0) of a solid or liquid target where

$$N_e \leqslant \frac{\omega_0^2 m}{4\pi e^2} \sim 10^{29} \lambda_0^{-2}\ [\text{Å}]. \tag{19}$$

Now using the relation between the charge z of a hydrogenlike ion and the wavelength of the $4 \rightarrow 3$ transition, $\lambda \sim 2 \cdot 10^4 z^{-2}$, we find the relationship between the laser wavelength λ and the wavelength λ_0 of the pump laser:

$$\lambda \sim 0.27 \lambda_0^{4/7}. \tag{20}$$

Also useful is the equation

$$z \sim 2.7 \cdot 10^2\ \lambda_0^{-2/7}\ [\text{Å}]. \tag{21}$$

For example, when $\lambda_0 \sim 10^4$ Å, these equations yield $z \sim 20$ and $\lambda \sim 50$ Å.

Estimates (20) and (21) refer to extremely high free-electron densities. If, however, we examine an expanding plasma with N_e 10^{α} times smaller than the initial value, then Eqs. (20)

†In [43] the prospect of an inverted population over a 4−3 transition of the lithiumlike ion of Al XI in a plasma expanding into a vacuum was considered. We note that for this choice of working ion the population of the state with n = 3 is already sensitive to radiative reabsorption. Thus, in general, there is no certainty that due to absorption at the 3p−2s line the population inversion over the states with n = 4 and n = 3 will not disappear.

and (21) become

$$z \sim 2.7 \cdot 10^2 \cdot 10^{-\alpha/7} \lambda_0^{-2/7}, \tag{22}$$

$$\lambda \sim 10^{2\alpha/7} \cdot 0.27 \lambda_0^{4/7}. \tag{23}$$

In particular, for $\alpha = 2$ and $\lambda_0 = 10^4$ Å, we obtain $z = 11$ and $\lambda \sim 200$ Å.

Using Eq. (14) for the minimal recombination flux and taking $N_+ \sim N_e/2z$, we estimate the minimum recombination rate for achieving $\varkappa \gtrless 1$ to be

$$\tau^{-1} \gtrsim 3 \cdot 10^7 \cdot 10^\alpha. \tag{24}$$

Inequality (24) is an indication of the possibility of a scheme operating in a quasistationary regime ($\tau^{-1} \ll |K_{33}|$). It is also important that this does not require extremely rapid supercooling of the plasma. It is sufficient that the characteristic time of the temperature drop be $\lessgtr 10^{-8}$–10^{-9} sec, which is fully possible even for ordinary expansion of a laser plasma into a vacuum.

The impression may be created that in this particular problem recombination conditions must exist in the plasma even at the initial moment. This is not so. At the initial moment of expansion the balance of ionization and recombination might not be upset, or might even be shifted toward ionization. However, as the plasma is cooled, a time occurs when recombination becomes dominant. This time is determined by the initial conditions in the plasma and by the characteristic supercooling time.

One inadequacy of quasistationary schemes for lasing on transitions between excited states of hydrogenlike ions is the large energy needed to create a highly ionized plasma. Neglecting the kinetic energy of particles with $z \geq 10$ compared to the energy spent in ionizing them, we can extrapolate this energy (for $z = 10$–40) to

$$E \sim 12 \; z^{2.5} N \quad \text{eV/cm}^3 \tag{25}$$

or, with the aid of Eqs. (19) and (21), to

$$E \sim 10^{33.5} \lambda_0^{-2.5}. \tag{26}$$

However, despite the large expenditures of energy (of the same order of magnitude as in ionization schemes for producing population inversions) the recombination approach to filling the upper working state has a substantial advantage in the power of the pump laser. In fact, the characteristic time in an ionization laser is A_{nm}^{-1}, while in a recombination laser, it is much longer at τ. It has already been noted that $A_{nn} \propto z^4$ while $\tau \sim$ const. For a 4−3 transition of a hydrogenlike ion we have $A_{33} \sim 10^8 z^4$ and $\tau \sim 10^{-8}$–10^{-9} sec. The divergence in these times for $z \sim 20$ ($\lambda \sim 50$ Å) reaches five or six orders of magnitude. This constitutes the advantage of the recombination principle for lasing in the x-ray region.

The pump laser power must be

$$W\,[\text{watts/cm}^3] \sim 10^{22} \lambda_0^{-2.5}\,[\text{Å}] \sim 10^{20} \lambda^{-4.5}\,[\text{Å}], \tag{27}$$

that is, to create an active medium in the form of a cylinder with diameter $D \sim 0.1$ cm and length $L \sim 10$ cm with a laser at $\lambda_0 \sim 10^4$ Å will require $E \sim 10^{10.5}$ ergs and $W \sim 10^{11}$ watts/cm^3.

Therefore, simple estimates indicate the possibility of obtaining effective amplification on a 4 → 3 transition of a hydrogenlike ion in an expanding, fully ionized, laser-produced plasma up to wavelengths $\lambda \sim 10$ Å.

Inversions with Respect to the Ground State in Hydrogenlike Ions

We now discuss the possibility of obtaining population inversions over transitions to the ground state of a hydrogenlike ion which is depopulated by recombination into a heliumlike ion. The interest in transitions spanning a wide part of the energy range of the bound states is due both to the relatively high energy yield of such an active medium and to the associated prospects for entering the short-wavelength region using transitions of ions which still have relatively low charge states.

Choosing the upper working level, as before, to be the most populated state n* [see the discussion of Eq. (17)], in the intense recombination approximation we have

$$N_{n^*} \sim \frac{\tau^{-1} N_+}{A_{n^*n^*}}, \qquad N_1 \sim \frac{\tau^{-1} N_+}{\tau_1^{-1}}, \tag{28}$$

where τ_1^{-1} is the recombination rate of the ground state of the hydrogenlike ion. The condition for a population inversion over the $n^* \rightarrow 1$ transition is

$$\frac{g_1}{g_{n^*}} \frac{\tau_1^{-1}}{A_{n^*n^*}} = \tau. \tag{29}$$

Using Eqs. (15) and (16), this condition yields

$$\tau_1^{-1} > 10^5 z^{1.8} N_e^{5/16}. \tag{30}$$

Condition (30) depends weakly on N_e and fairly strongly on z. For z = 2 ($\lambda \sim 300$ Å) and $N_e \sim 10^{16}$ cm^{-3}, Eq. (30) yields $\tau_1^{-1} \sim 3 \cdot 10^{10}$ sec^{-1}; for z = 3 ($\lambda \sim 100$ Å) and $N_e \sim 10^{16}$ cm^{-3} we have $\tau_1^{-1} \sim 7 \cdot 10^{10}$ sec^{-1} ($n^* \sim z$).

In these cases condition (14) is fulfilled more than adequately. The recombination rates indicate that the proposed scheme for an inversion is unrealistic without some mechanisms for extremely rapid supercooling of the free electrons in the plasma, even for the hydrogenlike ion of helium. The possible means for efficient cooling of the free electrons will now be analyzed.

We first note that the possibility of intensively supercooling a plasma contradicts its idealness since an ideal plasma is characterized by a high temperature and low free-electron density. The condition for idealness is derived by comparing the kinetic energy of the free particles in the plasma with their mean interaction energy. It has the form [44]

$$N_e \ll 10^{20.5} (T_e/z^2)^3. \tag{31}$$

When this condition is violated the role of collective effects among the charged particles is enhanced and the properties of the plasma approach those of a liquid. The picture of free motion of the particles with relatively rare collisions is then lost and the role of nonradiative transitions between states with large energy defects becomes greater. A comparison of the previous discussions with these new properties clearly shows that it would be unjustified to use the previous description of the relaxation processes under such conditions. The high probability of nonradiative deexcitation into the lower energy states makes the very possibility of creating an inversion in a nonideal plasma generally doubtful, particularly for short-wavelength transitions.

A functional relation similar to Eq. (31) between the parameters N_e and T_e follows from another customary assumption of plasma theory, namely, that with a sphere of radius equal to the Debye screening distance there must be a large number of free electrons. Violation of this much stricter condition

$$N_e \ll 10^{17} (T_e/z^2)^3 \tag{32}$$

leads to "weak nonidealness" of the plasma, a condition in which the role of field fluctuations increases and the frequency of collisional transitions with small energy defects becomes greater. It seems, however, that the limitation (32) does not determine the character of the relaxation kinetics; that is, in the case of "weak nonidealness" the basic ideas of recombination in a dense supercooled plasma are retained. Thus, we shall only take Eq. (31) into account, as violation of it will certainly lead to serious consequences. To do this we rewrite the condition for intense pumping in the form

$$J = \tau^{-1}N_+ \sim \beta N_e^3 \frac{1}{2z} \gtrsim 10^{32}\lambda^{-3}. \tag{33}$$

Assuming $\lambda \sim 10^3/z^2$ and $\beta \sim 10^{-27}z^3T_e^{-9/2}$, we obtain, after some simple transformations,

$$N_e \gtrsim 10^{20.5} (T_e/z^2)^3 q, \quad q \equiv 10^{2.5} \lambda^{-13/6} (T_e/z^2)^{-3/2}. \tag{34}$$

The inequalities (31) and (34) are equivalent for $q \gg 1$ which implies $\lambda^{13/6} \gg 10^{2.5}(T_e/z^2)^{-3/2}$. Assuming $T_e/z^2 \sim 0.1$ we find that $\lambda \gg 10^2$ Å.

Thus, strictly speaking, the x-ray range is not accessible to schemes using recombination depopulation of the ground state of a hydrogenlike ion. Estimates for other kinds of ions yield the same order of magnitude for the limitation of λ. We may point out, however, that it is equally difficult to obtain a strongly nonideal plasma as it is to achieve extremely small ($\lesssim 10^{-11}$ sec) cooling times for it.

A Collisional Depopulation Mechanism with Lithiumlike Ions as an Example

Another prospect for entering the short-wavelength range, without expending as much energy as in lasing on transitions between excited states of hydrogenlike ions, is to use depopulation of the lower working level by means of deexciting impacts by the free electrons of the plasma. The basic features of this scheme may be traced in a plasma of lithiumlike ions of various charge states. The working transition is the 3s−2p transition as with LiI and BeII. Calculations were done for BeII, BIII, CIV, NV, OVI, FVII and NeVIII, respectively.

The states with principal quantum numbers n = 2, 3, and 4 were included in the population balance. The populations of the energetically close 3p and 3d sublevels as well as of all the states with n = 4 were assumed to be related to one another according to a Boltzmann law with temperature T_e. In the framework of the intense recombination approximation we may write the following for the total population N_4 of all states of a lithiumlike ion with n = 4:

$$|K_{44}|\, N_4 = \beta N_e^2 N_{(z+1)}. \tag{35}$$

(For the values of T_e and N_e under consideration the "bottleneck" in the relaxation flux certainly lies "above" the states with n = 4.)

Here β is the three-body recombination coefficient for the (z + 1)-fold heliumlike ion into the z-fold lithiumlike ion we are discussing. The populations of the remaining (n < 4) states were found by solving a system of the form

$$K_{2p2p} N_{2p} + K_{2p3s} N_{3s} + K_{2p3pd} N_{3pd} + K_{2p,4} N_4 = -K_{2p2s} N_{2s},$$
$$K_{3s3s} N_{3s} + K_{3s3pd} N_{3pd} + K_{3s4} N_4 = 0,$$
$$K_{3pd3s} N_{3s} + K_{3pd3pd} N_{3pd} + K_{3pd4} N_4 = 0.$$

All downward transitions and excitation of the 2s state into the 2p state and of states with n = 3 into states with n = 4 have been taken into account. The coefficient β has been evaluated using the formula

$$\beta = 10^{-27} z^3 T_e^{-9/2} \tag{36}$$

obtained by analyzing the results of detailed calculations [22] for a hydrogen plasma. In this approximation the populations of the levels are proportional to β, so our conclusions about the inverted populations are independent of β. We also note that the errors produced by our use of a cruder relaxation scheme are clearly less than the uncertainties in our current knowledge of the collisional transition probabilities.

In computing the transition probabilities we have approximated the ion characteristics by simple functions of its spectroscopic charge z. The energies of the ionic levels divided by z^2 are given fairly accurately by two-parameter formulas of the type

$$\frac{E_m(z)}{z^2} \simeq a_m + \frac{b_m}{z}. \tag{37}$$

In determining the coefficients a_m and b_m our aim was to ensure maximal accuracy just for z = 2–8 rather than for the entire periodic chart. Their values are given in Table 2. In the probabilities $K_{nm} = V_{nm}N_e + A_{nm}$ of m → n transitions the Einstein coefficients A_{nm} were calculated using the well-known formula $A_{nm} = 4.3 \cdot 10^7 \Delta E_{nm}^2 f_{nm}$. The values of the oscillator strengths $f_{mm'}$ were approximated from the data of [30] by the expression

$$f_{mm'}(z) \simeq \alpha_{mm'} + \frac{\beta_{mm'}}{z}, \tag{38}$$

for which the coefficients are given in Table 3.

We note that for the plasma parameters being considered here the radiative rates for (n = 4) → (n = 3) transitions are already negligible compared to the collisional rates.

TABLE 2

m	2s	2p	3s	3p	3d	4s	4p	4d	4f
a_m	3.50	3.50	1.58	1.55	1.51	0.89	0.88	0.86	0.86
b_m	2.00	0.12	0.48	0.03	0	0.18	0.01	0	0

TABLE 3

Transition	2p — 2s	3p — 2s	4p — 2s	3s — 2p	4s — 2p	3d — 2p	4d — 2p
$\alpha_{mm'}$	0	0.108	0.0296	0.044	0.016	0.400	0.073
$\beta_{mm'}$	0.370	—0.160	—0.0384	0.300	0.028	0	0

The collisional transition rates are caclulated using Eqs. (8).

To describe the z dependence of the parameters A in Eq. (8) it was necessary to use the three-parameter expressions

$$A_{mm'} \simeq \begin{cases} \gamma_{mm'} + \dfrac{\delta_{mm'}}{z+\eta_{ll'}}, & n' \neq n, \\ \dfrac{1}{z-1}\left(\gamma_{mm'} + \dfrac{\delta_{mm'}}{z+\eta_{ll'}}\right), & n' = n, \end{cases} \tag{39}$$

where $\eta_{ll'}$ depends only on l and l'. For sufficiently low temperatures $\Delta E_{nm'} > T_e$ for each transition being considered. Then the values of $\chi_{mm'}$ depend weakly on z and have comparatively little effect on $V_{mm'}$. Thus, the functions $\chi_{mm'}(z)$ were approximated here by the constants $\chi_{mm'}(z) \simeq \bar{\chi}_{mm'}$. We note that for a number of transitions ($n' \neq n$) and sufficiently low temperatures T_e the values of $\Delta E_{mm'}/T_e$ go beyond the interval within which Vainshtein et al. [32] guarantee the accuracy of Eqs. (8). However, these calculations mainly used deexciting transitions, so it is not the values of the $V_{mm'}$ themselves which have the principal effect on the results, but their ratios.

In the analysis of the results it is convenient to use quantities reduced to z, i.e., $\overline{T}_e \equiv T_e/z^2$, $\overline{N}_e \equiv N_e/z^7$, $\overline{N}_m \equiv N_m/z''$, together with normalization of the populations in the form

$$\mathcal{N} = \overline{N}_m/g_m\overline{N}_e\overline{N}_{(z+1)}. \tag{40}$$

Our calculations showed that a population inversion over the 3s and 2p levels corresponds only to a rather low reduced temperature, $\overline{T}_e \gtrless 3 \cdot 0.2$ eV, and to substantial reduced electron densities, $\overline{N}_e \gtrless 3 \cdot 10^{12}$ cm^{-3}. On going to rather high densities N_e it must be kept in mind that in a chemically pure plasma the ion density also rises. In such situations (denoted above by the symbol $N_e \to \infty$) it was assumed that $A_{mm'} = 0$ in the population equations. This case corresponds to reabsorption of the emission from strong radiative transitions with a large energy gap. A population inversion exists over the 4s–3p transition only when $\overline{T}_e \lessgtr 0.05$ eV for not too high z (when $\overline{T}_e = 0.05$ eV and $\overline{N}_e = 10^{13}$ cm^{-3} for $z \leq 4$) and over the 4p–3d and 4d–3p transitions in the same temperature range at still smaller z (when $\overline{T} = 0.05$ eV only for z = 2). It should be noted that even when $\overline{T}_e = 0.05$ eV the populations of all states with n = 3 (not only 3p and 3d, but also 3s) obey a Boltzmann distribution with temperature T_e to satisfactory accuracy. Thus, in the parameter range under discussion it is possible to simplify the relaxation scheme still further.

At these $\overline{T}_e$ the population of the ground state, N_{2s}, can affect only N_{2p}. The dependence of $\mathcal{N}_{2p}$ on $\mathcal{N}_{2s}$ may be expressed by

$$\mathcal{N}_{2p} = \mathcal{N}^0_{2p}(\overline{T}_e, \overline{N}_e, z) + Q(\overline{T}_e, \overline{N}_e, z)\frac{g_{2s}}{g_{2p}}N_{2s}, \tag{41}$$

where

$$Q \equiv V_{2p2s}\overline{N}_e/|K_{2p2p}|.$$

The sharp dependence of Q on z and $\overline{T}_e$ makes it possible to determine the region in which the population of the ground state $\mathcal{N}_{2s}$ has a strong effect on $\mathcal{N}_{2p}$. In a multiply charged ion plasma $N_{2s} < N_e$ and at best is a fraction 1/z of N_e. As an estimate we take $N_{2s} \sim 0.1N_e$. Then it is possible to say that N_{2s} begins to affect N_{2p} when

$$0.1\,\frac{Q\overline{N}_e g_{2s}}{g_{2p}} \sim \mathcal{N}^0_{2p}. \tag{42}$$

We now evaluate the unsaturated gain coefficient $\varkappa$. Using the known formula (e.g., in [45]) we may write

$$\bar{\varkappa} \sim 2 \cdot 10^{-12} z^{-3} (\mathcal{N}_{3s} - \mathcal{N}_{2p}), \quad \bar{\varkappa} \equiv \varkappa / z'' \bar{N}_e \bar{N}_{(z)}. \tag{43}$$

The basic reason for the appearance of a population inversion over the 3s–2p transition is the buildup of electrons in the 3s level due to an intense flux from the 3p and 3d levels while the rate of depopulation of this level due to electron collisions is low. The 2p level, on the other hand, is emptied into the ground state of the ion by electron collisions comparatively quickly. As opposed to a plasma that is recombining into hydrogenlike ions, reabsorption of radiation in this case does not interfere with the population inversions. Furthermore, high free-electron densities facilitate a population inversion over the level pair 3s–2p. Here the limit on the values of N_e delineates excessively small (for a population inversion among the ions) rather than excessively large values of the free-electron densities, as in the case of lasing on hydrogenlike ions. A population inversion over the 3s–2p level pair begins only in a strongly supercooled dense plasma. For a decaying BeIII plasma with $T_e \sim 0.6$ eV and $N_e \geq 10^{16}$ cm^{-3} we have $\varkappa \gtrsim 1$ cm^{-1} and an acceptable decay time of $\tau \sim 10^{-7}$. As z increases the requirement on T_e becomes stricter; in the case of complete reabsorption of the characteristic radiation the corresponding estimate may be given by the formula

$$T_e \lesssim 0.23 \quad z^{3/2}. \tag{44}$$

Besides increasing the demands on N_e this leads to a sharp decrease in the decay time of the plasma. Thus, the threshold values of N_e and T_e for a multiply ionized neon plasma (λ = 103 Å) already correspond to a recombination time of $\tau \sim 10^{-12}$ sec, which is unrealistic [see the discussion of Eq. (30)]. Evidently, the wavelength region in which lasing may be realized by this scheme is limited to $\lambda \gtrsim$ 400 Å.

A Scheme Using an Intermediate Level

Let us assume that the working ion has a state c with the following properties.

1. Spontaneous radiative decay from it is difficult.
2. The energy separation between the lower working level and this level is a substantial fraction of the ionization potential I of the ion, and no other states lie between these levels.
3. The first state above c is a state b, transitions from which to the lower working state are forbidden, and which is located above c with such an energy gap that for $T_e \lesssim 0.01$T the transition c → b may be neglected. Then state c acts as a reservoir in which some of the electrons are trapped during the recombination process as they "drain" over the levels of the working ion. Practically the only disintegration channel for such a state is transitions to higher excited states. As T_e is lowered, a Boltzmann relation will be maintained between the populations N_c and N_b until some time. Then, in general, this Boltzmann distribution is disrupted in the direction of increased N_c/N_b due to the decay of b and inhibition of the c → b transition. It is proposed to create plasma decay conditions such that the intermediate metastable state c is more highly populated than the lower working state a. It is possible to realize this population inversion by transferring a large fraction of the electrons stored in state c through the short-lived state b lying above it by means of a supplementary pulse from a comparatively long-wavelength "illumination" laser.

These requirements are satisfied, in particular, by two-electron (heliumlike) multiply charged ions, whose energy spectra and level schemes are similar in many ways to those of the helium atom.†

Heliumlike ions have two long-lived metastable states, 2^3S and 2^1S, which are the "lowest" of the group of excited states with a principal quantum number $n = 2$. They have little likelihood of radiative decay because such transitions are strongly forbidden or of collisional de-excitation due to the large energy defect. Penning reactions are not effective in a multiply ionized plasma because of the repulsive Coulomb forces among the ions.‡ Their decay rate due to formation of autoionizing states is also small [46]. Thus the 2^1S and, especially, the 2^3S states may be populated to much more than equilibrium values under certain conditions.

In a plasma of heliumlike ions the only state lying "below" the metastable states by a significant energy and, therefore, appearing as a lower working level is the ground 1^1S state. In direct recombination pumping schemes, in order to produce an inversion between an excited state and the ground state, it was necessary to require a recombination rate that was comparable to the rate of decay of the upper working level. This led to unrealistic plasma parameters. However, in a scheme with an "accumulator" it is possible to reduce substantially the requirements on the rate of cooling of the free electrons in the plasma.

First of all, in a supercooled lithiumlike ion plasma the lifetime of the 2^3S and 2^1S metastable states is determined by the rate of "upward" transitions due to collisions with free electrons and may be much greater than the rate of radiative decay of the upper working level (for example, 2^1P).

Secondly, the fact that it is not necessary to worry about an inversion over the working states for a certain time makes it possible to efficiently use the reabsorption of resonance radiation. In fact, the "downward" decay of states with $n = 2$ takes place mainly through the 2^1P level whose radiative lifetime (for reasonable plasma parameters) is several orders of magnitude less than the characteristic times for decay to the ground state of the other states with $n = 2$. However, in a sufficiently dense plasma the lifetime of the 2^1P state is effectively increased due to multiple reabsorption of resonance radiation. Accordingly, the requirements on the rate of supercooling the plasma for producing an inversion between the states with $n = 2$ and the ground state (1^1S) of a heliumlike ion become less stringent.

We have made some estimates of the prospects for using this scheme to obtain lasing on short-wavelength transitions of CV and FVIII.

For the conditions $N_e \gtrsim 10^{16}$ cm^{-3} and $T_e/z^2 > 0.02$ eV of interest here the distribution of relative populations in the group of states with $n = 2$ may be regarded as a Boltzmann distribution.

In a two-level approximation we have

$$K_{12} N(2^1P) = \tau^{-1}_{(z+1)} N_{(z+1)}, \tag{45}$$

where $K_{12} = V_{1^1S2^1P} N_e + \bar{A}_{1^1S2^1P}, \bar{A}_{mm'}$ is the effective (including reabsorption) Einstein coefficient, and $N_{(z+1)}$ is the hydrogenlike ion density.

† We recall that the ground state of such ions, as opposed to He, may decay into a lithiumlike ion in a supercooled plasma.

‡ It is assumed that in this plasma there are almost no neutral atoms.

According to the local Boltzmann distribution among the states with n = 2, for the population N_c of the metastable state we have

$$N_c = \frac{g_c}{g_{2^1P}} N(2^1P) \exp\left(\frac{\Delta E_{2^1P,\,c}}{T_e}\right). \tag{46}$$

The condition for an inversion between states c and 1^1S takes the form

$$\frac{g_{2^1P}}{g_{(z)}} \frac{N_{(z)}}{N_{(z+1)}} \frac{K_{12}}{\tau^{-1}_{(z+1)}} \exp\left(-\frac{\Delta E_{2^1P,\,c}}{T_e}\right) < 1. \tag{47}$$

Here $N_{(z)}$ denotes the density of heliumlike ions in the ground state. Since with these estimates it is impossible to follow the time variation of the ratio $N_{(z)}/N_{(z+1)}$, we have examined two cases: (a) $N_{(z)} = N_{(z+1)}$ and (b) $N_{(z)} = 0.1N_{(z+1)}$. The recombination rate τ^{-1} was evaluated according to [22]. It was assumed everywhere that the plasma was expanding into a vacuum from a state of thermodynamic equilibrium. The initial values of T_e and N_e for cases (a) and (b) were estimated using the approximations of [44]. The conditions for onset of inversion are more favorable the greater the initial plasma density. We note that to produce nonequilibrium conditions in an expanding plasma the characteristic expansion time t_0 must have the following relation to the characteristic recombination time τ:

$$t_0 \lesssim \tau.$$

This condition limits the possibilities of this proposal in the case of plasmas with $z \gtrsim 10$.

An estimate of the characteristic saturation fields needed to transfer a substantial portion of the metastable ions to the upper working state yields [47]

$$E(2^1P,\ 2^1S) \sim 5 \cdot 10^4 \text{ V/cm} \quad \text{and} \quad E(2^1P,\ 2^3S) \sim 2.5 \cdot 10^7 \text{ V/cm}.$$

The values of the unsaturated gain coefficients for a carbon plasma with initial $N_e \sim 10^{19}$ cm^{-3} are

$$\varkappa(2^1P,\ 2^1S) \sim 10^2 \text{ cm}^{-1} \quad \text{and} \quad \varkappa(2^1P,\ 2^3S) \sim 10^3 \text{ cm}^{-1}.$$

Literature Cited

1. L. I. Gudzenko, L. A. Shelepin, and S. I. Yakovlenko, Usp. Fiz. Nauk, 114:457 (1974).
2. L. I. Gudzenko and V. S. Marchenko, Trudy FIAN, 90:90 (1967) [this issue].
3. A. G. Molchanov, Usp. Fiz. Nauk, 106:165 (1972).
4. V. A. Bushuev and R. N. Kuz'min, Usp. Fiz. Nauk, 114:677 (1975).
5. B. M. Smirnov, Pis'ma Zh. Éksp. Teor. Fiz., 6:565 (1967).
6. A. V. Vinogradov and I. I. Sobel'man, Zh. Éksp. Teor. Fiz., 63:2113 (1972).
7. L. I. Gudzenko, Yu. I. Syts'ko, S. S. Filippov, and S. I. Yakovlenko, Preprint IPM AN SSSR, No. 37 (1973).
8. L. I. Gudzenko, Yu. I. Syts'ko, and S. I. Yakovlenko, Preprint FIAN, No. 70 (1973).
9. H. Griem, Plasma Spectroscopy [Russian translation], Atomizdat, Moscow (1969).
10. B. F. Gordiets, L. I. Gudzenko, and L. A. Shelepin, J. Quant. Spectrosc. Radiat. Transfer, 8:971 (1968).
11. V. V. Ivanov, Radiation Transfer and the Spectra of Astronomical Objects [in Russian], Nauka, Moscow (1969).
12. T. Holstein, Phys Rev., 72:1212 (1947); 82:1159 (1951).
13. L. I. Gudzenko, V. V. Evstigneev, Yu. I. Syts'ko, S. S. Filippov, and S. I. Yakovlenko, Preprint IPM AN SSSR, No. 63 (1971).

14. D. R. Bates, A. E. Kingston, and R. W. P. McWhirter, Proc. Roy. Soc., 267:297 (1962); 270:155 (1962).
15. R. W. P. McWhirter and A. G. Hearn, Proc. Phys. Soc., 82:641 (1963).
16. V. V. Evstigneev, Candidate's Dissertation, Moscow Engineering Physics Institute (1975).
17. Yu. K. Zemtsov, Candidate's Dissertation, Nuclear Physics Institute, Moscow State University (1971).
18. V. S. Vorob'ev, Zh. Éksp. Teor. Fiz., 51:327 (1966).
19. S. I. Yakovlenko, Candidate's Dissertation, Moscow State University (1973).
20. L. I. Gudzenko, V. V. Evstigneev, and S. I. Yakovlenko, Kratk. Soobshch. Fiz., No. 9, p. 23 (1973).
21. L. M. Biberman, V. S. Vorob'ev, and I. T. Yakubov, Usp. Fiz. Nauk, 107:353 (1972).
22. L. G. Johnson and E. Hinnov, J. Quant. Spectrosc. Radiat. Transfer, 13:333 (1973).
23. R. W. Motley and D. L. Jassby, Phys. Rev., 187:314 (1969); A1:265 (1970).
24. B. P. Curry, Phys. Rev., A1:166 (1970).
25. D. Ya. Dudko, Yu. P. Korchevoi, and V. I. Lukashenko, Opt. Spektrosk., 34:33 (1973).
26. B. F. Gordiets, L. I. Gudzenko, and L. A. Shelepin, Zh. Éksp. Teor. Fiz., 55:942 (1968).
27. V. A. Abramov, Teplofiz. Vys. Temp., 3:18 (1965).
28. É. E. Son, Teplofiz. Vys. Temp., 8:1128 (1970).
29. L. I. Gudzenko, V. V. Evstigneev, S. S. Filippov, and S. I. Yakovlenko, Preprint IPM AN SSSR, No. 36 (1973); Teplofiz. Vys. Temp., 12:964 (1974).
30. W. L. Wiese, M. W. Smith, and B. M. Clennon, NSRDS-NBS (1969), p. 22.
31. É. M. Anderson and V. A. Zilitis, Opt. Spektrosk., 16:177 (1964).
32. L. A. Vainshtein, I. I. Sobel'man, and E. A. Yukov, Electron Impact Excitation Cross Sections of Atoms and Ions [in Russian], Nauka, Moscow (1973).
33. I. L. Beigman and L. A. Vainshtein, Izv. Akad. Nauk SSSR, Ser. Fiz. 27:1018 (1963).
34. E. E. Antonov, Yu. P. Korchevoy, and V. I. Lukashenko, Proc. 11th Int. Conf. on Phenomena in Ionized Gases, Prague (1973), Contributed papers, p. 33.
35. E. L. Latush and M. F. Sém, Zh. Éksp. Teor. Fiz., 64:2017 (1973).
36. E. L. Latush, V. S. Mikhalevskii, and M. F. Sém, Opt. Spektrosk., 34:214 (1973).
37. E. L. Latush and M. F. Sém, Kvant. Élektron., No. 3 (15), p. 66 (1973).
38. L. I. Gudzenko and S. I. Yakovlenko, Kratk. Soobshch. Fiz., No. 7, p. 3 (1970).
39. I. L. Beigman and L. A. Vainshtein, Trudy FIAN, 51:8 (1970).
40. V. V. Zhukov, V. G. Il'yushenko, E. L. Latush, and M. F. Sém, Kvant. Élektron., 2:1409 (1975).
41. B. F. Gordiets, L. I. Gudzenko, and L. A. Shelepin, Zh. Prikl. Mekh. Tekh. Fiz., 5:115 (1966).
42. A. Jacobs, J. Quant. Spectrosc. Radiat. Transfer, 12:243 (1972).
43. É. Ya. Kononov, and K. N. Koshelev, Kvant. Élektron., 1:2411 (1974).
44. Ya. B. Zel'dovich and Yu. P. Raizer, The Physics of Shock Waves and High-Temperature Hydrodynamic Phenomena [in Russian], Nauka, Moscow (1966).
45. I. I. Sobel'man, Introduction to the Theory of Atomic Spectra [in Russian], Fizmatgiz, Moscow (1963).
46. R. L. Kauffmann et al., J. Phys., B6:2197 (1973).
47. L. I. Gudzenko, V. V. Evstigneev, and S. I. Yakovlenko, Preprint FIAN, No. 4 (1975).

CHAPTER II

PRODUCTION AND KINETIC ANALYSIS OF SUPERCOOLED PLASMAS

The General Problem

Two problems must be solved to create the active medium for a plasma laser: (1) to obtain an intensely recombining plasma (this automatically ensures filling of the upper working level), and (2) to efficiently depopulate the lower working level. Here we shall discuss the questions associated with the first problem.

To produce a supercooled plasma it is desirable that energy be introduced into the gas in such a way that most of it goes into ionizing the gas, and, in addition, it is necessary to ensure cooling of the free electrons. In a pulsed regime the energy is first delivered to a comparatively small number of free electrons which get heated and ionize the medium. Then heating is sharply curtailed and the free electrons are rapidly cooled and undergo recombination. This way of obtaining a supercooled plasma "in the afterglow of an ionization pulse" was analyzed theoretically in [1] and has been realized in laser experiments [2-7]. A stationary method of obtaining a supercooled plasma would require a nontraditional type of gaseous discharge since in ordinary discharges the plasma is superheated, that is, the free-electron temperature T_e exceeds the thermodynamic equilibrium value T_{eq}. For stationary maintenance of a supercooled plasma in which the mean electron energy (temperature) obeys the inequality $T_e < T_{eq}$ the electron distribution must differ from Maxwellian in such a way that there is a sufficient density of fast electrons to produce the specified degree of ionization.

The following mechanisms for cooling the free electrons may be considered during the theoretical analysis of both pulsed and steady-state plasma laser schemes: (1) wall cooling, (2) cooling during elastic collisions with heavy particles, (3) cooling during inelastic collisions due to loss of energy by the electrons in excitation of vibrational degrees of freedom and dissociation of the molecules of an impurity gas, and (4) cooling of the plasma during adiabatic expansion into a vacuum. We indicate here that the wall-cooling channel is ineffective in a dense medium (this case is analyzed in [8]), that the effect of elastic collisions on the free-electron temperature is discussed briefly in the next two sections of this chapter, and that the third channel will be discussed later (at the beginning of the last section of this chapter).

The fundamental problems in calculating the electronic relaxation and in the theoretical analysis of the amplification properties of a stationary supercooled plasma are practically independent of the auxiliary ionization source [whether it is a beam of charged particles (longitudinal or transverse, electrons or ions), x radiation from an outside source, or showers of fission products formed in the medium during nuclear reactions]. In any case, since ionization is usually accompanied by the continual excitation of levels, including the ground state, it is necessary to examine anew the conditions for a population inversion.

An Analysis of the Afterglow of a Helium Plasma

The study of nonstationary processes in a dense recombining nonequilibrium plasma requires keeping track of the time variations of a large number of parameters which are interrelated by nonlinear equations. In this regard we shall analyze the afterglow of a helium plasma by numerically solving the level kinetic equations together with the heat balance equations. The basic goal of the numerical analysis is not so much to obtain specific data as to clarify the role played by various processes in this problem. Hence we consider the time variation of the level populations and plasma parameters together with the time variation of other physical characteristics of the relaxation process.

Statement of the Problem

We shall analyze the relaxation stage of a dense plasma following the instantaneous cutoff of an ionizing pulse. We limit ourselves to a spatially homogeneous model. To simplify the problem it must be remembered, when setting up the kinetic equations, that the relaxation times of different characteristics differ strongly. Thus, when considering the relaxation of comparatively slowly varying characteristics it is possible to regard more rapid processes as already established and to describe them by a quasistationary approximation. Hence, the difference in the masses of an electron and an atom implies that the translational velocity distributions of the electrons and heavy particles may be characterized by different temperatures T_e and T. The characteristic times for establishing each of these distributions are much less than the characteristic time for energy exchange between these plasma components.

We shall tentatively assume an electron density of $N_e \sim 10^{16}$–10^{19} cm^{-3} and heavy-particle and free-electron temperatures of $T \sim T_e \lesssim 1$ eV. In addition to elastic collisions leading to energy exchange between the electron and ion components, the following inelastic processes play an important role in the relaxation of such plasmas:

$$X(p) + e \rightleftarrows X^+ + e + e \tag{1}$$

the ionization of an X atom in state p by an electron e with formation of a singly charged ion (the inverse process is three-body recombination);

$$X(p) + e \rightleftarrows X(q) + e \tag{2}$$

transitions between excited states of an X atom due to collisions of the first and second kinds;

$$X + Y \rightleftarrows X^+ + Y + e \tag{3}$$

ionization due to collisions with heavy particles Y (the inverse process is three-body recombination, with the third particle being a heavy particle); and

$$X(p) + Y \rightleftarrows X(q) + Y \tag{4}$$

a process analogous to (2).

Important processes involving photons include

$$X + \hbar\omega \rightleftarrows X^+ + e \tag{5}$$

and

$$X(p) + \hbar\omega_{pq} \rightleftarrows X(q), \tag{6}$$

that is, photoionization (photorecombination) and photoabsorption (radiative decay).

Collisional recombination (and, by analogy, ionization) takes place in a stepwise fashion. At first an electron is captured by an ion and forms a highly excited atom with an ionization energy of the order of the free-electron temperature. Then the atom is deexcited due to process (2).

The main processes which control relaxation in a dense low-temperature plasma are processes (1), (2), and (6), and these will be taken into account in the following. These processes dominate since, because of the smallness of the cross sections for inelastic collisions among heavy particles, processes (3) and (4) can be important only at rather low degrees of ionization $\alpha \lesssim 10^{-6}$ [9-12].† Similarly, photorecombination is important only in a rarefied gas in which the condition

$$\alpha < \frac{3 \cdot 10^{13} T_e^{3,75}\ [\mathrm{eV}]}{N\ [\mathrm{cm}^{-3}]}, \tag{7}$$

where N is the heavy-particle density (cf. [9]), is satisfied.

We now discuss the hierarchy of relaxation times in the plasma under consideration below. As already noted, the fastest process with a characteristic time of the order of a few electron–electron collision times is the "Maxwellianization" of the free-electron (and, analogously, of the heavy particle) velocity distribution, that is, the evolution of the electron and gas temperatures. Over practically the same times a Boltzmann distribution develops in common with the continuum over the more highly excited discrete atomic levels with ionization potentials that are less than or of the order of T_e. (In a supercooled plasma, $T_e \ll J$, where J is the ionization potential of the atom.) We shall say that these levels belong to the quasiequilibrium spectrum. The next, slower, process is the establishment of a stationary flow of electrons down over the lower discrete levels. A quasistationary nonequilibrium distribution of the populations of the lower excited levels is formed which is determined by the instantaneous values of T_e and N_e. This process is slower because during an exchange of energy between a bound electron and a plasma electron the latter gains an energy equal to the energy of an atomic transition from one state to another and greater than T_e for the lower excited states. Equalization of the free-electron and heavy-particle temperatures may take a characteristic time of the same order as this. The slowest process is the reduction in the free-electron density due to recombination and filling of the ground states of the atoms. This finally leads to the establishment of an equilibrium electron distribution over all the levels if there are no steady-state sources to maintain the disequilibrium. The combination of atoms to form molecules and the relaxation of excited molecules will not be treated here.

If at the initial time the plasma was far from equilibrium, then as it relaxes it passes through a series of stages corresponding to the relaxation processes listed above. The characteristic time for each stage of relaxation depends strongly on the plasma parameters.

Taking the preceding qualitative discussion into account, we may write the system of relaxation equations for the afterglow of a supercooled helium plasma in the following form [14-15]:

$$\frac{dN_m}{dt} \sum_{i=1}^{n_1} (K_{mi} + 2q_n \delta_{mi} \delta_{m,\,2^3S}) N_i + D_m + q_n N_{2^3S}^2 \delta_{m,\,1^1S} \equiv \Gamma_m; \tag{8}$$

$$m = 1^1S,\quad 2^3S,\quad 2^1S,\quad 2^3P,\quad 2^1P,\quad n = 3,\quad n = 4,\ldots,\quad n = n_1 - 3 = 10;$$

$$N = N_+ + \sum_{m=1}^{n_1} N_m, \qquad N_e = N_+; \tag{9}$$

$$\frac{3}{2} \frac{d}{dt} (N_e T_e) = Q_{\mathrm{inel}} - Q_{\Delta T} + q_n N_{2^3S}^2 (J - 2E_{2^3S}); \tag{10}$$

$$\frac{3}{2} \frac{d}{dt} (NT) = Q_{\Delta T}. \tag{11}$$

Here the N_m are the populations of the discrete levels; $n_1 = 13$ is the number of the last discrete level to be considered (i.e., that is not included in the continuous spectrum); $N_e = N_+$ is the density of free electrons and singly charged helium ions; T_e and T are the temperatures of

---

†An exception is resonance processes, of which we must point out ionization of one of two colliding metastable atoms in the processes $X(p) + X(p) \to X^+ + X + e$.

the free electrons and heavy particles; q_n is the rate of Penning ionization,

$$\mathrm{He}\,(2^3S) + \mathrm{He}\,(2^3S) \to \mathrm{He}\,(1^1S) + \mathrm{He}^+ + e$$

and

$$\delta_{m_i} = \begin{cases} 0, & m \neq i, \\ 1, & m = i. \end{cases}$$

The relaxation matrix $\{K_{mi}\}$ includes only spontaneous radiative transitions and collisions with electrons, i.e.,

$$K_{mi} = V_{mi}N_e + A_{mi} \quad (m \neq i),$$
$$K_{mm} = -(\sum_{i \neq m} K_{im} + V_{em}N_e), \tag{12}$$

where V_{mi} is the rate of the $i \to m$ transition due to electron impact and A_{mi} is the spontaneous radiative transition rate. The free term in Eq. (8) has the form

$$D_m = V_{me}N_e^+N_+, \tag{13}$$

where V_{me} is the rate of three-body recombination to level m. In the heat balance equations the term

$$Q_{\mathrm{inel}} = N_e \sum_{m=1}^{n_1} \left[N_m\left(\sum_{i=1}^{n_1} E_{mi}V_{im}\right) + E_m(V_{me}N_e^2 - V_{em}N_m)\right] \tag{14}$$

gives the heat released to the electron gas due to inelastic collisions of electrons with excited atoms (where $E_{mi} = E_i - E_m$ and E_m is the ionization energy of the m-th level) and

$$Q_{\Delta T} = 3\frac{m}{M}\nu_{\mathrm{el}}N_e(T_e - T) \tag{15}$$

characterizes the exchange of energy between electrons and heavy particles in elastic collisions. (Here ν_{el} is the sum of the elastic collision frequencies of electrons with ions and unexcited atoms.)

In solving these equations numerically we have included the reabsorption of radiation by introducing effective probabilities [13] and assuming the line profiles to be Doppler broadened.

Discussion of Results

In solving the system of equations (8)-(11) we have studied both the time variation of the plasma parameters N_e, T_e, and T, and of the populations, and the behavior of various relaxation characteristics. In particular, we analyzed the recombination coefficient

$$\beta = \Gamma_e/(N_e^2N_+), \tag{16}$$

the recombination heating of the electrons Q_{inel}, the heat exchange between the electrons and gas $Q_{\Delta T}$, and the average energy released per electron recombination,

$$E^* = Q_{\mathrm{inel}}/\Gamma_e. \tag{17}$$

Examining the variation of these quantities makes it possible to analyze the interrelations among the various relaxation mechanisms in detail.

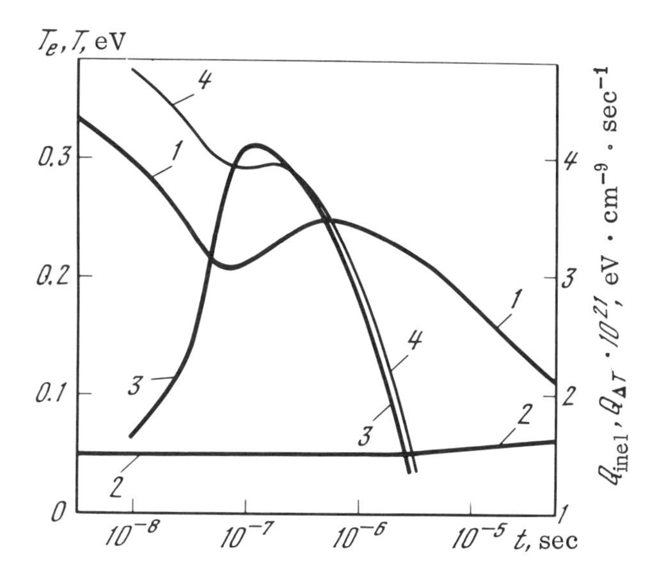

Fig. 1. The temperature relaxation: 1) T_e ; 2) T; 3) Q_{inel}; 4) $Q_{\Delta T}$.

Figures 1-3 show the results of the relaxation calculations for a stationary optically thick helium plasma with R = 1 cm. The initial conditions for the calculations were $T_0 = 0.05$ eV, $T_{e0} = 0.33$ eV, $N_{e0} = 6.4 \cdot 10^{14}$ cm^{-3}, and $N = 10^{18}$ cm^{-3}. The initial populations N_{m0} of the excited levels were chosen to be the stationary-sink populations $\tilde{N}_m(T_{e0}, N_{e0})$, corresponding to the initial values of N_e and T_e. These initial conditions were obtained by solving the steady-state problem of finding the plasma parameters created by an electron beam with a current density j = 10 A/cm^2 and electron energy V = 10 keV. (See the last section of this chapter.)

We note the somewhat unusual behavior of the electron temperature (curve 1 of Figs. 1 and 4). It falls, then rises, and later decreases again. Other relaxation characteristics also

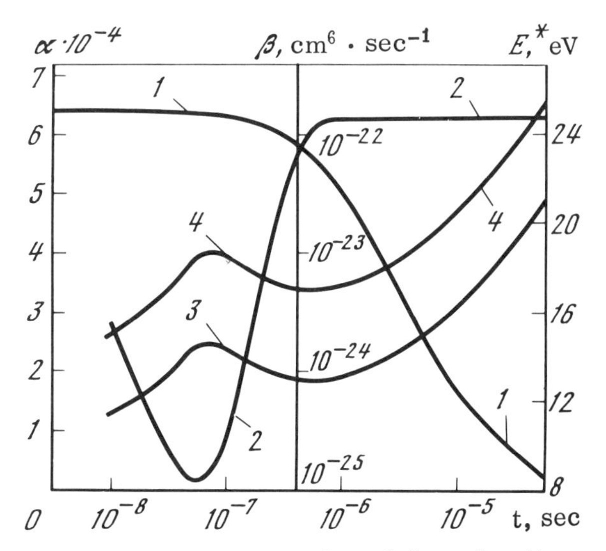

Fig. 2. The time variation of the relaxation characteristics: 1) degree of ionization α; 2) heat release E* ; 3, 4) recombination coefficient β.

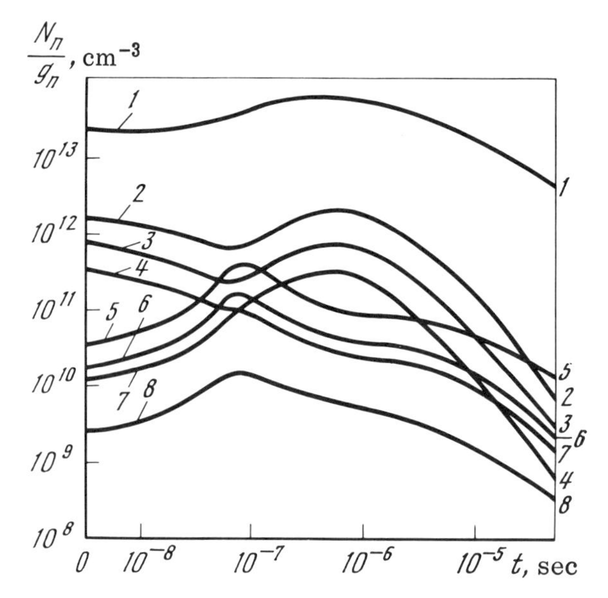

Fig. 3. The relaxation of the level populations. The numbers on the curves correspond to the following states: 1) 2^3S; 2) 2^1S; 3) 2^3P; 4) 2^1P; 5) 3^3S; 6) 3^1S; 7) 3^3P; 8) average population of the states with n = 4.

behave nonmonotonically. Of particular interest is the time behavior of E* (curve 2, Fig. 2). This effect is due to the existence of the metastables and to the small relaxation time of the electron temperature. The quantity T_e falls rapidly and the states with n = 2 cannot be filled up to their quasistationary values. Here the heat released per act of recombination decreases (as the electrons only "reach" the first excited state) although the recombination coefficient is large. When the populations of the states with n = 2 reach their quasistationary values E* is practically equal to the ionization energy of helium (24.6 eV). The total heat release into the electron gas due to recombination increases during the rapid relaxation stage and achieves values exceeding $Q_{\Delta T}$ (Fig. 1, curves 3 and 4), so the electron temperature increases slightly.

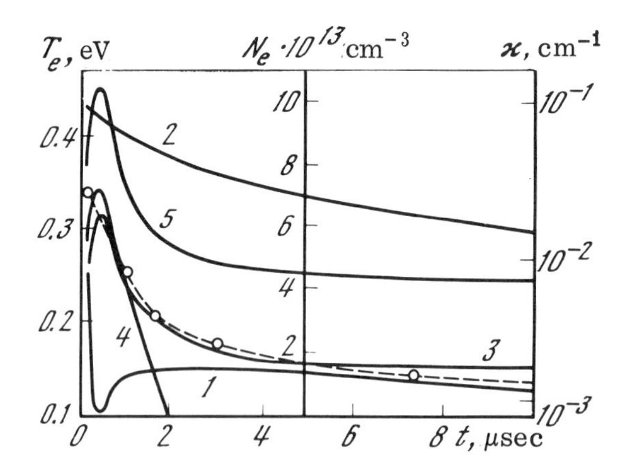

Fig. 4. The afterglow of a helium plasma. 1) Electron temperature; 2) electron density; 3-5) gain coefficient for the $3^1S \rightarrow 2^1P$, $3^1D \rightarrow 2^1P$, and $3^3S \rightarrow 2^3P$.

Curve 4 of Fig. 2 was obtained by calculating the recombination coefficient with Pitaevskii's formula

$$\beta\,[\mathrm{cm}^6 \cdot \mathrm{sec}^{-1}] = C \cdot 10^{-27} T_e^{-9/2}\,[\mathrm{eV}], \quad (18)$$

where C = 8.75.

An exact calculation of the relaxation using Eq. (16) indicates that the recombination coefficient is well described by Eq. (18) with C = 1.4. This value is close to that obtained for helium in 1968 by Mansbach and Keck who found C = 2.0 and four times smaller than the value (C = 5.6) found by Hinnov and Hirshberg (in 1962) and by Bates and Kingston (in 1964) [17].

Figure 3 shows the time variation of the reduced populations N_n/g_m of the excited levels. The sublevels with n = 3 were assumed to be populated according to a Boltzmann law. It is clear that initially ($t \leqslant 10^{-7}$ sec) the electrons "accumulate" in states with n = 3 because of an "upward flow," as estimates of the transition rates demonstrated. The states with n = 2 are then emptied due to transitions into the ground state. Then (at $t \sim 10^{-6}$ sec) there is a "return flow" of electrons into states with n = 2. Consequently, the populations of all states begin to decrease due to the reduction in the free-electron density. We note that an inversion begins to appear over the $3^3S \rightarrow 2^3P$ transition when the electron temperature has fallen enough.

Figure 4 compares the computed time variation of the free-electron temperature for $N = 4 \cdot 10^{17}$ cm^{-3}, $T_{e0} = 0.43$ eV, $T_0 = 0.026$ eV, and $N_{e0} = 10^{14}$ cm^{-3} with probe measurements made under the same initial conditions in [18]. From the figure it is clear that the computed results (curve 1) are in good agreement with the measured values of T_e (dashed curve) at times of the order of 10 μsec. An estimate of the characteristic temperature relaxation time yields $\tau_{T_e} \sim 2 \cdot 10^{-7}$ sec, in agreement with the calculated results but not in accordance with the experimentally measured variation of T_e. This difference is apparently due to a smearing out of the leading front of the heating-field pulse or to inaccurate measurement of T_e in the early stage of relaxation. Curves 2-5 of Fig. 4 describe the time behavior of the free-electron density (curve 2) and gain coefficients for the $3^1S \rightarrow 2^1P$, $3^1D \rightarrow 2^1P$, and $3^3S \rightarrow 2^3P$ transitions (curves 3, 4, and 5, respectively). From these graphs it is clear that substantial population inversions may be observed in the afterglow of helium plasmas over time intervals $\Delta t \sim 10$ μsec.

Production of a Supercooled Plasma by Means of an Electron Beam

We shall consider the formation of a supercooled plasma by a hard source. Later we shall use two specific ways of ionizing a gas as examples: first, a beam of fairly hard (high-energy) electrons delivered from without to a dense cold gas through a foil (or through a window in the gas-pumpout diaphragms and, second, nuclear fission products inside a gas reactor [19]. We shall first examine qualitatively the formation of the electron energy distribution. Fast charged particles almost exclusively interact inelastically with the gas, exciting and ionizing it. For the secondary electrons knocked out of the atoms with sufficient energy to excite and ionize the medium only inelastic collisions are significant as well. We shall include these fast electrons in the "ionization cascade." Secondary electrons with energies below the excitation threshold $\varepsilon < E_{thr}$ will be referred to as "subthreshold" electrons. They are cooled by elastic collisions. The cooled electrons form a Maxwellian (with temperature T_e) distribution of recombining "plasma" electrons.† It is these which predominate in the cold dense plasma and determine the kinetics of the excited levels. Thus, the relaxation equations for the

† The different regions of the energy spectrum are illustrated in Fig. 5.

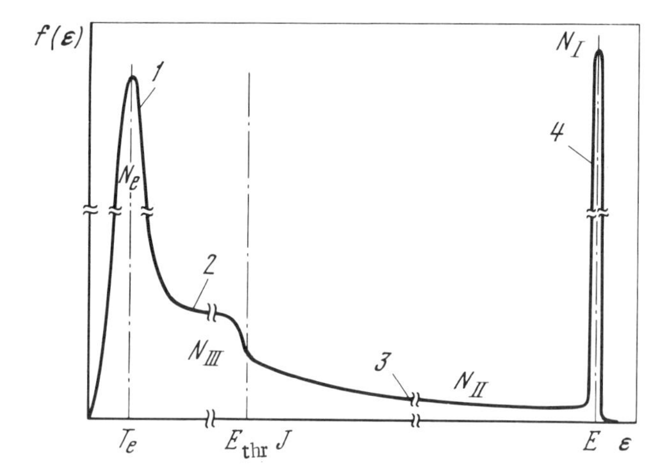

Fig. 5. The energy distribution of the free electrons in a dense gas ionized by a hard source: 1) plasma electrons; 2) subthreshold electrons; 3) cascade electrons; 4) primary particles (in the case of an electron beam as ionizer).

excited-level populations have the usual form [Eqs. (8)-(11)], and the contribution of the hard source reduces to supplementing these equations with terms which describe the ionization and excitation of atoms by electrons belonging to the ionization cascade.

The Degradation Spectrum

In discussing the excitation and ionization kinetics of the gas it is convenient to use the concept of a "degradation spectrum" $Z(\varepsilon)$ [20].† This refers to the energy distribution (density) function of the electrons formed in the energy interval $\varepsilon, \varepsilon + d\varepsilon$ due to the slowing down of a single fast particle. For simplicity we shall assume the gas to be atomic. Then electrons with an energy $\varepsilon + E_m$ fall into the interval $\varepsilon, \varepsilon + d\varepsilon$ with an energy loss E_m due to exciting an atom. Electrons with energies $\varepsilon' > \varepsilon + J$ (where J is the ionization potential of the atom) which have lost energy $\varepsilon' - \varepsilon$ due to ionization and to electrons knocked from atoms during ionization by either initial fast particles or by the electron cascade will also enter this energy interval. Thus,

$$Z(\varepsilon)|_{0<\varepsilon<E} = \sum_m Z(\varepsilon + E_m) P_m(\varepsilon + E_m) + \int_{\varepsilon+J}^{\infty} Z(\varepsilon') P_i(\varepsilon', \varepsilon' - \varepsilon)\, d\varepsilon'. \tag{19}$$

Here E is the energy of a fast particle which generates the cascade. The quantities P_α determine the relative probability of the transition $P_m(\varepsilon)$ during an inelastic collision; thus, α is the probability that an electron with energy ε will excite an atomic state m, and $P_i(\varepsilon', \varepsilon)$ is the probability density for ionization of an atom by an electron with energy ε' such that the electron losses energy ε. In the following we shall also use the relative probability of ionization by an electron with energy ε': $P_i(\varepsilon') = \int_J^{0.5(\varepsilon'+J)} P_i(\varepsilon', \varepsilon)\, d\varepsilon$. All these probabilities are simply

† Some related questions have been discussed in detail by G. D. Alkhazov [21] in calculating an ionization cascade in helium.

related to the cross sections for the corresponding transitions:

$$P_m(\varepsilon)=\frac{\sigma_m(\varepsilon)}{\sigma_{\text{inel}}(\varepsilon)},\qquad P_{\text{i}}(\varepsilon',\varepsilon)=\frac{\sigma_{\text{i}}(\varepsilon',\varepsilon)}{\sigma_{\text{inel}}(\varepsilon')},\qquad P_{\text{i}}(\varepsilon)=\frac{\sigma_{\text{i}}(\varepsilon)}{\sigma_{\text{inel}}(\varepsilon)}$$

where $\sigma_{\text{inel}}(\varepsilon)$ is the total cross section for inelastic interactions between an electron with energy ε and an atom, $\sigma_{\text{inel}}=\sum_m \sigma_m+\sigma_{\text{i}}$, $\sigma_{\text{i}}=\int_J^{0.5(\varepsilon+J)}\sigma_{\text{i}}(\varepsilon,\varepsilon')\,d\varepsilon'$. In view of the indistinguishability of the incident and ejected electrons the ionization cross sections must be symmetrized, i.e., $\sigma_{\text{i}}(\varepsilon',\varepsilon'-\varepsilon)=\sigma_{\text{i}}(\varepsilon',\varepsilon+J)$. The boundary condition for the degradation spectrum is chosen to correspond to a single electron incident on the medium, i.e.,

$$Z(\varepsilon>E-E_{\text{thr}})=\delta(\varepsilon-E). \tag{20}$$

Here $E_{\text{thr}}=\min_m(E_m)$ is the minimum excitation energy of the atom. If the initial electrons are not monoenergetic and have some distribution $g(E)$, then the degradation spectrum is the sum of the corresponding spectra $Z_E(\varepsilon)$, taking into account the initial electron distribution

$$Z(\varepsilon)=\int_0^\infty Z_E(\varepsilon)\,g(E)\,dE. \tag{21}$$

The number of ionizations Z and excitations Z_m which take place as an electron is slowed down and some related quantities are defined in terms of the degradation spectrum as follows:

$$Z_{\text{i}}=\int_J^E P_{\text{i}}(\varepsilon)\,Z(\varepsilon)\,d\varepsilon,\qquad Z_m=\int_{E_m}^E P_m(\varepsilon)\,Z(\varepsilon)\,d\varepsilon,\qquad Z_{\text{inel}}(E)=Z_{\text{i}}+\sum_m Z_m, \tag{22}$$

$$U_{\text{i}}=\frac{E}{Z_{\text{i}}(E)},\qquad U_m=\frac{E}{E_m(E)},\qquad U_{\text{inel}}=\frac{E}{Z_{\text{inel}}(E)}, \tag{23}$$

where U_{inel} denoted the average energy expended in a single inelastic interaction; U_{i}, U_m, and U_{inel} depend weakly on the energy of the incident electron and are thus already convenient to use. The Z_m are basically the values of the degradation spectrum in the negative discrete energy region.

Equation (19) for $\varepsilon < E_{\text{thr}}$ gives the spectrum of the subthreshold electrons formed due to slowing down. Using the function $Z(\varepsilon)|_{\varepsilon<E_{\text{thr}}}$ it is possible to calculate the number of subthreshold electrons generated, as well as the average energy expended in the formation of such an electron,

$$\begin{aligned} Z_{\text{thr}}&=\int_J^E\left[\int_J^{J+E_{\text{thr}}} P_{\text{i}}(\varepsilon',\varepsilon)\,d\varepsilon\right]Z(\varepsilon')\,d\varepsilon'+\sum_{m}\int_{E_m}^{E_m+E_{\text{thr}}} P_m(\varepsilon)\,Z(\varepsilon)\,d\varepsilon,\\ U_{\text{thr}}&=\frac{E}{Z_{\text{thr}}(E)}. \end{aligned} \tag{24}$$

From the standpoint of producing a plasma the subthreshold energy of the electrons in an atomic gas is wasted.

Summing up the preceding remarks, we may define the degradation spectrum $Z(\varepsilon)$ as a function given over the entire energy axis $-\infty<\varepsilon<\infty$ such that for $0<\varepsilon<\infty$ it is the solution of the integral equation (19) with the boundary condition (20), while for $-\infty<\varepsilon<0$, $Z(-E_m)=Z_m$ and $Z(-J)=Z_{\text{i}}$, and $Z(\varepsilon)=0$ in the remaining points of the nagative energy half-axis.

We note that Eq. (19) for the degradation spectrum in the region $\varepsilon > E_{thr}$ is equivalent to the balance equation for a stationary free-electron distribution function when inelastic collisions predominate. In fact, if G fast electrons are generated in unit volume per unit time then the distribution function of the cascade electrons has the form

$$f(\varepsilon) = GZ(\varepsilon) \Big/ \left[\sqrt{\frac{2\varepsilon}{m}}\, \sigma_{inel}(\varepsilon) N\right]. \tag{25}$$

Then substituting $f(\varepsilon)\, \nu_{inel}(\varepsilon) \equiv f(\varepsilon) \sqrt{\frac{2\varepsilon}{m}}\, \sigma_{inel}(\varepsilon)\, N$ in Eq. (19) in place of $Z(\varepsilon)$ leads to a balance equation which takes only inelastic processes into account. The distribution function of the subthreshold electrons is not described by Eq. (25) since it is not produced solely by inelastic collisions. However, the degradation spectrum makes it possible to find the rate $GZ(\varepsilon)$ at which electrons with those energies appear, in order to then include this rate in the corresponding balance equation. This also applies to the discrete spectrum; the expressions $G_i \equiv GZ_i$ and $G_m = GZ_m$ may be used in the kinetic equations for the populations of the excited states and for the ion densities.

The Subthreshold Electrons

The subthreshold electrons formed as a result of cascade ionization are cooled further due to excitation of electronic states. When there is a negligibly small density of molecules in the gas, cooling takes place only during elastic collisions with atoms, ions, and plasma electrons, the last of which already have a comparatively low-temperature Maxwellian distribution. When a significant amount of molecules are present, significant energy losses also occur due to electronic excitation of vibrational states and dissociative attachment of electrons. We shall not include the effects of these processes here as we assume the plasma to be effectively made up of atoms.

To describe the collisional relaxation of the distribution function $f(\varepsilon)|_{\varepsilon<E_{thr}}$ of the subthreshold electrons we shall use the Fokker–Planck form of the collision integral. Including the production of electrons by the ionization cascade we have

$$-\frac{d}{d\varepsilon}\left\{\varepsilon\nu_\varepsilon\left[f(\varepsilon) + T_\varepsilon \frac{df}{d\varepsilon}\right]\right\} = GZ(\varepsilon), \tag{26}$$

where $\nu_\varepsilon = \nu_{ei} + \nu_{ea} + \nu_{ee}$ is the elastic collisional energy exchange frequency (ν_{ei} is the frequency for electron–ion collisions; ν_{ea}, for electron–atom collisions; and ν_{ee}, for subthreshold electrons with plasma electrons); and $T_\varepsilon = [(\nu_{ei} + \nu_{ea})\, T + \nu_{ee} T_e]/\nu_\varepsilon$ is the temperature averaged over the collision frequencies. Integrating Eq. (26) over ε we obtain

$$\varepsilon\nu_\varepsilon\left[f(\varepsilon) + T_\varepsilon \frac{df}{d\varepsilon}\right] = \int_\varepsilon^{E_{thr}} GZ(\varepsilon')\, d\varepsilon', \tag{27}$$

the right-hand side of which gives the diffusion flux along the energy axis. In choosing the constants of integration we have used the fact that for $\varepsilon > E_{thr}$ this flux is practically zero since inelastic processes predominate here. For the same reason we may assume $f(\varepsilon)|_{\varepsilon=E_{thr}+0} = 0$. Thus, we obtain

$$f(\varepsilon) = \int_\varepsilon^{E_{thr}} \varphi(\varepsilon') \exp\left(\int_\varepsilon^{\varepsilon'} \frac{d\varepsilon''}{T_{\varepsilon''}}\right) d\varepsilon', \tag{28}$$

where

$$\varphi(\varepsilon) = \frac{G}{\varepsilon v_\varepsilon} \int\limits_{\varepsilon}^{E_{thr}} Z(\varepsilon')\, d\varepsilon' \tag{29}$$

is the function into which $f(\varepsilon)$ transforms for $\varepsilon \gg T_\varepsilon$ and $|\varepsilon - E_{thr}| \gg T_\varepsilon$. Usually, $T_\varepsilon \ll E_{thr}$, so $f(\varepsilon) = \varphi(\varepsilon)$ for $\varepsilon \sim E_{thr}$.

Generally speaking the function (28) describes the electron distribution inaccurately near the threshold. In order to find the distribution in this region it is necessary to match the function (28) and the flux (27) to the distribution function of the electron cascade and its flux. By linearizing the inelastic cross sections in the prethreshold region with respect to the energy, this may be done analytically and it may be shown that this region is rather narrow, i.e., $\Delta\varepsilon_{thr} \sim E_{thr}[(N_e/N) + (2m_e/M)]$, and in fact has no effect on the plasma kinetics.

The Effect of an Electron Beam on a Gas

It was assumed above that the passage of fast particles through the medium is characterized by a quantity [or distribution g(E)] which is independent of position and time, and which gives the number of ionizer particles appearing in unit volume of gas per unit time. This description corresponds to a particle source which is uniformly mixed with the medium being ionized (such as, for example, in a gaseous-phase nuclear reactor [19, 22]). If the plasma is produced by a directed particle beam, this description is not completely adequate for the problem. On the other hand, it is useful for estimating the average effective characteristics even in this case.

Let us suppose that a broad, monoenergetic beam of charged particles enters a gas which uniformly fills the half-space $x > 0$ in the Ox direction. The beam is characterized by the particle flux† j and the energy V of the particles entering the medium. We shall examine how the slowing-down parameters are related to the variables which characterize the kinetics of the beam–gas interaction. We begin with the fact that the effective slowing-down length (mean free path [23]) of the beam electrons is given by

$$l = \frac{V}{U_{inel}} \frac{1}{N\sigma_{inel}(V)}, \tag{30}$$

where N is the gas (number) density. Since jZ_m/l is the number of atoms excited per unit volume, the probability of exciting a single atom per unit time can be written in the form

$$v_m = j\frac{Z_m}{lN} = j\frac{U_{inel}}{V} Z_m \sigma_{inel}(V) \equiv j\sigma_m(V)\beta_m, \tag{31}$$

where

$$\beta_m = \frac{U_{inel}}{V} Z_m \frac{\sigma_{inel}(V)}{\sigma_m(V)} \tag{32}$$

are coefficients which characterize the role of secondary processes. For m = u Eq. (31) gives the ionization frequency.

† The idea of current density as opposed to flux density is usually used for an electron beam; then $1\ A/cm^2 = 6.25 \cdot 10^{18}\ cm^{-2} \cdot sec^{-1}$.

We note that for an electron beam Eq. (31) describes the excitation and ionization frequencies adequately over only a comparatively small depth of penetration of the beam into the gas,

$$l \gg x \gg \frac{1}{N\sigma_{\text{inel}}(\varepsilon \sim J)},$$

in which the monoenergeticity and directionality of the beam particles are still practically intact and the particles have not yet lost much energy but the secondary electrons with $\varepsilon \sim J$ are already completely slowed down in the medium. Equation (31) gives values of the collision frequencies that are accurate only as the order of magnitude for a thick layer. This is due to the importance of multiple scattering. In fact, in each act of ionization a fast electron loses the momentum gained by a "newly born" electron [i.e., an amount of the order of $(mJ)^{1/2}$] and is deflected by an angle

$$\Delta\theta_e \sim \sqrt{mJ}/\sqrt{mV}. \tag{33}$$

Since the deflection in each act is small and fluctuates, the total deflection θ_e over the path length is proportional to the square root of the number of ionizations, $(Z_i)^{1/2} \sim (V/J)^{1/2}$, that is,

$$\theta_e \sim 1. \tag{34}$$

Therefore, the depth to which the electrons penetrate the gas is equal to the slowing-down length only in order of magnitude. An electron beam may be regarded as broad only if its transverse dimensions are much greater than the slowing-down length or, perhaps, if the electrons penetrate the gas in a fairly strong longitudinal magnetic field. In the case of heavy particles of mass M their momentum is $(MV)^{1/2}$ times greater than that of electrons with the same energy, and the deflection is small, i.e.,

$$\Delta\theta_M \sim \sqrt{m/M}\sqrt{J/V}, \qquad \theta_M \sim \sqrt{m/M}, \tag{35}$$

so Eq. (31) applies more rigorously to them.

Despite the fact that the average quantities derived using Eq. (31) do not fully describe the ionization kinetics, their use is fully justified, for instance, in laser applications. Indeed, from the standpoint of finding whether population inversions can be obtained, the main interest is in the ratios $\nu_m/\nu_i = Z_m/Z_i$ which are practically the same at different points of space and therefore may be found by the degradation spectrum method. This is because the equations for these relative quantities involve the ratios of inelastic cross sections which depend on the energy as a ratio of logarithms over nearly the entire slowing-down length.

Up to now we have spoken only of inelastic collisions between beam particles and gas atoms. However, in analyzing the passage of a beam through a gas there is still another means of energy dissipation to be considered – the interaction of the beam with the collective oscillations of the plasma it has created. The excitation of plasma instabilities leads to losses due to heating of the free electrons of the plasma [24-26] which reduces the energy put into ionization and is therefore undesirable for producing a strongly supercooled plasma. Thus it is best to settle for parameters in the medium such that plasma-beam instabilities are suppressed. Usually the most important instability which is excited is Langmuir (plasma) oscillations due to the interaction with fast charged particles of those waves whose phase velocity is the same as the velocity of the beam particles. The reciprocal of the characteristic time (growth rate) for excitation of this mechanism is given by the equation (page 20 of [24]) $\gamma_L \approx (N_I/N_e)^{1/3}\omega_L$, where N_I is the charge density in the beam, N_e is the plasma electron density, and $\omega_L = (4\pi N_e^2 e^2/m)^{1/2}$ is the plasma frequency. The excitation of Langmuir waves is limited by collisions, which randomize the motion of the electrons. If the elastic collision frequency exceeds the growth rate of the Langmuir oscillations, i.e.,

$$\nu_{el} > \gamma_L, \tag{36}$$

then it may be assumed that these oscillations have been suppressed. Condition (36) is not strict in general and is satisfied at moderate gas densities ($N \gtrsim 10^{18}$ cm^{-3} for $j \sim 100$ A/cm^2, $V \sim 10$ keV). In [27] and elsewhere the collective effects which may develop when a beam is sent into a dense gas are discussed. However, such effects are possible only for monoenergetic beams (with a small spread in velocity) and they may also be suppressed, for example, by making the beam distribution function sufficiently nonmonoenergetic.

The Electron Densities in the Various Groups

As already noted (Fig. 5) the ionized gas contains the primary fast particles, the cascade electrons, and the slow (subthreshold) electrons. The electrons with a Maxwellian distribution, i.e., the plasma electrons, form a special group. We now estimate the relative densities of the free electrons in these groups. Let N_I be the density of fast particles slowed down in the medium. If the ionizer is an electron beam, then in the region where they have not yet lost their energy we have $N_I \sim j/ev_{beam} = 6.25 \cdot 10^9\ J[\mathrm{A/cm^2}]/(V[\mathrm{keV}])^{1/2}$, where j is the current density of the beam, v_{beam} is the velocity of the beam electrons, and $V = \frac{1}{2} m v_{beam}^2$ is the energy of a beam electron. If the plasma is formed by uranium "daughters" then $N_I \sim \frac{G}{\langle\sigma_I v_I\rangle N} \equiv \frac{G}{\nu_I}$, where G is the number of fissions per unit volume per unit time, v_I is the velocity of the nuclear fragments, σ_I is the cross section for inelastic interactions between them and the medium, and ν_I is the ionization frequency.

We now estimate the density N_{II} of the cascade electrons. From our results on the degradation spectrum (in this chapter) it follows that

$$N_{II} = \int_{E_{thr}}^{V} N_I f(\varepsilon)\, d\varepsilon > \frac{G}{\nu_I} Z_i = N_I Z_i. \tag{37}$$

Since $Z_i \gg 1$ for $V \gg J$, it is always true that $N_{II} \gg N_I$. The density N_{III} of the subthreshold electrons may be evaluated using Eq. (29) and is

$$N_{III} = \int_0^{E_{thr}} \varphi(\varepsilon)\, d\varepsilon \sim \frac{G}{\nu_\varepsilon} Z_i \sim N_I \frac{\nu_I}{\nu_\varepsilon} Z_i \sim N_{II} \frac{\nu_I}{\nu_\varepsilon}. \tag{38}$$

The plasma electron density N_e in the steady-state case is given by

$$N_e \sim \frac{GZ_i}{\nu_{rec}} \sim \frac{\nu_\varepsilon}{\nu_{rec}} N_{III}. \tag{39}$$

Here $\nu_{rec} = 1/\tau_{rec}$ is the recombination frequency. In the case of three-body recombination $\nu_{rec} = \beta N_e^2$. We now write Eqs. (37)-(39) together:

$$N_e \sim \frac{\nu_\varepsilon}{\nu_{rec}} N_{III} \sim \frac{\nu_I}{\nu_{rec}} N_{II} \sim \frac{\nu_I}{\nu_{rec}} Z_i N_I.$$

If

$$\nu_{rec} \ll \min(\nu_\varepsilon,\ \nu_I) \tag{40}$$

(in other words, if the rate of formation and cooling of the free electrons is much greater than their recombination rate), then the plasma electrons are more numerous than all the others. As a rule, $\nu_I \gg \nu_\varepsilon$ [since $(2m/M)\langle\sigma_{el}v\rangle \ll \sigma_H v_I$ due to the small mass ratio]; hence using Eq. (40) it is possible to write the following hierarchy of densities:

$$N_e \gg N_{III} \gg N_{II} \gg N_I, \tag{41}$$

which yields the reverse hierarchy of characteristic relaxation times for the electrons in the corresponding portions of the energy spectrum,

$$\tau_{rec} \ll \tau_{III} \ll \tau_{II} \ll \tau_I. \tag{42}$$

Under ordinary conditions in a low-temperature dense plasma (cf., e.g., [28-30]) condition (40) is satisfied with a large margin and the plasma electron density exceeds all the others by several orders of magnitude.

The Kinetic Equations of a Beam Plasma

We shall write a simplified system of equations for the parameters of a thermally unstable plasma produced by an electron beam. At first we shall discuss a gas made up of a single chemical element. We assume the problem to be homogeneous. With these assumptions the system of equations for N_e, T_e, and T takes the form

$$\frac{dN_e}{dt} = \nu_i N - \beta N_e^3, \tag{43}$$

$$\frac{3}{2} N_e \frac{dT_e}{dt} = [(U_i - J)\nu_i] N + \beta N_e^3 J - Q_{\Delta T}, \tag{43a}$$

$$\frac{3}{2} N \frac{dT}{dt} = Q_{\Delta T}. \tag{43b}$$

This system of equations is not stationary; however, N_e and T_e rapidly enter a quasi-stationary regime. In fact, the characteristic relaxation time τ_{N_e} for N_e is

$$\tau_{N_e} = \left| \frac{1}{N_e} \nu_i N \right|^{-1} = \alpha \nu_i^{-1}, \tag{44}$$

where $\alpha = N_e/N$ is the degree of ionization.

Since, as has been shown, $\alpha \lesssim 10^{-3} \ll 1$ at the beam and plasma parameters of interest to us, $\tau_{N_e} \ll \tau_i \equiv \nu_i^{-1}$ (where τ_i is the characteristic time for ionization of the medium by the beam). This means that, as we are interested in solving the system (43) for times $\tau_{N_e} \ll t \ll \tau_i$ it is possible to set $dN_e/dt = 0$ when solving it. Then

$$N_e = \left(\frac{\nu_i N}{\beta} \right)^{1/3}. \tag{45}$$

Using this we may rewrite Eq. (43) in the form

$$\frac{3}{2} N_e \frac{dT_e}{dt} = U_i \nu_i N - Q_{\Delta T},$$

and for the characteristic time to establish T_e we have

$$\tau_{T_e} = \left| \frac{1}{T_e} \frac{U_i \nu_i N}{\frac{3}{2} N_e} \right|^{-1} = \frac{T_e}{U_i} \frac{3}{2} \frac{N_e}{N} \nu_i^{-1} \sim \frac{T_e}{U_i} \tau_{N_e} \ll \tau_{N_e}. \tag{46}$$

Thus, T_e enters a quasistationary regime even sooner than N_e and obeys the relation

$$T_e = T + \frac{U_i \nu_i}{\alpha \frac{2m}{M} \nu_e}. \tag{45a}$$

In addition, the characteristic time for relaxation of the gas temperature is

$$\tau_T = \left| \frac{1}{T} Q_{\Delta T} \right|^{-1} = \left| \frac{1}{T} U_i \nu_i \right|^{-1} = \frac{T}{U_i} \nu_i^{-1}, \tag{47}$$

and we finally establish the following hierarchy of characteristic times:

$$\tau_i \gg \tau_T \gg \tau_{N_e} \gg \tau_{T_e}. \tag{48}$$

These considerations about quasistationary regimes were tested directly by numerical calculations on a computer. The system of nonstationary equations (43) with β evaluated using Eq. (18) was solved. The characteristics of the elementary processes were taken to be those of hydrogen: U_i = 36 eV, σ_i (50 keV) = $1.5 \cdot 10^{-18}$ cm^2, ν_e [sec^{-1}] = $3.14 \cdot 10^{-7}$ N [cm^{-3}] × (T_e [eV])$^{1/2}$ + $2.9 \cdot 10^{-6} N_e T_e^{-3/2}$ (23.4 − 1.15 $\log_{10} N_e$ + 3.45 $\log_{10} T_e$).

The current density was taken to be j = 100 A/cm^2 and N was assumed constant. The system of equations was solved by the Runge–Kutta method with automatic selection of step size for a number of densities in the range N = 10^{16}–10^{19} cm^{-3}. The results of the solution are shown in Figs. 6 and 7. As can be seen from the figures, the above qualitative discussion is confirmed by the calculation: The temperature and density of the electrons relax rapidly from arbitrarily specified initial values to quasistationary values and then parametrically follow the temperature of the heavy particles. Here it should also be noted that in the case of hydrogen the relative difference in T_e and T is small. This is due to the rather large cross section for

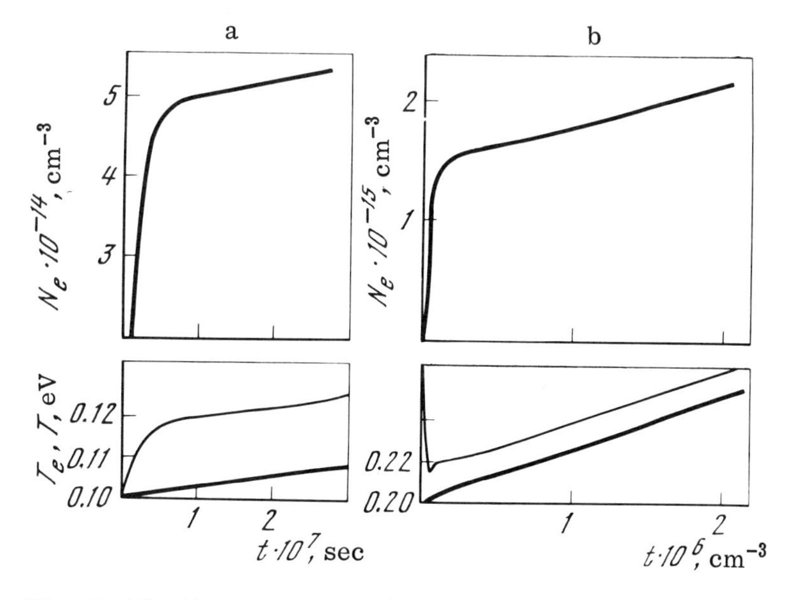

Fig. 6. The time variation of the free-electron density and temperature for a gas density of N = 10^{19} cm^{-3}: a) $T_e(0)$ = T(0); b) $T_e(0)$ > T(0).

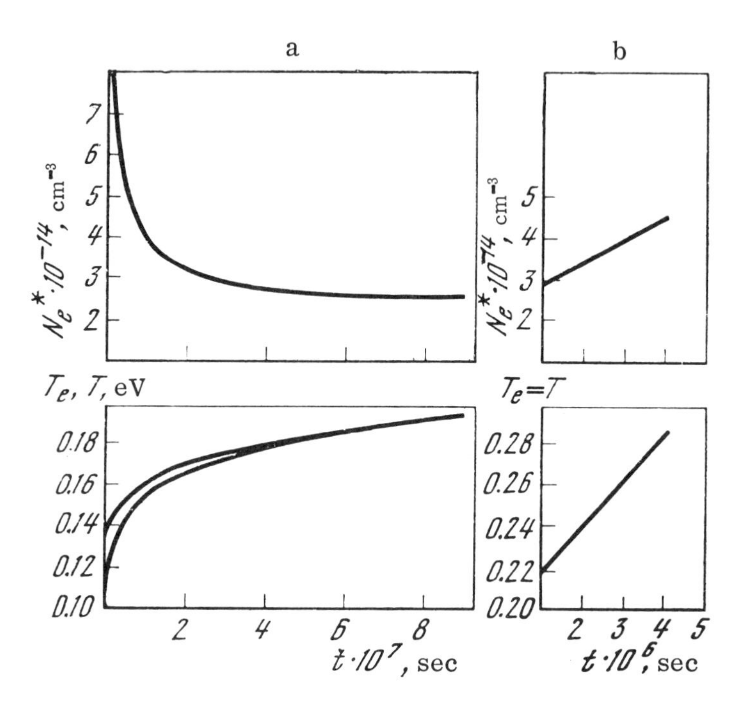

Fig. 7. The time variation of the electron density and temperature and of the gas temperature after instantaneous switching on of the beam. The gas density is $N = 10^{17}$ cm^{-3}. a) For different initial gas and electron temperatures with $T_e(0) > T(0)$; b) for equal initial temperatures $T_e(0) = T(0)$.

elastic collisions of electrons with hydrogen atoms, $\sigma_{el}(0) \approx 40\pi a_0^2$, and the relatively low mass of these atoms.

Population Inversions in a Beam Plasma

The stationary action of an auxiliary hard ionization source may complicate depopulation of the lower working level. Thus, for example, in a number of cases (Li [1]; Be^+ [31]; Mg^+, Ca^+, Sr [2]) a low-lying resonance state which is emptied by collisions with the cooled plasma electrons may serve as a lower laser level in a freely decaying plasma. In the stationary case fast electron collisions may excite this state (from the heavily populated ground state) much more effectively than the large cross section deexcitation collisions, with the more numerous plasma electrons, can depopulate it. Stationary pumping can hardly occur in this problem if the lower working level is not a resonance level. Thus, we here consider situations in which a stationary recombining plasma is made up of a mixture of chemically nonreacting atoms X and Y and lasing takes place on a transition X(b) → X(a), where the lower level X(a) is a resonance level. We assume the basic mechanism for its depopulation to be collisions of the type

$$X(a) + Y(1) \rightarrow X(1) + Y^+ + e. \tag{49}$$

We shall consider a simplified model. Assume that there are no intermediate energy levels between a and b and that the recombination flux of the electrons reaches this level in a single step, that is, transition from levels above b to level a or below and transitions from level b to levels below a may be neglected. Assuming that the conditions for intense decay of the plasma are fulfilled, we shall also neglect upward transitions from levels a and b. Both working levels are depopulated by reactions (49) which we shall assume to be of no importance for levels lying above b. Keeping in mind the sufficient conditions for efficient lasing which we have derived, we shall neglect depopulation of the lower working level by all mechanisms except (49). Some estimates of the populations of the n = 1, 2, 3, and 4 levels of the hydrogen atom (neglecting excitation by the fast electrons) made on the basis of this "open two-level model" were reasonably close to numerical calculations done on a computer including the entire transition matrix [32]. In the case of a steady-state beam-produced plasma from this model we have

$$(q_b N^{\rm Y} + K_b)\, N_b = (\nu_{\rm i} + \nu_b)\, N^{\rm X}, \qquad q_a N^{\rm Y} N_a = \nu_a N^{\rm X} + K_b N_b. \tag{50}$$

Here $q_a = \langle\sigma_a v\rangle$ and $q_b = \langle\sigma_b v\rangle$ are the rates of reactions (49), v is the thermal speed of the atoms, $K_b = A_b + V_{ab} N_e$ is the frequency of X(b) → X(a) transitions, $\nu_{a,b}$ are the excitation frequencies for states a and b, and N^X and N^Y are the densities of the unexcited atoms. We thus find the linear unsaturated gain coefficient to be

$$\varkappa_{ab} = \sigma_{ab}^{\rm ph} g_b \frac{(\nu_a + \nu_b)\, N^{\rm X}}{K_b}\, \Phi_{ab}(N^{\rm Y}), \tag{51}$$

where

$$\Phi_{ab}(N^{\rm Y}) = \frac{K_b}{g_b} \frac{1}{q_b N^{\rm Y} + K_b} \left[1 - \frac{g_b}{g_a} \frac{\nu_a}{\nu_a + \nu_b} \frac{q_b}{q_a} - \frac{K_b(\nu_{\rm i} + \nu_b + \nu_a)}{\nu_{\rm i} q_a N^{\rm Y}} \right], \tag{52}$$

σ_{ab}^{ph} is the photoabsorption cross section at the center of the X(b) → X(a) line, and g_a and g_b are the statistical weights of the working levels. The function Φ_{ab} may have positive values only if

$$q_a/q_b > \nu_a g_b / (\nu_a + \nu_b)\, g_a. \tag{53}$$

If this condition is fulfilled $\Phi_{ab}(N^Y)$ is negative for small values of N^Y; here it increases with N^Y, crossing zero at $N^Y = N_0^Y$. At the point $N^{\rm Y} = N_m^{\rm Y}$, where $N_m^{\rm Y} \gtrsim 2N_0^{\rm Y}$, this function reaches a maximum and then approaches zero as N^Y is increased further. The significance of this behavior in $\Phi_{ab}(N^Y)$ (and the coefficient $\varkappa_{ab}$) is clear. An inversion may be realized over the transition X(b) → X(a) only when the ratio of the rates at which electrons leave the lower and upper states exceeds the ratio of the rates at which they are filled (taking the statistical weights into account) by the beam. If the concentration of the impurity is too large, then it will noticeably depopulate the upper working level as well as the lower level.

Condition (53) differs qualitatively from the inversion criterion derived for a freely decaying supercooled plasma, for which (in the two-level model) it is always possible to choose a sufficiently high N^Y that $\varkappa_{ab} > 0$. Here the very possibility of obtaining a population inversion is determined by the ratio of the excitation and deexçitation cross sections. The right-hand side of Eq. (53) depends on the contribution of secondary electrons, which requires a special analysis including the multiplication kinetics of the fast electrons. It should be noted that condition (53) is greatly weakened in cases where levels a and b are split into sublevels in the energy interval $\gtrsim T_e$. This is because the working states of an atom with split levels are s and p states as a rule. Since the s state is the "lowest" sublevel of the upper working level, its population is greater than the "average population" of the level. Similarly, p states are usually the "highest" sublevels of the lower working level and their population is less than the "average." Thus, under certain circumstances the population of an upper working s state might be greater than that of a lower working p state although the inversion criterion (53) for the "average" populations of the working levels might not be satisfied.

To conclude this section we consider the role of secondary electrons whose effect on the sublevel kinetics has been neglected in deriving the criterion (53). We note that the cross section for a transition between excited states of an atom decreases as the ratio of ΔE (the energy difference between the states) to E (the energy of a free electron) decreases. This indicates that transitions between levels with $\Delta E \sim T_e$ are mainly caused by plasma electrons, whose average energy is T_e, rather than by secondary electrons for which $E \gg T_e$. In addition, as demonstrated above [cf. Eq. (41)] the plasma electron density N_e is much greater than the density N_{II} of fast secondary electrons; hence, the secondary electrons cannot affect the recombination coefficient and the recombination flux going to the upper working level of the atom, if the plasma electron temperature is sufficiently low,

$$T_e < E_{i.u.}$$

where $E_{i.u.}$ is the ionization potential of the upper working level. Thus, the secondary electrons may complicate the attainment of a population inversion over the working levels only by exciting atoms from the ground state, but this effect was taken into account in deriving the criterion (53).

Some numerical model calculations were done to verify these qualitative considerations. As examples we analyzed the atomic mixtures H + Li and H + Xe. In the calculations hydrogen was assumed to be atomic. The gas temperature was assumed to be given. A detailed scheme for these calculations and the quantitative results are given in [33]. The calculations showed that in the mixture H + Li an inversion is obtained over the $H(3) \rightarrow H(2)$ transition despite the fact that the ionization frequency for the hydrogen atom and the excitation frequency for its resonance state H(2) are the same. When xenon was added an inversion was also realized on the $H(4) \rightarrow H(3)$ transition.

This examination was mainly illustrative as it is rather difficult to obtain a dense, atomic hydrogen plasma and maintain it in a steady state. Indeed, at low temperatures and high densities hydrogen is molecular. When an H_2 molecule is ionized by fast electrons a molecular ion H_2^+ is generally formed in the reaction

$$H_2 + e \rightarrow (H_2^+)^* + 2e.$$

The removal of two electrons,

$$H_2 + e \rightarrow 2H^+ + 3e,$$

or dissociative ionization,

$$H_2 + e \rightarrow H^* + H^+ + 2e,$$

is unlikely [34]. The formation of atomic hydrogen ions at low degrees of ionization is also improbable since the reaction to form the H_3^+ molecular ion is highly efficient in dense hydrogen:

$$(H_2^+)^* + H_2 \rightarrow H_3^+ + H.$$

The cross section for this reaction at the thermal velocities of the colliding molecules ($T \lesssim 1$ eV) is extremely large, $\sigma \sim (2\text{-}5) \times 10^{-15}$ cm^2 [35]; hence a collision between H_2^+ and H_2 has a higher probability of resulting in the formation of H_3^+ than in the dissociation of H_2^+. Thus, atomic hydrogen, the simplest medium from the computational standpoint, is not optimal for experiments on lasing media. In addition, the levels of hydrogen are not split, so a population inversion is harder to obtain. These calculations have demonstrated the possibility in principle of making a plasma laser with an electron beam in a steady-state regime using an impurity to depopulate the lower working level, even in such an apparently disadvantageous medium as hydrogen.

The Gain Properties of a Helium Beam Plasma with an Admixture of Molecular Hydrogen

We shall now discuss the possibility of operating a quasistationary beam plasma laser using helium with an admixture. We compute the parameters of a plasma created by an electron beam ($j = 10$–100 A/cm^2, $V \sim 10$ keV) injected into a comparatively dense gas ($N \sim 10^{18}$–10^{19} cm^{-3}) made up of a mixture of helium and molecular hydrogen. The populations of the excited states of helium and the gain coefficients for transitions between states with $n = 2$ and $n = 3$ are calculated.

A simplified scheme for obtaining a population inversion in such a laser is the following. The beam electrons spend their energy ionizing and exciting the medium. The ionization cascade generated by the beam electrons produces electrons with energies below the excitation energy of the gas. These electrons are cooled due to elastic collisions with heavy particles and form a recombining plasma. The recombination flux, which fills the excited states of helium (in particular, the upper working state with $n = 3$), is mainly due to three-body recombination. The impurity deexcites the lower working states with $n = 2$ by means of Penning ionization (ionization during transfer of excitation), so a population inversion is realized.

Upon closer examination the beam laser scheme seems much more complicated. It is necessary, for example, to take into account the following factors which have a negative effect on the possibility of obtaining a population inversion:

(a) excitation of the lower working level by beam electrons and secondary electrons from the ionization cascade;
(b) Penning reactions which disrupt the upper working state;
(c) conversion of helium ions into molecular ions followed by dissociative recombination.

At the same time, an important factor which makes it easier to obtain a population inversion is the splitting of the $n = 2$ and 3 levels. In addition, if the impurity is a molecular gas, then there is an additional mechanism for cooling the electrons by excitation of vibrational degrees of freedom of the gas which (as will be seen from the following) favors the achievement of high gain coefficients.

Helium in particular was chosen for detailed examination primarily because elementary processes have been studied in more detail in it than in any other gas [17]. Calculating the relaxation of helium is not much more difficult than hydrogen and, at the same time, as the preliminary calculations of [36] have shown, lasing in helium should be easier than in hydrogen due to the favorable ratio of the excitation and ionization cross sections† as well as to the level splitting. Also important is the fact that lasing has been observed experimentally on $n = 3 \rightarrow n = 2$ transitions in the afterglow of helium plasmas [37, 38]. Finally, a steady-state, recombining, supercooled plasma has been obtained in beam experiments [39] for the purpose of constructing a steady-state plasma laser.

†The most important quantity characterizing the feasibility of a population inversion is the ratio of the excitation rate of the level with principal quantum number $n = 2$ to the sum of the excitation rates of all the other levels including ionization,

$$\delta = \frac{\nu_2}{\nu_i + \sum\limits_{n>2} \nu_n} = \frac{\sum\limits_{l} \nu_{2,\,l}}{\nu_i + \sum\limits_{n>2} \sum\limits_{l} \nu_{n,\,l}} .$$

Here l is the orbital quantum number; δ was calculated for various concentrations of neon as an impurity. The calculations showed that $\delta < 0.5$ and, to within the accuracy to which the cross sections are known, it is independent of the impurity concentration for partial impurity concentrations of less than 100%.

The Parameters of an Electron Beam Plasma

In this section we analyze the relationship between the beam and gas parameters and the characteristics of the plasma produced by the beam. The calculations are for a mixture of helium and molecular hydrogen. The choice of hydrogen as an impurity was due mainly to the following reasons:

(a) hydrogen has the lowest (after helium) cross section for ionization by fast electrons; this is favorable from the standpoint of minimizing the energy loss due to ionization of the impurity;
(b) excitation of vibrational degrees of freedom of the hydrogen molecule is an additional mechanism for cooling the plasma electrons;
(c) hydrogen is the simplest molecular gas and its kinetics are well known.

The equations for the density of helium ions, N_+, and of He_2^+ molecular ions, N_{2+}, may be written in the form

$$\frac{dN_+}{dt} = \nu_i N - \beta N_e^2 N_+ - K_c N_+ N^2 = 0, \tag{54}$$

$$\frac{dN_{2+}}{dt} = K_c N^2 N_+ - K_d N_e N_{2+} = 0, \tag{54a}$$

$$N_e = N_+ + N_{2+}. \tag{54b}$$

Here N is the helium atom density; β is the three-body recombination rate for helium, $He^+ + e + e \xrightarrow{\beta} He^* + e$; K_c is the rate of conversion in $He^+ + He + He \xrightarrow{K_c} He_2^+ + He$; and K_d is the rate of dissociative recombination, $He_2^+ + e \xrightarrow{K_d} He^* + He$. The contribution to N_e from ionization of hydrogen is assumed to be negligibly small. This is because ionization of the H_2 molecule results in the formation of the molecular ion H_2^+ which rapidly recombines (cf. the discussion at the end of the last section).

The system of equations (54) must be supplemented by the heat balance equation for the electron temperature,

$$\frac{3}{2} N_e \frac{dT_e}{dt} = (U_i - J)\nu_i N + J(\beta N_e^2 N_+ + K_d N_{2+} N_e) - Q_{\Delta T} - \nu_{vib} E_{vib} N_e = 0,$$

which with the aid of Eq. (54) reduces to

$$U_i \nu_i N - \nu_{vib} E_{vib} N_e - Q_{\Delta T} = 0. \tag{55}$$

Here ν_{vib} is the rate of inelastic collisions between the plasma electrons and molecular hydro-

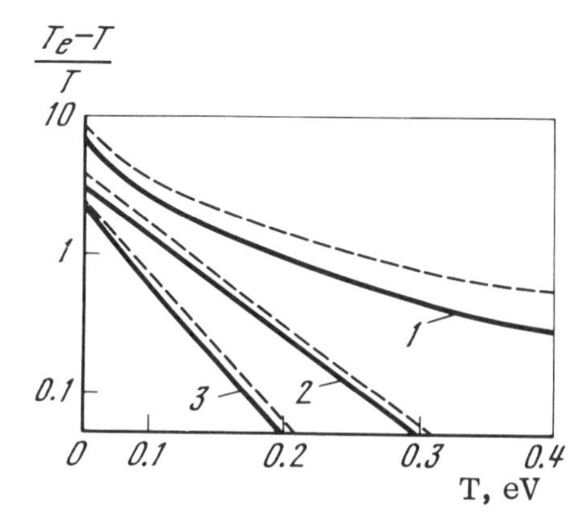

Fig. 8. The divergence between the electron and gas temperatures for various beam and gas parameters. The smooth curves are for a beam parameter $\nu_i = 10\ sec^{-1}$ and a helium density $N = 10^{18}\ cm^{-3}$; the dashed curves are for $\nu_i = 10^3\ sec^{-1}$ and $N = 10^{19}\ cm^{-3}$. Curves 1-3 correspond to molecular hydrogen densities $N_{H_2} = 0$, N/10, N.

gen. These collisions are accompanied by the transfer of an effective energy E_{vib} to excitation of vibrational degrees of freedom. The quantity $Q_{\Delta T}$ describes the cooling of the plasma electrons due to elastic collisions with atomic particles,

$$Q_{\Delta T} = (T_e - T) \sum_z \frac{3m}{M_z} \nu_{ez} N_e. \tag{56}$$

Here the summation is over the particle types z = He, He^+, He_2^+, and H_2; m is the electron mass; M_z is the mass of an atomic particle; and ν_{ez} is the rate of elastic collisions between electrons and particles of type z.

Some results of the calculations are shown in Figs. 8-10.

Figure 8 illustrates the effect of the hydrogen impurity on the divergence between the electron (T_e) and gas (T) temperatures. It is clear that for small T the divergence may be very great and the effect of the impurity increases as the gas temperature rises. Figure 9 illustrates the effect of the hydrogen impurity on the free electron temperature and density. As the amount of hydrogen increases and the gas temperature falls, three-body recombination speeds up and, therefore, the electron temperature falls.

Figure 10 illustates the role of helium ion conversion (into molecular ions) and dissociative recombination of molecular ions. As the gas density is increased to $N > 10^{18}$ cm^{-3} molecular ions begin to predominate in the plasma. This, however, still does not signify substantial dominance of molecular over atomic recombination. The ratio of the flux of helium ions undergoing three-body recombination, $J_e = \beta N_e^2 N_+$, to the flux of ions undergoing conver-

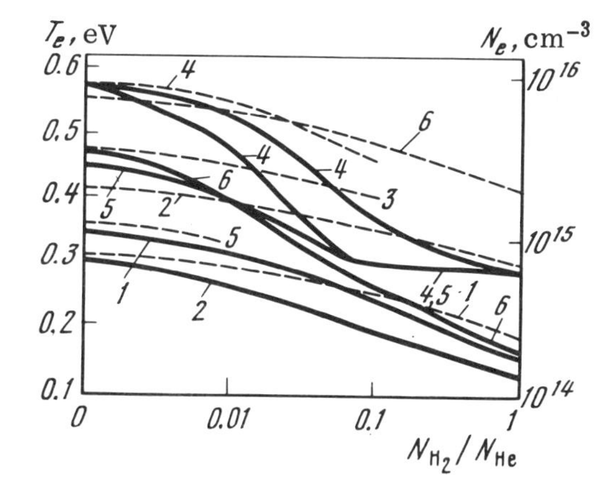

Fig. 9. The dependence of the density (dashed curves) and temperature (smooth curves) of the electrons on the relative amount of hydrogen impurity, N_{H_2}/N. The curves correspond to the parameters ν_i (sec^{-1}), T_e (eV), and N (cm^{-3}) (the resulting ratio N_+/N_e is given in parentheses): 1) 100, 0.1, 10^{18} (0.76); 2) 100, 0.1, 10^{19} (0.10); 3) 1000, 0.3, 10^{18} (0.94); 4) 1000, 0.3, 10^{19} (0.31); 5) 100, 0.3, 10^{18} (0.87); 6) 1000, 0.1, 10^{19} (0.18).

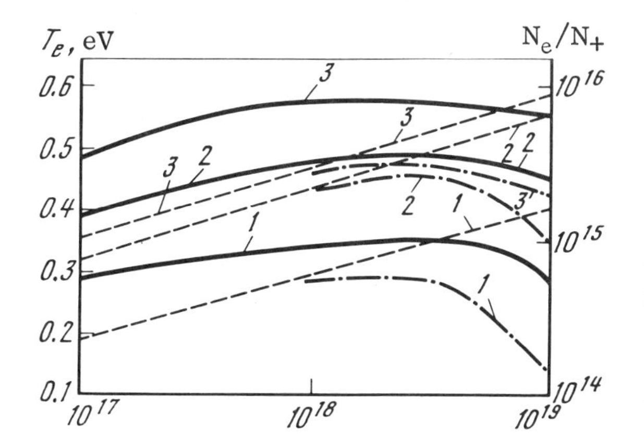

Fig. 10. The dependence of the electron temperature and density and the atomic helium ion density on the helium density N. The smooth curves are the electron temperature, the dashed curves are the electron density, and the dot-dashed curves are the atomic ion density as functions of N. The numbers on the curves correspond to the parameters T (eV), ν_i (sec^{-1}), and N_{H_2} (cm^{-3}): 1) 0.1, 100, 0; 2) 0.1, 1000, 0; 3) 0.3, 1000, 0.

sion, $J_e = K_c N^2 N_+$, can be written in the form

$$\frac{J_e}{J_c} = \frac{\beta N_e^2 N_+}{K_c N^2 N_+} = 4 \cdot 10^5 T_e^{-9/2} T^{3/4}\,[\mathrm{eV}] \left(\frac{N_e}{N}\right)^2. \tag{56a}$$

From this it is clear that although for $N = 10^{19}$ the molecular ion density is 5-10 times the atomic ion density, the flux J_e dominates J_c for small divergences $T_e - T$ (between the electron and gas temperatures).

The Populations of Excited Helium Levels.

Gain Coefficients

To determine the gain coefficients for various transitions between the n = 2 and n = 3 states we have computed the populations of the helium states for given ν_i, N, N_{H_2}, and T_e. Along with Eqs. (54) we have solved the following population balance equations:

$$\frac{dN_\gamma}{dt} = \sum_{\gamma'} K_{\gamma\gamma'} N_{\gamma'} + D_\gamma = 0. \tag{57}$$

Here

$$K_{\gamma\gamma'} = A_{\gamma\gamma'} + V_{\gamma\gamma'} N_e \tag{58}$$

is the relaxation matrix including spontaneous radiative transitions ($A_{\gamma\gamma'}$) and transitions due to electron impacts ($V_{\gamma\gamma'} = \langle \sigma_{\gamma\gamma'} v \rangle$). The diagonal elements of the relaxation matrix have the form

$$K_{\gamma\gamma} = -\sum_{\gamma' \neq \gamma} K_{\gamma'\gamma} - V_{e\gamma} - q_\gamma N_{H_2},$$

where $q_\gamma = \langle \sigma_\gamma v \rangle$ is the rate of Penning ionization, $He(\gamma) + H_2 \rightarrow He(1) + H_2^+ + e$. The quantity D_γ gives the rate at which atoms arrive at level γ due to recombination from the continuum and excitation from the ground state; $D_\gamma = V_{\gamma e} N_e^2 N_+ + \nu_\gamma N$, where $V_{\gamma e}$ is the rate of three-body recombination to level γ. The values of $V_{\gamma\gamma}$, q_γ, and $A_{\gamma\gamma'}$ in the relaxation matrix have been taken from [14]. The values of the ratios used in the numerical calculations were effectively overestimated, so they were done with $\delta = 0.7$, and this scheme is probably somewhat more restrictive than the actual amplification conditions would be. In addition, to verify the reliability of our conclusions about the possibility of steady-state lasing in a helium–hydrogen mixture we have checked a scheme in which the entire flux of He^+ ions, J_c, undergoes conversion and ends up in the helium 2^1P state. This scheme is very much stricter than the real situation.

Some results of the gain-coefficient calculation are shown in Fig. 11. The results for the $3^3S \rightarrow 2^3P$, $3^1S \rightarrow 2^1P$, and $3^1D \rightarrow 2^1P$ transitions are shown. These data characterize quite well the dependence of the gain coefficient on the amount of impurity and the level splitting. The gains on other transitions may be easily estimated [22] on the basis of these results.

From Figs. 11 and 12 it is apparent that the main factor affecting the feasibility of a population inversion and the gain coefficient is the plasma electron temperature. For $T_e \leq 0.1$ eV no further emptying of the lower state by Penning ionization is even necessary. As T_e increases for a constant rate of pumping, $\nu_i N$, there is both an increase in the amount of impurity required for lasing to occur and a decrease in the maximum gain.

Discussion of the Results

First we note that the calculations done here fully confirm the preliminary estimates [36]. (See also [22] and the references there.) In fact, "parasitic" processes (excitation of the lower working level n = 2 during the ionization cascade, formation of molecular ions, and depopulation of the upper working level by the impurity), within the parameter ranges $\nu_i = 10^2$-10^3 sec^{-1} (which corresponds, for example, to a current density j = 10-100 A/cm^2 at an

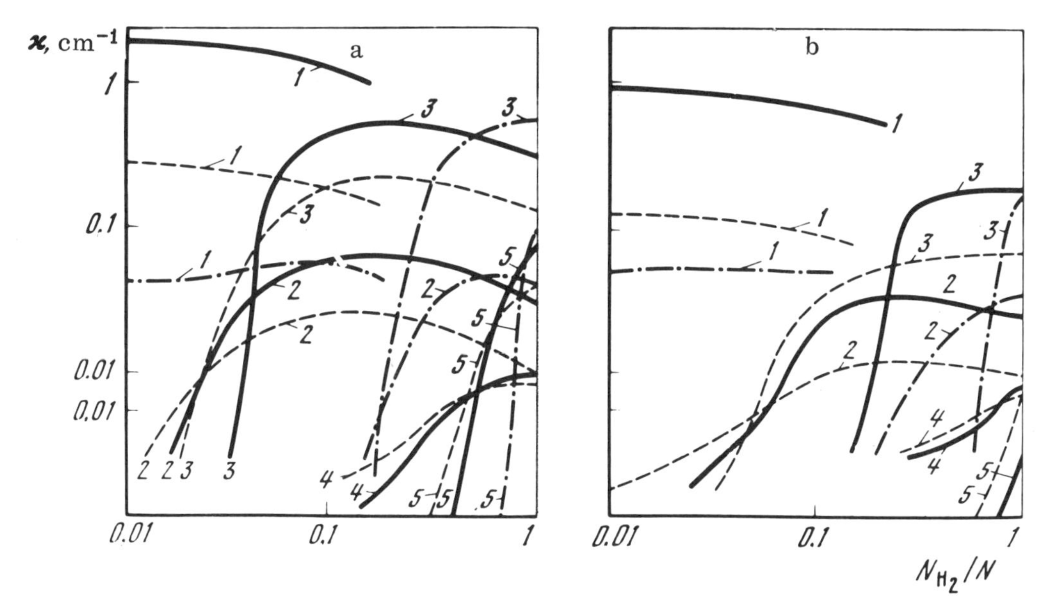

Fig. 11. The dependence of the gain coefficient for the $3^3S \rightarrow 2^3P$ (continuous curves), $3^1S \rightarrow 2^1P$ (dashed), and $3^1D \rightarrow 2^1P$ (dot-dashed) transitions on the fractional amount of hydrogen impurity: a) helium density 10^{19} cm^{-3}; b) helium density 10^{18} cm^{-3}. The numbers on the curves correspond to the parameters T_e (eV), ν_i (sec^{-1}) and (in parentheses) the free-electron density N_e (cm^{-3}) for case (a) (the first number) and for case (b) (second number): 1) 0.1, 100 ($4.9 \cdot 10^{14}$, $1.3 \cdot 10^{14}$); 2) 0.2, 100 ($1.2 \cdot 10^{15}$, $3.7 \cdot 10^{14}$); 3) 0.2, 1000 ($2.4 \cdot 10^{15}$, $7.9 \cdot 10^{14}$); 4) 0.3, 100 ($1.9 \cdot 10^{15}$, $6.7 \cdot 10^{14}$); 5) 0.3, 1000 ($4.2 \cdot 10^{15}$, $1.5 \cdot 10^{15}$).

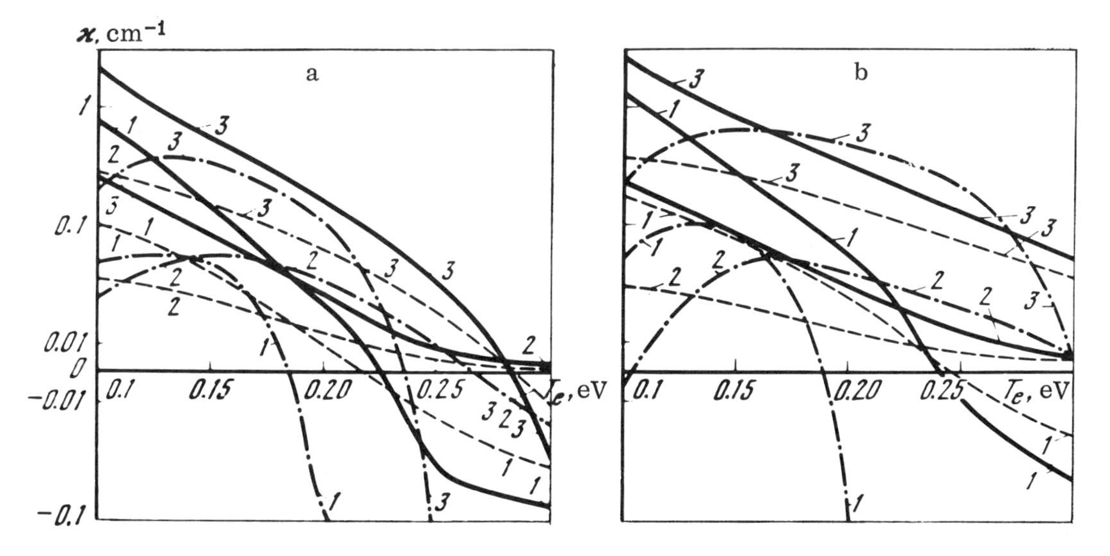

Fig. 12. The dependence of the gain coefficient on the electron temperature: a) for N = 10^{18} cm^{-3}; b) N = 10^{19} cm^{-3}; 1) $\nu_i = 10^2$, $N_{H_2}/N = 0.1$; 2) $\nu_i = 10^2$, $N_{H_2}/N = 1$; 3) $\nu_i = 10^3$, $N_{H_2}/N = 1$.

electron energy V of about 10 keV) and $N = 10^{18}$-10^{19} cm^{-3}, do not cause significant deterioration in the conditions for lasing. This is due first of all to the favorable (for amplification) fact that the level with n = 2 is strongly split. Actually, if we neglect splitting then an inversion is possible only if criterion (53) is fulfilled; it is satisfied with a small margin for a helium–hydrogen mixture. However, at low electron temperatures the population of the 2^1P and 2^3P states is much less than the average population of the level with n = 2, so criterion (53) is greatly moderated. As has already been noted, during the calculations we made some assumptions which should lead to stricter conditions for lasing than actually exist. Nevertheless, the conditions for lasing at low electron temperatures $T_e < 0.3$ eV which follow from the calculation are even somewhat weaker than those given by the estimates of [36].

It follows from the calculations that low temperatures are favorable both for achieving a population inversion and for obtaining high average populations in the upper working levels (that is, for obtaining large $\varkappa$). This is due both to splitting of the levels and to a reduction in the electron density as the electron temperature falls. In fact, for a constant pump rate $\nu_i N$ the average population of the upper working level $N_3 \sim \nu_i N / |K_{33}|$ increases as N_e is decreased since $|K_{33}| = |V_{33}| N_e + |A_{33}| + q_3 N_{H_2}$. Also partially associated with the reduction in the electron density is a reduction in the impurity density needed for an inversion.

Therefore, our estimates and calculations imply that in an experimentally realizable electron beam plasma laser it is necessary to avoid overheating of the gas. The characteristic time for heating the gas to temperature T may be estimated using the relation

$$\tau_T \approx T/(J\nu_i) \sim 10^{-2}\nu_i^{-1},$$

so for $\nu_i = 10^2$ sec^{-1} and $\nu_i = 10^3$ sec^{-1} the gas remains cold over hundreds and tens of microseconds, respectively.

We note that our calculations of the plasma parameters agree with experiment [39]. Unfortunately, the measurements described in [39] were made at a time when the gas had already been heated and the electron temperature was fairly high. As can be seen from curves 4 and 5 of Fig. 11, the conditions are already unfavorable for lasing in that parameter range.

Literature Cited

1. B. F. Gordiets, L. I. Gudzenko, and L. A. Shelepin, Zh. Éksp. Teor. Fiz., 55:942 (1968).
2. E. L. Latush and M. F. Sém, Zh. Éksp. Teor. Fiz., 64:2017 (1973).
3. E. L. Latush, V. S. Mikhailovskii, and M. F. Sém, Opt. Spektrosk., 34:214 (1973).
4. E. L. Latush and M. F. Sém, Abstracts of Reports to the Vth All-Union Conference on the Physics of Electronic and Atomic Collisions, Uzhgorod (1972), p. 165.
5. E. L. Latush, Izv. Severo-Kavkazsk. Nauchn. Tsentra Vyssh. Shkoly Ser. Estestv. Nauk, No. 2, p. 101 (1973).
6. E. L. Latush and M. F. Sém, Pis'ma Zh. Éksp. Teor. Fiz., 15:645 (1972).
7. E. L. Latush and M. F. Sém, Kvant. Élektron., No. 3 (15), 66 (1973).
8. B. F. Gordiets, L. I. Gudzenko, and L. A. Shelepin, Zh. Éksp. Teor. Fiz., 36:1662 (1966).
9. Ya. B. Zel'dovich and Yu. P. Raizer, The Physics of Shock Waves and High-Temperature Hydrodynamic Phenomena [in Russian], Nauka, Moscow (1966).
10. H. W. Dravin and F. Emard, J. Phys. 254:202 (1972).
11. H. W. Dravin, Naturforscher, 25a:145 (1973).
12. E. M. Pavlovskaya and I. V. Podmoshenskii, Opt. Spektrosk., 34:19 (1973).
13. L. M. Biberman, V. S. Vorob'ev, and I. T. Yakubov, Usp. Fiz. Nauk, 107:353 (1972).
14. V. V. Evstigneev and S. S. Filippov, Preprint IPM AN SSSR, No. 5 (1974).
15. Yu. I. Syts'ko and S. I. Yakovlenko, Kvant. Élektron., 2:657 (1975); Preprint of IAÉ, No. 2424 (1974).

16. I. Ya. Fugol', O. N. Grigorashchenko, and D. A. Myshkis, Zh. Éksp. Teor. Fiz., 60:423 (1971).
17. J. F. Delpech, Low-Temperature Helium Plasmas, 11th Int. Conf. on Phenomena in Ionized Gases, Prague (1973), Invited Papers.
18. E. L. Latush, Candidate's Dissertation, Rostov State University (1974).
19. L. I. Gudzenko, I. S. Slesarev, and S. I. Yakovlenko, Preprint FIAN, No. 109 (1974).
20. L. V. Spenser and V. Fano, Phys. Rev., 93:1172 (1954).
21. G. D. Alkhazov, Zh. Tekh. Fiz., 41:2513 (1971).
22. L. I. Gudzenko, L. A. Shelepin, and S. I. Yakovlenko, Usp. Fiz. Nauk, 114:457 (1974).
23. A. Dalgarno, in: Atomic and Molecular Processes [Russian translation], Mir, Moscow (1964), p. 540.
24. A. B. Mikhailovskii, Theory of Plasma Instabilities, Vol. 1, Consultants Bureau, New York (1974).
25. Ya. B. Fainberg, Usp. Fiz. Nauk, 93:617 (1967).
26. Yu. V. Tkach, Ya. B. Fainberg, L. I. Bolotin, Ya. Ya. Bessarab, N. P. Gadetskii, I. I. Magda, and A. V. Sidel'nikova, Zh. Éksp. Teor. Fiz., 62:1702 (1972).
27. A. A. Ivanov, V. V. Parail, and T. K. Soboleva, Zh. Éksp. Teor. Fiz., 63:1678 (1972).
28. G. D. Alkhazov, Zh. Éksp. Teor. Fiz., 40:97 (1970).
29. B. M. Smirnov, Atomic Collisional and Elementary Processes in Plasmas [in Russian], Atomizdat, Moscow (1968).
30. H. Massey and E. Burhop, Electronic and Ionic Impact Phenomena, Oxford University Press (1971).
31. L. I. Gudzenko and S. I. Yakovlenko, Zh. Éksp. Teor. Fiz., 59:1863 (1970).
32. L. I. Gudzenko, V. V. Evstigneev, Yu. I. Syts'ko, S. S. Filippov, and S. I. Yakovlenko, Material from the First All-Union Conference "The Thermodynamics of Irreversible Processes and Its Applications," Chernovtsy (1972); Preprint IMP AN SSSR, No. 62 (1971).
33. L. I. Gudzenko, Yu. I. Syts'ko, and S. I. Yakovlenko, Preprint FIAN SSSR, No. 70 (1973).
34. J. McDaniel, Collision Phenomena in Ionized Gases, Wiley (1964).
35. V. V. Ivanov, V. A. Kalinskii, M. V. Tikhomirov, and I. N. Tunitskii, Khim. Vys. Énerg., 3:539 (1969).
36. L. I. Gudzenko, M. V. Nezlin, and S. I. Yakovlenko, Zh. Tekh. Fiz., 43:1931 (1973).
37. R. M. Pixton and G. R. Fowles, Phys. Lett., 29A:654 (1969).
38. C. B. Collins, A. J. Cunningham, and B. W. Johnson, 11th Int. Conf. on Phenomena in Ionized Gases, Prague (1973), contributed papers.
39. S. V. Antipov, M. V. Nezlin, E. N. Snezhkin, and A. S. Trubnikov, Zh. Éksp. Teor. Fiz., 65:1866 (1973).

CHAPTER III

PLASMA LASERS USING TRANSITIONS OF DIATOMIC DISSOCIATING MOLECULES

Introduction

Attention was first brought to the use of a recombination plasma as a laser medium by Gudzenko and Shelepin [1]. Lasers using a recombining plasma were subsequently called plasma lasers. An alternative concept is the gas laser in which lasing takes place during ionization of the gas. It may be said that the fundamental qualitative difference between plasma and gas lasers is that their active media deviate from thermodynamic equilibrium in opposite senses: In a gas laser the electrons are overheated (that is, the temperature T_e of the free electrons in the plasma is greater than the effective equilibrium temperature T_{eq} corresponding to the actual degree of ionization [2]) while in a plasma laser the electrons are supercooled (that is, $T_e < T_{eq}$). This basic difference determines the methods for creating the lasing medium.

Any pumping regime must be matched to the characteristic parameters of the specific laser scheme involved. The activation mechanisms which result in a population inversion produced mainly by a recombination flux over the excited states of atoms, ions, or molecules are most often realized in a weakly ionized plasma with a high density N and a sufficiently low heavy particle temperature T. On the other hand, when there is a low gas temperature ($T \lessapprox 1$ eV) and the medium is dense, the atoms and atomic ions tend to form more complex molecules. This overlap of the regions where a recombination regime can be realized and where molecules can exist serves as a preliminary indication that the plasma laser principle is more effective whem amplification is on transitions between molecular states. In the following we shall discuss schemes for whose realization the achievement of recombination conditions is of decisive importance.

In a figurative sense there is a mutually favorable relationship between the plasma principle and lasers using molecular transitions. In fact, the presence of molecules in an ionized medium speeds up recombination since the plasma electrons then have another important cooling channel (inelastic collisions accompanied by the excitation of vibrational and rotational states of the molecules). In turn, the use of decaying plasmas as laser media has not only made it possible to expand the list of "molecular" lasers but also to construct lasers with record-breaking characteristics. (See the sections in this chapter on the comparison of the model with experiment and on the monohalides of the inert gases, as well as the next section.)

The variety in chemical composition and structural features of molecules makes it reasonable to expect lasing in any part of the spectrum from microwave to vacuum ultraviolet wavelengths. An equally important advantage of molecules for lasing on an electronic transition, for example, is that their internal degrees of freedom greatly increase the statistical weight of a given electronic state. At the characteristic densities of a recombination regime this may ensure a very high absolute population inversion unattainable with atoms. In enumerating the general advantages of molecular active media over atomic, we must also note that in many cases the natural linewidth of a molecular transition is several orders of magnitude greater than the corresponding atomic width. Although this reduces the gain coefficient and, therefore, makes more powerful pumping sources necessary, there is the possibility of tuning the frequency of the laser output over fairly wide limits and of producing ultrashort pulses.

At present the group of molecular species on whose transitions (especially electronic) lasing has been achieved or attempted is limited, with rare exceptions, to diatomic particles. A fundamental reason for this is the objective complexity of excitation relaxation and plasma chemical kinetics in mutiatomic molecules. For many diatomic (as opposed to multiatomic) molecules substantial quantitative data have already been accumulated on term structures, radiative transition probabilities, and the rates of elementary collisional processes. The following discussion of so-called dissociating molecules as active media for lasers is also conveniently based on the example of diatomic molecules although the specifics of this type of molecule is not determined by the number of chemically bound atoms.

We shall take the basic parameter characterizing a transition between two electronic states of a diatomic molecule to be the ratio of the dissociation energies of the upper (D_2) and lower (D_1) combination terms. Then the set of molecular electronic transitions may be arbitrarily divided into three classes: (a) $D_2/D_1 \ll 1$ (Fig. 1a), (b) $D_2/D_1 \gg 1$ (Fig. 1b), and (c) $D_2/D_1 \sim 1$ (Fig. 1c). This classification may be extended to molecules as well if we consider (for definiteness) transitions to the ground state (1). Then we must assign the alkali metal halides and molecules of monovalent thallium to class (a) as their dissociation en-

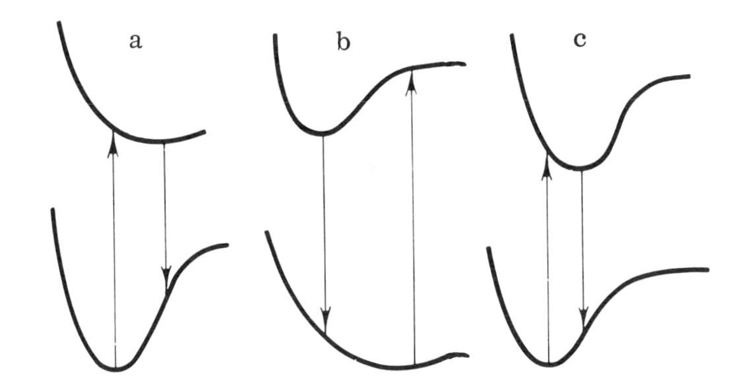

Fig. 1. The classes of radiative transitions in diatomic molecules.

ergies tend to be reduced significantly by electronic excitation. Molecules whose transitions from bound electronically excited states to a repulsive ground state belong to class (b) are referred to as dissociating molecules. They include the dimers of mercury (Hg_2) and the inert gases as well as compounds of the inert gases (LiXe, ArO, etc.).

Various processes in class (a) and (b) molecules take place in different ways. As an example, if a molecule XY belongs to class (a) (abbreviated $XY \in a$) then the effective rate of chemical processes of the type $X + Y \rightarrow XY$ into the ground state is much greater than into the excited states. In class (b) the situation is the opposite. Second, when creating an active medium in the case $XY \in a$, it is natural to consider an inverted population of the atomic levels of X and Y controlled by depopulation of the lower levels by entry of the atoms into a chemical reaction. If, on the other hand, $XY \in b$, it is necessary to use a population inversion over molecular states since the lower term of XY decays rapidly. Third, when excited $XY \in$ a molecules are concentrated in a liquid medium a liquid-to-gas phase transition is possible. When $XY \in b$ a gas-to-liquid or a liquid-to-crystal transition is possible.

Here we shall analyze fundamental questions about lasing in a gas containing dissociating molecules from the standpoint of the recombination principle. Lasing on dissociating molecules is of current interest for the following reasons: (a) the closeness of the upper working level to the ionization energy for such molecules makes it possible to expect high output energies per unit volume, high laser powers, and high efficiencies; (2) the working transition for many dissociating molecules lies in the ultraviolet; and (3) the unusually wide line (for an active medium) of the corresponding spontaneous transition makes it possible to obtain quasi-monochromatic radiation with a tunable frequency.

The first part of this chapter is devoted to a discussion of the mechanisms and kinetics of relaxation of a dense, weakly ionized plasma leading to a population inversion. The balance of radiation at the working frequency is examined. Model calculations of the parameters of the time regimes for pumping are made. The results of experimental work are analyzed briefly. In the second part the most important properties of a laser scheme using dissociating molecules are used in a qualitative attempt to optimize the choice of combining molecular electronic states. This is then used to propose a number of promising chemical compounds for lasing. The results of some preliminary experiments studying the resonance fluorescence spectra of these compounds are reported.

Historical Background

The idea of using dipole transitions of molecules from bound electronically excited states into dissociating (photodissociation) states to obtain stimulated emission was discussed many years ago by F. Houtermans [3]. He obtained (starting with the technically possible pumping sources of that time) a gain coefficient for the $a^3\Sigma_g^+ \to b^3\Sigma_u^+$ transition of molecular hydrogen H_2 that was too small, $\varkappa \sim 10^{-8}$ cm^{-1}. Houtermans first pointed out that the great difficulty in obtaining effective amplification is due to the large linewidth of the photodissociating transition and also proposed using a gaseous discharge plasma of inert gases and elements from the II B group (diatomic molecules of which are stable only in electronically excited states) as possible active media.

The first experimental attempts to obtain lasing on the $a\,(^31_u{}^3O_u^-) \to X^1\Sigma_g^+$ transition of the mercury Hg_2 molecule in a pulsed arc were unsuccessful [4-6]. Some attempts to excite hydrogen under similar conditions also failed [7]. The way to success was discovered after reports appeared on the development of directionality [8] and narrowing of the luminescence spectrum line [9] of liquid xenon irradiated with a relativistic electron beam.

The use of a high-current electron beam injected into a dense gas as a pumping source led to lasing in the vacuum ultraviolet on the $^3\Sigma_u^+ \to X^1\Sigma_g^+$ transitions of the xenon [10-25], krypton [17], and argon [11, 21] molecules at wavelengths of 173, 143, and 126 nm, respectively. It should be noted that this method has been used to produce the most powerful laser at wavelengths shorter than the red [22] and to construct a tunable laser [24].

The theoretical discussion of lasing on dissociating molecules has mainly been devoted to producing a sufficient population in the upper working state in the ionization regime [26, 27]. The first laser scheme based on the recombination ideology was [28]. A direct experimental verification of the nature of the disequilibrium in the lasing medium has not been reported in any of the published work on this type of lasing. Schemes have also been proposed for single-photon [29-31] and multiphoton [32] populating of the upper working level of dissociating molecules. The search continues for other laser-active compounds with repulsive ground terms.

Production of a Population Inversion in a Dense Plasma of Pure Inert Gases

The Term Structure of Dissociating Molecules

It is expedient to begin our analysis of the basic features of lasing on transitions of dissociating molecules with the simplest and most typical case of pure inert gases. All of the inert elements form very unstable diatomic molecules in the ground electronic state $R_2(1)^1\Sigma_g^+$ and have potential curves characterized by a small van der Waals minimum at relatively large internuclear distances r. As the internuclear distance is reduced the growth in the repulsive force between R atoms is faster the lower the atomic number of the element. The lowest electronically excited states of the molecules, $R_2(2)^3\Sigma_u^+$, are bound and as the nuclei are moved infinitely apart they correlate with the metastable 2^3S atomic state of helium and the m^3P_2 $(3 \leq m \leq 6)$† state of the other inert elements. The next bound electronically excited state $^1\Sigma_u^+$ correlates with the lowest resonance‡ level of the atoms and, as a rule, has a potential energy curve [33-35] with a structure similar to the $^3\Sigma_u^+$ state. The energy splitting between the $^3\Sigma_u^+$ and $^1\Sigma_u^+$ terms is much less than the first excitation energy of the atom, $E_2 \approx {}^3/_4 J$, where J is the ionization energy of the atom. A term diagram for the lowest states of a dimer molecule R_2 and the molecular ion R_2^+ is shown in Fig. 2. Other dissociating molecules XY with unlike nuclei also have similar curves [35].

The dipole radiative transitions $^3\Sigma_u^+ \to X^1\Sigma_g^+$ are forbidden by the selection rules. Although the prohibition becomes weaker with increasing atomic number, the opinion exists [36-

† Radon and its chemical derivatives are evidently of no practical interest.

‡ Except for He, in which the 2^1S level is also metastable.

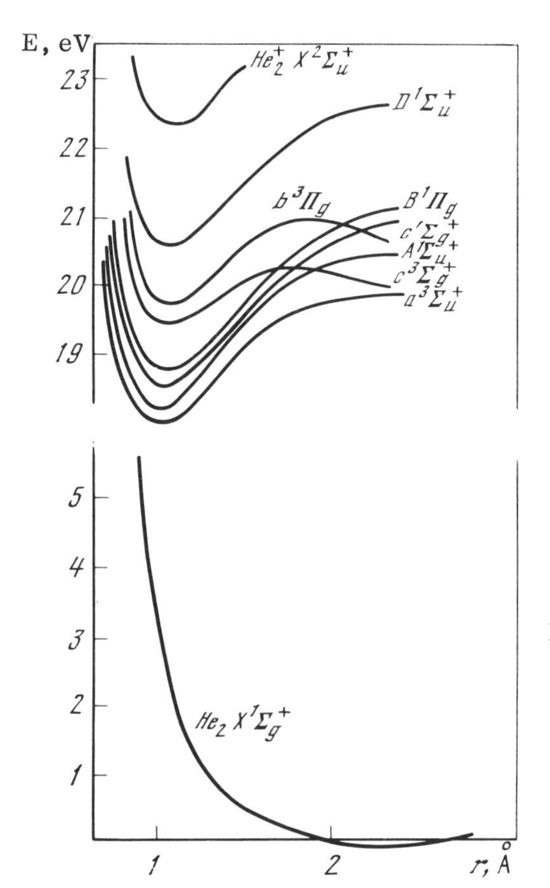

Fig. 2. Potential curves for several states of the He_2 molecule.

39], and is apparently confirmed by experiment [11, 17, 18, 21], that for a high density of particles in an active medium a radiative transition into the ground dissociating state takes place from both the lower electronically excited states, the metastable $^3\Sigma_u^+$ and "resonance" $^1\Sigma_u^+$ state, with an average effective probability $A \approx (A_r N_r + A_M N_M)/(N_r + N_M)$, where A is the Einstein coefficient for a spontaneous radiative transition and N is the population.† (The indices r and m refer to the resonance and metastable states, respectively.) Taking mixing of the energetically close $^3\Sigma_u^+$ and $^1\Sigma_u^+$ states in a dense plasma into account, we shall assume in the following that the working transition is between a generalized electronically excited state $R_2(2, n, l)$ of the molecule and a repulsive ground state $R_2(1, U(r))$:

$$R_2(2, n, l) - \hbar\omega_0 \rightarrow R_2(1, U(r)) \equiv 2R(1) + E_{\text{dissoc}}, \tag{1}$$

where n and l are the vibrational and rotational quantum numbers, ω_0 is the transition frequency, and $E_{\text{dissoc}} \approx U(r)$ is the kinetic energy of dissociation of the atoms R(1), which depends generally on n and l (Fig. 3).

The characteristic decay time of the $R_2(1)$ molecule, $\tau_{\text{dissoc}} \sim |dU/dr\,(r_{02})|^{-1}\sqrt{ME_{\text{eq}}}$, is not more than 10^{-11} sec in order of magnitude. Here M is the mass of the atom; $E_{\text{eq}} \equiv E_{\text{dissoc}}(r_{02})$, where r_{02} is the equilibrium internuclear distance of the $R_2(2)$ term; and $dU/dr(r_{02})$ is the curvature of the ground-state term in the neighborhood of r_{02}. Some characteristics of the R_2

† This structure for the upper working level is due to a number of subtle effects that are of interest in the spectroscopy of the laser output, for example, a small displacement of the spectrum toward longer wavelengths [17, 21] and a linear growth in the rate constants for damping of fluorescence, R_2^* [37], as the gas density is increased.

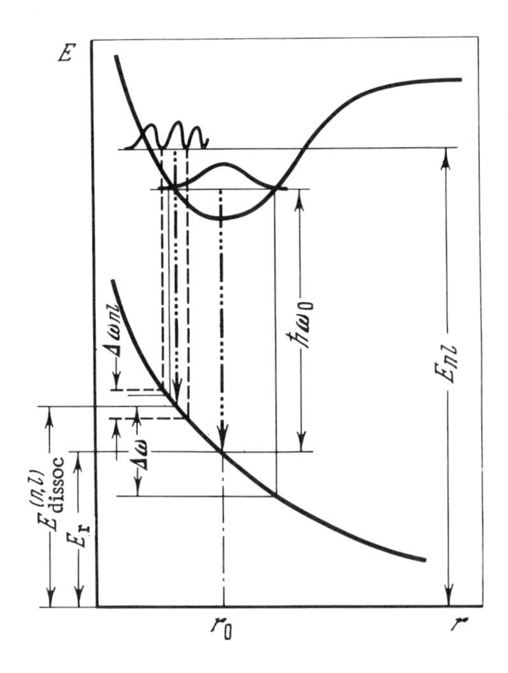

Fig. 3. The structure of the parameters of a photodissociative transition.

TABLE 1

	He	Ne	Ar	Kr	Xe
J, eV	24.59	21.56	15.76	14.00	12.13
		$^3\Sigma_i^+$			
E_{2M}, eV	19.82	16.62	11.55	9.92	8.32
D_{2M}, eV	1.9	—	1.2	0.9	0.6
r_{02}, Å	1.05	2.1	2.6	2.8	3.12
τ_{rad}, sec	10^{-2}	10^{-5}	$3\cdot10^{-6}$	$3\cdot10^{-7}$	$2\cdot10^{-8}$
$\hbar\omega_{vib}$, eV	0.2	0.09	0.036	0.022	0.017
		$^1\Sigma_i^+$			
E_{2P}, eV	20.61†	16.67	11.62	10.03	8.44
D_{2P}, eV	2.3	‡	0.8	‡	0.47
τ_{rad}, sec	$2\cdot10^{-9}$	$3\cdot10^{-9}$	§	§	$4\cdot10^{-9}$
		$^1\Sigma_g^+$			
D_1, cm^{-1}	24	30	77	138	190
r_{01}, Å	2.8	3.1	3.7	4.0	4.42
E_d, eV	3.0	—	0.78	0.63	0.54
λ_0, mm	—	—	126	147	173
$\Delta\omega$, eV	—	—	0.5	0.55	0.65
$\sigma_{pd}\cdot10^{18}$, cm^2	—	—	0.04	0.5	4.0

† See second footnote on p. 63.

‡ It is assumed [33-34] that $D_{^3\Sigma_i^+} \approx D_{^1\Sigma_i^+}$. To the same degree of accuracy the equation holds for r_0 and $\hbar\omega_{vib}$.

§ The radiative lifetimes of $^1\Sigma_i^+$ and the correlated atomic level are close.

molecule that are important for the following discussion are collected in Table 1 which was compiled from data in a large number of original papers.

Basic Mechanisms for Relaxation of the Plasma

Following [40] we shall examine the case of a pure single-component (chemically) gas. Let an auxiliary source create a weakly ionized plasma which contains practically no multiply ionized ions. We assume the medium to be spatially uniform and of sufficient extent that the transport of particles out of the active zone may be neglected. It is known that as the density of the medium is increased the concentration of multiatomic complexes rises. Here we shall limit ourselves to processes which involve species with no more than two atoms.

In a dense recombining plasma two electron sinks are important: (a) over the excited states of atoms, and (b) over the excited states of molecules. It is important that both channels populate the upper working level $R_2(2)$.

The first relaxation mechanism appears as follows. Due to three-body recombination of atomic ions,†

$$R^+ + e + Z \rightarrow R(m) + Z \tag{2}$$

atoms $R(m)$ in highly excited states are formed. A flow over the levels is ensured by electron impacts of the second kind, ‡

$$R(m') + e \rightarrow R(m) + e, \qquad E_{m'} > E_m. \tag{3}$$

Relaxation over the excited states of the atom continues all the way to the first excited state $R(2)$. The energy of this state is sufficiently great that deexcitation by electrons,

$$R(2) + e \rightarrow R(1) + e, \tag{3a}$$

is unlikely. Since in a dense gas radiation on the $R(2) \rightarrow R(1)$ transitions is forbidden, relaxation over the excited atomic states "takes a turn" toward formation of excited molecules $R_2(2)$; collisional association,

$$R(2) + R(1) + R(1) \rightarrow R_2(2) + R(1), \tag{4}$$

takes place. Relaxation over the excited states of the molecule $R_2(m)$, i.e.,

$$R_2(m') + e \rightarrow R_2(m) + e, \tag{3b}$$

begins with formation of a molecular ion R_2^+ due to conversion of an atomic ion,

$$R^+ + R(1) + R(1) \rightarrow R_2^+ + R(1). \tag{5}$$

The atomic and molecular relaxation channels are mixed, for example, during collisional association and dissociation,

$$R(m) + R(1) + R(1) \rightleftarrows R_2(m) + R(1) \tag{6}$$

† This process is most efficient if the third particle Z is an electron. The excited atoms $R(m)$ appear not during the course of recombination but also as a result of direct excitation of atoms $R(1)$.

‡ The ratio of the rates of the direct (3) and inverse processes is determined by the factor $\exp\{-|E_{m'}-E_m|/T_e\}$, where T_e is the free electron temperature.

or when a molecule is broken up by electron impact, in particular, in a transition to one of the highly excited dissociating terms,

$$R_2(m') + e \rightarrow R(m) + R(1) + e. \tag{7}$$

In a number of cases the ratio of the relaxation fluxes is controlled by dissociative recombination,

$$R_2^+ + e \rightarrow R(m) + R(1), \tag{8}$$

so that at a certain stage the main flow of electrons shifts from the molecular channel to the atomic. The remaining portion of the flow over the excited states of the $R_2(m)$ molecules also ends up in the $R_2(2)$ state without appreciable further losses.†

We now consider the processes which disrupt the upper working level. One of them is spontaneous photodissociation (1) into a dissociating ground state. We shall refer to the situation in which each act of ionization (and excitation) of an R(1) atom is accompanied by formation of an $R_2(2)$ molecule and followed by its successive radiative decay as ideal. This case corresponds to the most complete transfer of the energy of the recombining plasma into laser radiation. The real situation usually differs from the ideal in that the decay of the upper working level may not only be radiative. Deexcitation by an electron,

$$R_2(2) + e \rightarrow R_2(1) + e, \tag{7'}$$

is unlikely if the excited state is sufficiently far from the ground state, as is process (3a). Nevertheless, its contribution cannot be neglected at high free-electron densities. Perhaps the most important losses of excited particles are in reactions which take place during inelastic collisions of two excited neutrals. Because of these reactions one of the particles is deactivated to the ground state while the other goes into a more excited state or into the continuum. In the case of the inert gases, for which $2(E_2 - D_2) > J$, the most important reaction of this type is a variant of the Penning effect (ionization during collisions of excited molecules),

$$R_2(2) + R_2(2) \rightarrow \begin{cases} R_2^+ + 2R(1) + e, & (9) \\ R^+ + 3R(1) + e. & (9a) \end{cases}$$

Summarizing the above remarks we may apply the following model: In practice the recombination flux passes through the upper working level of the molecule $R_2(2)$, which is destroyed due to radiative decay (1) or processes (3c) and (9). This relaxation scheme for the plasma has the important advantage of recombination filling of the $R_2(2)$ level over an ionization regime. In fact, in an ionization regime the flow, directed upward over the excited states of atoms and molecules, "sticks" badly at the states R(2) and $R_2(2)$ which lie near the continuum; that is, there is an additional mechanism for their becoming depopulated due to further excitation and ionization.

† These losses may be due primarily to radiative decay of highly excited molecules into the ground state. Considering the convergence of the sum of the oscillator strengths, the high rate of collisional processes of the type (3b), and the relatively low population of the states $R_2(m)$ for $m \geq 3$, these losses may be neglected. For the atomic channel this loss is still less important because of the almost complete reabsorption of the resonance emission of the atoms in a dense medium.

The Conditions for Lasing

In the laser radiation, field-stimulated processes are important, generally affecting the population kinetics of the particles in the active medium. However, at least at the stage when this field is being formed and its amplitude is still small, the role of stimulated (induced) photoprocesses in the population balance of the upper working level may be neglected. In an established lasing regime some of these processes, such as induced photodissociation [the inverse reaction of (1)], which is accompanied by the formation of an $R_2(2, n, l)$ molecule due to three-body collisions of two R(1) atoms and a photon, and photoionization of excited atoms and molecules, control the quantum yield at the working frequency and the lasing efficiency.

The problem of light amplification on photodissociating transitions including the effect of photoassociation was discussed in [41] where

$$\varkappa = \sum_{n,\, l} \sigma_{pd}(n, l; E_{dissoc}) \left[M_{2nl} - (2l+1) \left(\frac{2\pi\hbar^2}{\mu T} \right)^{3/2} \exp\{-(E_{nl} - \hbar\omega_0)/T\} N_1^2 \right] \tag{10}$$

was derived for the gain coefficient $\varkappa$. Here $\sigma_{pd}(n, l; E_{dissoc})$ is the cross section for photodissociation in processes (1), E_{nl} is the energy and M_{2nl} is the population of the $R_2(2, n, l)$ state, $\mu = M/2$ is the reduced mass of the molecule, N_1 is the ground-state population of the atom R(1), and T is the translational temperature of the gas particles. In Eq. (10) the first term in the square brackets describes the photodissociative processes and the second term, the photoassociative processes.

Assuming the excited state $R_2(2)$ to be more populated, we shall consider that the losses of output radiation due to photoionization are controlled by the processes

$$R_2(2) + \hbar\omega_0 \rightarrow \begin{cases} R_2^+ + e, \\ R^+ + R(1) + e. \end{cases} \tag{11}$$

The absorption of light by these processes could be simply accounted for by putting the factor $(1 - \sigma_{pi}/\sigma_{pd})$, where $\sigma_{pi}(n, l)$ is the effective cross section for (11), in front of the first term in the square brackets of Eq. (10). It is clear that lasing can occur only if $\sigma_{pi} < \sigma_{pd}$, which, as a rule, holds with some margin.† In the following we shall set $\sigma_{pi} \ll \sigma_{pd}$ and neglect the contribution of photoionization to the laser radiation balance.

We now consider the role of photoassociation. We assume that process (4) dominates in the kinetics of formation of $R_2(2)$ molecules. Then the maximum allowed population of the $R_2(2, n, l)$ state is given by the Saha–Boltzmann equation

$$M_{2nl} = (2l+1) \left(\frac{2\pi\hbar^2}{\mu T} \right)^{3/2} \exp\left\{ \frac{E_2 - E_{nl}}{T} \right\} N_2 N_1, \tag{12}$$

where N_2 is the population of the R(2) atomic level. We note that for this to be true it is necessary that the rate of vibrational relaxation of the $R_2(2)$ state should be much greater than the rate of all other processes of formation and disintegration of molecules. This evidently can be so for xenon (with a rate coefficient for vibrational relaxation $k_{vib} \approx 6 \cdot 10^{-11}$ cm$^3 \cdot$ sec^{-1} [42] at T ≈ 300°K), but is clearly impossible for helium (with $k_{vib} \approx 6 \cdot 10^{-16}$ cm$^3 \cdot$ sec^{-1} [43]). Substituting Eq. (12) in (10), we obtain

$$\varkappa = \left(N_2 - N_1 \exp\left\{ -\frac{\Delta E}{T} \right\} \right) N_1 \left(\frac{2\pi\hbar^2}{\mu T} \right)^{3/2} \sum_{n,\, l} \sigma_{pd}(n, l; E_{dissoc})(2l+1), \tag{13}$$

† This is because the wave function of the resulting photoelectron is smeared over a larger energy interval than the wave function of the nuclei.

where $\Delta E \equiv E_2 - \hbar\omega_0$. It follows from this that amplification on photodissociative transitions is possible only for sufficiently low heavy-particle temperatures,

$$T < T_{cr}, \quad T_{cr} = \Delta E / \ln (N_1/N_2). \tag{14}$$

Since the population of the atomic ground state is usually several orders of magnitude greater than N_2, $T \ll \Delta E$ must hold. From this follows a requirement for a sufficient rate of heat removal if we desire cw operation or that there be a limit on the amount of input energy for pulsed operation. Condition (14) indicates that the decay (dissociation) time of the ground state $R_2(1, E_{dissoc})$ is not small enough compared to the (radiative!) lifetime of the excited $R_2(2)$ state to achieve lasing on the transition (1), as proposed in [3, 26, 27]. The statement in [27] that acts of emission and absorption of photons during mutual passes by molecules may contribute positively to the gain is also incorrect. In fact, as shown in [44], amplification is possible during radiative collisions of atoms if $N_2 > g_2/g_1 N_1$. Since the ratio of statistical weights $g_2/g_1 \sim 1$, this condition, much stricter than (14), can scarcely be satisfied in a dense, weakly ionized, inert-gas plasma.

The cross section for photodissociation may be written in the form [41] $\sigma_{pd}(n, l; E_{dissoc}) = 1/4\lambda AQ_l(n, E_{dissoc})$, where $\lambda_0 = c/2\pi\omega_0$ is the wavelength of the laser light; $A = 4\omega_0^3 d_{12}/3\hbar c^2$ is the Einstein coefficient for spontaneous decay and is related to the matrix element of the transition d_{12} found from the electronic wave functions; and $Q_l(n, E_{dissoc})$ is the square of the overlap integral of the nuclear wave functions. The dependence of Q_l on E_{dissoc} determines the lineshape.† At the line center $Q_l(n) \approx 1/\Delta\omega_{nl}$, where $\Delta\omega_l$ is the width (projected on the dissociating term and then onto the energy axis; cf. Fig. 3) of the wave function of the excited state $R_2(2, n, l)$. Below, in our estimates of the gain coefficient we shall assume that condition (14) is satisfied and use the expression

$$\varkappa \approx {}^1/_4\lambda_0^2 A M_2/\Delta\omega, \tag{15}$$

where $M_2 \equiv \sum_{nl} M_{2nl}$ is the population of the $R_2(2)$ state and $\Delta\omega$ is the effective linewidth.

Minimal heating of the gas is typical of schemes with optical pumping of the upper working state. In [29] the use was proposed of optical pumping of xenon by photoassociation of atoms in the $^1\Sigma_u^+$ state, which quickly relaxes collisionally to the $^3\Sigma_u^+$ state from which the laser transition is expected to originate. This idea is difficult to realize because of the low probability of such a process. Thus, according to the estimates of the authors (of [29]) themselves, to obtain a gain coefficient of $\varkappa \sim 10^{-3}$ cm^{-1} it is necessary to have an absorption coefficient for the photoassociative transition of $\varkappa_{ac} \sim 1$-10 cm^{-1}. At the same time, since the photodissociation cross sections for the $R_2(2)^1\Sigma_u^+ \to R_2(1)^3\Sigma_g^+$ and $R_2(2)^3\Sigma_u^+ \to R_2(1)^1\Sigma_g^+$ transitions differ by roughly an order of magnitude, it follows from Eq. (10) that for this to occur at room temperature the gas density must exceed 10^{22} cm^{-3}. The absorption of a photon by an atom is several orders of magnitude more probable. Thus another way is more realistic: optical pumping of the excited state R(2) of an atom so that later, due to collisional association [reaction (4)], a molecule will be formed [31]. The prevalence of this mechanism may explain the reported gain coefficients $\varkappa \sim 10^{-2}$ cm^{-1} in [30].

The Role of Impurities

Modern techniques can ensure the purity of a gas to no better than 10^{-9}, but usually the fraction φ of impurity (Y) particles in gases considered to be pure is 10^{-4}-10^{-6}. At such im-

†A discussion of the shape of a photodissociative transition line (including photoassociation) is given in F. Mies, Mol. Phys, 26:1233 (1973).

purity concentrations the density of impurity particles is commensurate with the populations of the excited states of atoms and molecules in the working gas. The way the impurities affect the relaxation kinetics of the plasma in a specific case depends on their chemical properties; hence, it is appropriate here to limit ourselves to a consideration of this problem in general outline.

If the impurity is easily ionized ($J^{(Y)} + D_2^{(R)} < E_2^{(R)}$) then the loss of $R_2(2)$ molecules may be substantial. A parasitic channel for relaxation of excitation over the discrete states of the impurity then arises. The effectiveness of this channel is mainly determined by the rates of charge exchange

$$(R)^+ + (Y)(1) \rightarrow (Y)^+ + (R)(1) \tag{16}$$

(here the parentheses signify a particle which may be an atom or molecule) and Penning ionization

$$R_2(2) + (Y)(1) \rightarrow 2R(1) + (Y)^+. \tag{9b}$$

If the energy spectra of the states (R) and (Y) overlap substantially it may be assumed that reactions involving impurity particles lead only to a partial redistribution of the populations of the excited working particles.

The presence of impurities has a much greater effect on the balance of laser radiation. This must be especially included when the laser radiation is in the vacuum ultraviolet. In this spectral region the absorption cross sections of typical impurities, such as the basic components of air, may reach 10^{-17} cm^2 [45], which is usually several times greater than the photodissociation [reaction (1)] cross sections characteristic of inert gas dimers. For amplification it is necessary that $\varkappa > \varkappa_{abs}^{Y}$. This leads to a rather strict limitation

$$M_2 > \frac{\sigma_{abs}^{Y}}{\sigma_{pd}} \varphi N_1, \tag{17}$$

since the photodissociation scheme requires fairly large densities N_1. Thus, photoabsorption by the impurities is an important hazard for lasing on transitions of dissociating R_2 molecules, and the purity of the working gas must be monitored by spectroscopic methods.

In individual cases it is possible to find materials whose presence in the active medium poses no hazard. It is even possible to add them as a buffer gas for intensifying recombination in the plasma. (See the discussion of the steady-state regime in the next section of this chapter.) For example, for xenon (as shown in [16, 22]) such an additive might be the light inert gases (He, Ne, Ar). On the other hand, impurity amounts of xenon are critical for lasing on dimers of argon and krypton [46] since the energy of the lower excited states of the lighter gas is sufficient to ionize (or excite) the ground state of an atom of the heavier gas and the corresponding excitation transfer reactions have very high rates [47]. A buffer gas is also desirable for shifting the density of the medium at which substantial photoassociation occurs toward higher values.

Recombination Mechanisms for Lasing on Transitions of Dissociating Molecules

The Conditions for Obtaining a Supercooled Plasma

To create a supercooled, intensely recombining plasma in which the average energy of the free electrons obeys the inequality $T_e < T_{eq}$, it is necessary to ensure that their distribution differs from Maxwellian in such a way that there is a sufficient (to achieve a given degree of ionization α) density of fast electrons. Thus, it is necessary to point out the conditions for forming such a distribution, in particular the means for injecting energy into a dense medium. The basic problems in calculating the electron relaxation and in theoretically analyzing the prospects for lasing in a supercooled plasma are practically independent of the nature of the auxiliary ionization source. It may be, for example, a beam of charged particles (electrons or ions) injected into the medium, x radiation, or showers of nuclear fission products [12]. Of special interest for a broad range of applications is the transverse (relative to the direction of the laser output) breakdown of a gas by a specially formed pulsed electric field. Breakdown of a gas by a laser pulse that has been focused by cylindrical optics or other methods of optical pumping may be feasible.

We shall briefly describe the formation of the free-electron energy distribution in a dense atomic gas that is being ionized and excited by a hard external source. The slow electrons can be divided into fast (with energies greater than J) and plasma (cooled by collisions with heavy particles) groups. The collisions of the fast electrons with one another and with the plasma electrons may be neglected compared to their inelastic interactions with the atoms. Since the cross sections for Coulomb collisions between cold plasma electrons are large, it is possible to introduce the concept of their temperature T_e ; T_e is close to the relatively low temperature T of the particles of the dense gas, so the plasma electrons recombine intensely.

We shall consider how the parameters of the plasma created by the electron beam are related to the beam characteristics. The populations (N_m) of the atomic and molecular states at different times are given by the balance equations

$$\frac{d(N_m)}{dt} = \sum_{m \leqslant m'} K_{mm'}(N_{m'}) + D_m \equiv \Gamma_m. \tag{18}$$

The matrix K_{mm} is called the relaxation matrix. Its elements give the average number of transitions of a particle per unit time interval from state m' to state m. The diagonal element K_{mm} determines the total loss of particles (per unit time) from state m. The quantity D_m characterizes the influx from the continuum. We shall write the balance equation for the plasma electrons in the form

$$\frac{dN_e}{dt} = -\Gamma_e + \Gamma_i + \nu_i N_1. \tag{19}$$

Here ν_i is the rate of ionization of atoms by beam electrons and fast secondary electrons (usually $\nu_i \gtrsim \sigma_i j/e$, where j is the current density), σ_i is the cross section for ionization by the beam electrons, Γ_e is the total recombination flux over the electron states, and Γ_i is the ionization flux, including all processes which eject electrons into the continuum (including ionization by hot plasma electrons, Penning reactions, etc.) except direct ionization of R(1) atoms by beam electrons or fast secondary electrons. The ionization regime corresponds to the condition $dN_e/dt > 0$; however, already in the stationary state $(dN_e/dt = 0)$ the plasma electrons are supercooled. Actually, in thermal equilibrium, when the degree of ionization and the electron temperature are related by the Saha equation, the first two terms in Eq. (19) add out. From this it is clear that if the third term is nonzero the electron density must be above the equilibrium value.

To evaluate recombination in a nonequilibrium plasma it is convenient to introduce the parameter

$$\xi = \frac{\Gamma_e - \Gamma_i}{\nu_i N_1 + \Gamma_i}. \tag{20}$$

In the steady state it is appropriate to distinguish two limiting cases: dominant ($\xi \to 1$) and weak ($\xi \to 0$) recombination. When the source is rapidly turned off it is possible to have $\xi \gg 1$. It is natural to assume the existence of a correlation between ξ and the efficiency of conversion of pump energy into light.

We now estimate the characteristic times to establish the recombination regime. For this purpose, following [48], we consider the relaxation of the gas (T) and plasma electron (T_e) temperatures starting with the heat balance equations† for a spatially uniform medium together with the steady-state form of Eq. (19):

$$^3/_2 \frac{d}{dt}(N_e T_e) = Q_{\text{inel}} - Q_{\Delta T}, \tag{21}$$

$$^3/_2 N_1 \frac{dT}{dt} = Q_{\Delta T}, \tag{22}$$

$$\frac{dN_e}{dt} = -\Gamma_e + \Gamma_i + \nu_i N_1 = 0. \tag{19a}$$

Here Q_{inel} is the energy released to the electron-plasma component due to inelastic collisions; the quantity

$$Q_{\Delta T} = \frac{3m_e}{M} \nu_{\text{el}} N_e (T_e - T) \tag{23}$$

is the heat exchange between the electrons and the heavy particles in elastic collision, ν_{el} is the elastic collision frequency, and m_e/M is the ratio of the electronic and atomic masses.

In the dominant recombination regime‡ the second term of Eq. (19a) may be assumed small. Then Eqs. (19a) and (21) for N_e and T_e take the form§

$$\begin{aligned} \frac{dN_e}{dt} &= \nu_i N_1 - \Gamma_e = 0, \\ ^3/_2 N_e \frac{dT_e}{dt} &= E_0 \nu_{\text{inel}} N_1 - Q_{\Delta T} = 0, \end{aligned} \tag{24}$$

where E_0 is the mean energy transfer in inelastic collisions of atoms with fast electrons, $\nu_{\text{inel}} = \nu_i + \nu_{\text{ex}}$ is the inelastic collision frequency, and ν_{ex} is the effective excitation rate for R(1) atoms ($\nu_{\text{ex}} \sim \nu_i$). With Eqs. (24) and (22) it is possible to estimate the times for establishing the temperature and density of the electrons as well as the time to heat the gas:

$$\tau_{N_e} \sim \frac{\alpha}{\nu_i}, \quad \tau_{T_e} \sim \tau_{N_e} \frac{T_e}{E_0}, \quad \tau_T \sim \frac{T}{E_0} \nu_i^{-1}. \tag{25}$$

For a degree of ionization $N_e/N_1 = \alpha \ll T_e/E_0$ we have

$$\tau_{T_e} \ll \tau_{N_e} \ll \tau_T \ll \nu_i^{-1}. \tag{26}$$

Usually $T_e/E_0 \sim 10^{-2}$; therefore, the quasistationary approximation (24) is correct for $\alpha \ll 10^{-2}$.

† The variation in N_1 may be neglected. Elastic energy exchange with the electrons is not the only way of heating the gas. Hot atoms are formed, for example, by reactions (1) and (9). However, the role of these processes in the heat balance of a weakly ionized gas is, as a rule, small.

‡ The case $\xi \to 1$ can be realized only for sufficiently low T_e since the main contribution to Γ_i is from electronic collisions of the first kind with excited particles.

§ The possible heating of the plasma electrons by collective interactions with the beam electrons has been neglected here.

In the weak-recombination disequilibrium regime we may regard the third term of Eq. (19a) as small. This term determines a small deviation from equilibrium in the direction of recombination. Equations (25) remain valid if we replace ν_i by Γ_i/N_1. Then the recombination regime is established in two steps: first, after a characteristic time

$$\tau_{N_e} \sim N_e/\Gamma_i \tag{25a}$$

an electron density close to equilibrium is established and, second, after a characteristic time

$$\tau_{\delta N_e} \sim \delta N_e/N_e \ \alpha/\nu_i \tag{25b}$$

the recombination flux is established to compensate for the ionization due to the external source. Here δN_e is the deviation in the electron density from the equilibrium value.

The Steady-State Population Kinetics of the $R_2(2)$ State

At present there are a number of papers (cf. the review [48]) in which the system of balance equations (18), (21), and (22) has been solved for relatively simple models of atomic media. In such models the elements of the relaxation matrix have a fairly simple structure, and the transition probabilities often may be calculated theoretically. Right now in molecular problems it is possible to speak of a semiempirical approach to calculating the population kinetics although for a very limited number of species. A similar calculation was done in [38] for the relaxation of a xenon plasma in its afterglow. The authors included reactions (4), (5), (8), and (9). At the same time, they ignored the basic mechanism for the plasma decay, electronic collisions of the second kind (assuming that the Xe* atoms relax to the $Xe_2(2)$ state radiatively), while including in their considerations the much less probable processes of ionization of excited Xe(2) atoms and $Xe_2(2)$ molecules by the plasma electrons. The temperatures T_e and T were specified as parameters. Despite the obvious inadequacies of their kinetic model, they were able to obtain a result in qualitative agreement with experiment; more precisely, for a certain choice of parameters the $M_2(t)$ curve satisfactorily fit the experimental points for the laser pulse signal. Taking into account the lack of information on the probabilities of elementary processes, in evaluating the prospects for lasing on transitions of dissociating molecules it is appropriate to limit ourselves to the analysis of crude models corresponding to the limiting cases of the pumping regime.

Let us consider the case of quasistationary recombination of a strongly nonequilibrium plasma ($\xi \to 1$). Let the pumping conditions be chosen so that the time hierarchy (26) is satisfied. Then for $\tau_{T_e} < t < \tau_T$ it is possible to regard T_e and T as constants. The balance equations for the plasma electrons and the $R_2(2)$ molecules in the stationary-sink approximation [48] have the form

$$\begin{gathered}-\Gamma_e + \nu_i N_1 + qM_2^2 = 0.\\ \Gamma_e + \nu_{ex} N_1 - 2qM_2^2 - AM_2 = 0.\end{gathered} \tag{27}$$

Here q is the rate coefficient for reaction (9), state 2 is assumed to be energetically far removed from state 1, and processes (3a) and (3b) are neglected. From this we obtain

$$M_2 = \frac{A}{2q}\left[\left(1 + 4q\nu_{inel}\frac{N_1}{A^2}\right)^{1/2} - 1\right]. \tag{28}$$

For $4qN_1\nu_{inel} \ll A^2$ the situation is ideal and $AM_2 \approx \nu_{inel}N_1$. As the pumping intensity ($\nu_{inel}N_1$)

is increased the quantity M_2 varies almost linearly up to some critical value $M_2 \approx A/2q$, and, later, when $4qN_1\nu_{inel} \gg A^2$ the function $M_2(\nu_{inel}N_1)$ goes into a square-root dependence $M_2 \approx (\nu_{inel}N_1/q)^{1/2}$. Thus, the greatest interest is in molecules with high Einstein coefficients A since a reduction in the gain coefficient is difficult to compensate by increasing the population (due to a reduced pumping efficiency which appears sooner the smaller A is).

There is a yet more serious reason for limiting the pumping intensity from above. As can be seen from Eqs. (23) and (25), as ν_i increases τ_T decreases. This may lead to a cutoff of lasing due to overheating of the gas. Cooling times for the heavy particles $\tau_{cool} \lessgtr 10^{-6}$ sec cannot be realistically assured. This forces us to settle for quasistationary lasing in pulses of duration $\tau_{N_e} < \tau < \tau_T$.

We note that the conditions in this model may correspond to some extent to the case of lasing on argon molecules [21]: In the pressure range 30-70 atm a linear dependence of the peak laser output signal on N_1 has been observed. However, the data are not adequate for a more detailed comparison of experiment with the theory. In a xenon plasma (as shown by a preliminary calculation [40]) and possibly in a krypton plasma a weak recombination nonequilibrium develops.† This is due to the fact that the quasistationary divergence between the electron and gas temperatures ($T_e - T$) is in fact linearly dependent on the atomic mass of the element. Thus, the addition of a light buffer gas should speed up cooling of the electrons in elastic collisions. Experiments with xenon–neon and xenon–argon mixtures [22] show that roughly the same laser output is obtained from mixtures with lower densities of working gas atoms than in the case of pure xenon. When xenon is doped with helium [16] at partial density ratios of up to 10:1 of He and Xe the efficiency of conversion of beam energy into radiation is at least not lowered and the peak laser signal (for fixed equivalent pressure) in helium–xenon mixtures increases by more than twice compared with pure xenon. These results favor the recombination regime for lasing on photodissociative transitions of xenon dimers.

Pulsed Lasing

We now study the conditions for lasing on the photodissociative transition (1) in the afterglow of an ionization pulse with a short fall time. In accordance with the above discussion of the plasma relaxation mechanisms we shall write the balance equations for the electron (N_2) and excited (M_2) molecule $R_2(2)$ densities in the form

$$\frac{dN_e}{dt} = -k_r M_e + qM_2^2,$$
$$\frac{dM_2}{dt} = k_r N_e - 2qM_2^2 - AM_2 - V_{12}N_e M_2. \tag{29}$$

Here $k_r = \Gamma_e/N_e$ is a coefficient which characterizes the rate of recombination of electrons with atomic and molecular ions and is a complicated function of N_e, T_e, N_1, and T, and V_{12} is the rate coefficient of reaction (3b). In general these equations must be supplemented by the heat-balance equations (21) and (22). We assume that the degree of ionization of the medium is small ($\alpha \lessgtr 10^{-3}$) and that the energy supply is cut off sufficiently rapidly (more specifics below). Then the electrons can be cooled much more rapidly than the recombination flux (corresponding to the moment the pumping pulse begins to fall) can change, and therefore during the afterglow of the laser pulse T_e may be regarded as constant. We shall also assume that $T_{cr} > T \approx$ const.

Rapid cooling of the electrons leads to intensification of the recombination flux and a change in the initial ratio of N_e to M_2. It follows from the above plasma-relaxation scheme that

† It is clear that from the point of view of the efficiency with which the source energy is converted a weakly nonequilibrium plasma is not the ideal active medium.

the $R_2(2)$ state is usually a bottleneck for the recombination flux as in the absence of stimulated emission the element $K_{12}^{R_2}$ of the relaxation matrix is less than any of $(K_{mm'})^{R_1R_2}$ ($m \geq 2$, $m' > 2$). Thus, for sufficiently high initial densities $N_e(0) + M_2(0)$, $N_e(t)$ follows $M_2(t)$ in a quasistationary manner a short time $\Delta\tau$ following turnoff of the pumping pulse; that is, we may set $dN_e/dt \approx 0$ in Eq. (29) and consider the equations

$$\begin{aligned} dM_2/dt &= -qM_2^2 - AM_2 - V_{12}N_eM_2, \\ k_rN_e &= qM_2^2. \end{aligned} \tag{30}$$

To solve the nonstationary equation (30) we must refine the form of $k_r(N_e)$, which, besides depending on the plasma parameters, is determined by the structure of the energy terms of the atoms and molecules of the chemical element R. At high densities N_1 and low degrees of ionization the ion component of the plasma is mainly made up of R_2^+ particles which recombine with the electrons mainly (in the case of the inert gases) through the dissociative mechanism (8). Thus, we set $k_r = \beta_D N_e$ (where β_D is the effective dissociative recombination coefficient) in Eq. (30). After the substitution we obtain

$$dM_2/dt = -\hat{q}M_2^2 - AM_2, \tag{30a}$$

where $\hat{q} = q + V_{12}(q/\beta_D)^{1/2}$ (usually $q \gtrless V_{12}(q/\beta_D)^{1/2}$). The solution of Eq. (30a) has the form

$$M_2(t) = M_{20}e^{-At}\left[1 + \frac{\hat{q}}{A}M_{20}(1 - e^{-At})\right]^{-1}, \tag{31}$$

where† $M_{20} \approx N_e(0) + M_2(0)$. Then the characteristic time $\tau_2(t)$ increases monotonically:

$$\tau_2(t) \approx \left|\frac{1}{M_2}\frac{dM_2}{dt}\right|^{-1} = (\hat{q}M_2 + A)^{-1} = \begin{cases} (\hat{q}M_{20} + A)^{-1} & \text{for } t \to 0, \\ A^{-1} & \text{for } t \to \infty. \end{cases} \tag{32}$$

We now make some estimates. Let a part δW of the energy W delivered to the gas be spent in ionizing and exciting the particles of the medium. Then $M_{20} \approx \delta W/JSL$, where J is the ionization energy of an R atom, S is the perpendicular cross section, and L is length of the active zone. For amplification of light in a single transit of the resonator at $t \approx 0$ we have

$$I/I_0 = (1 - \gamma)\exp\{\varkappa L\} = (1 - \gamma)\exp\left\{\sigma_{pd}\frac{\delta W}{JS}\right\}. \tag{33}$$

Here I is the intensity of the radiation and γ is the coefficient of lumped energy losses in the resonator. The condition for self-excitation, $I/I_0 > 1$, therefore takes the form of a lower bound on the pump energy delivered over the perpendicular cross section, i.e.,

$$\frac{\delta W}{S} > U_p, \quad U_p = |\ln(1 - \gamma)|\frac{J}{\sigma_{pd}}. \tag{34}$$

An upper bound on the amount of energy input follows from Eq. (14):

$$\frac{(1 - \delta)W}{SL} < \frac{3}{2}N_1T_{cr}. \tag{35}$$

† The time is reckoned from a time $\Delta\tau$ after the pumping pulse begins to drop. For the quasistationary approximation (30) to apply it is necessary that $\tau_{N_e} \ll \tau_2$ and, therefore, that $N_{e0} \ll M_{20}$.

In combination with Eq. (34) we have

$$\frac{U_p}{\delta} < \frac{W}{S} < \frac{L}{1-\delta} N_1 T_{cr}, \quad L > \frac{1-\delta}{\delta} \frac{U_p}{N_1 T_{cr}}. \tag{36}$$

Equations (34)-(36) demonstrate the expediency of going to longer cavity lengths L so as to ensure a minimal input energy density U_p.

The condition for rapid cooling of the electrons, $\tau_{T_e} < \tau_2$, reduces to a limit on the density of the gas,

$$N_1 > \left(\hat{q} \frac{U_p}{JL} + A\right) \left(\frac{3m_e}{M} \langle \sigma_{el} v \rangle\right)^{-1}, \tag{37}$$

where $\langle \sigma_{el} v \rangle = \nu_{el}/N_1$ is the product of the cross section for elastic collisions between electrons and R atoms and the velocity averaged over a Maxwellian distribution.

In accordance with the above assumptions, the characteristic fall time (τ_f) of the pump pulse must at least satisfy the condition

$$\tau_f < \tau_2 \approx \left(\hat{q} \frac{U_p}{JL} + A\right)^{-1}, \tag{38}$$

that is, for any length of active medium the time τ_f must be less than A^{-1}. We note that condition (38) is easily satisfied. A more rigorous analysis of the limitations on the duration of the trailing edge of the pumping pulse in [49] did not lead to significantly more rigid requirements.

Comparison of the Model with Experiments

Despite the fact that no direct measurement was made of the characteristics of the active medium (such as N_e, T_e, T, and M_2) in the experiments [10-25] we have cited and the gain coefficient $\varkappa$ was not always evaluated (and if so, only approximately), several observed features of lasing on inert gas dimers confirm the basic properties (at least do not contradict them) of our simple scheme for lasing on electronic transitions of dissociating molecules to the ground state. This applies both to the possibility of lasers using photodissociating transitions and to the nature of the plasma disequilibrium under which effective lasing occurs. We have already noted the construction of a tunable laser [24] in the wavelength range 1690-1740 Å ($\Delta\hbar\omega_0 \approx 1500$ cm^{-1}) and the achievement of record high laser output powers at those wavelengths (75 MW [22]). We now consider the features of the recombination laser regime in more detail.

In [10-12] lasing was observed in xenon in the afterglow of a beam pulse ($W_{el} \sim 2$ MeV, $j \sim 1.5$ kA/cm^2, $\tau_{pulse} \sim 2$ nsec) with a very short fall time, $\tau_f \lessdot 1$ nsec. The observed dependence of the time delay of the output pulse peak ($\approx\Delta\tau$, see the discussion preceding Eq. (30)] on the gas density may be written in the form $\Delta\tau \sim (N_1)^{-\varepsilon}$, where $\varepsilon \approx 0.7$. This means that the delay is due to cooling and recombination of the plasma electrons. Indeed, from Eqs. (23), (25), and (37) it follows that

$$\tau_{N_e} \propto \tau_{T_e} \approx \left(\frac{3m_e}{M} \langle \sigma_{el} v \rangle N_1 \frac{T_e - T}{T_e}\right)^{-1}.$$

For $T \ll T_e$, if $\Delta\tau$ is proportional to τ_{T_e} or τ_{N_e} then the dependence $\Delta\tau \propto (N_1)^{-\varepsilon}$, where $\varepsilon \approx 1$, holds. In an ionization regime a possible delay could be due to loss of R(2) atoms in chemical reaction (4), but then we should expect another dependence $\Delta\tau\,(N_1)$ of the form

$$\Delta\tau \propto \tau_{R(2)} \propto (N_1)^{-2}.$$

We now estimate the characteristic times $\tau_{el} \sim [(3m_e/M)k\sigma_{el} v > n_1]^{-1}$ and τ_R (2) ~ $(k_2N_1^2)^{-2}$, where $k_2 \approx 3 \cdot 10^{-32}$ cm$^6 \cdot$ sec^{-1} [17,25] is the rate coefficient of process (4) at a xenon pressure of 115 psi ≈ 8 atm ($N_1 \approx 2.4 \cdot 10^{20}$ cm^{-3}). We use a value of $\langle\sigma_{el} v\rangle$ in the neighborhood of the Ramsauer minimum cross section σ_{el} [50], i.e., $\langle\sigma_{el}v\rangle$ – $5.4 \cdot 10^{-8}$ cm$^3 \cdot$ sec^{-1}, for $T_e \approx 0.6$ eV. We obtain $\tau_{T_e} \sim 6 \cdot 10^{-9}$ sec; $\tau_{R(2)} \approx 5 \cdot 10^{-10}$ sec. The observed value of $\Delta\tau$ at 115 psi [10] is 4.5 nsec, which is of the same order of magnitude as $\tau_{R(2)}$ and is much greater than the characteristic time for accumulation of excited $R_2(2)$ molecules which would correspond to the delay of lasing in an ionization regime.

The results of [13-16] are representative. When a dense gas (up to 25 atm) is ionized by a beam with an electron energy $W_{el} \sim 1.5$ MeV, a pulse duration $\tau_{pulse} \sim 50$ nsec, and a current density j ~ 200-300 A/cm^2 [14, 16], the medium begins to lase shortly before the beam is shut off. The maximum gain occurs during the afterglow, and the maximum spontaneous emission appears later. To achieve maximum energy output the current density of the beam was increased to 800 A/cm^2 [15]. The time variation in the intensity of stimulated emission obtained in this case is quite different: Lasing is cut off as the beam current is rising. This is explained by overheating of the gas due to delivery of a large specific energy and the consequent violation of criterion (14), and confirmed by the estimates of the time to heat the gas. Ault et al. [20] used a beam with the parameters $W_{el} \sim 700$ keV, j ~ 3 kA/cm^2, and $\tau_{pulse} \sim 60$ nsec and also obtained a very fast laser cutoff ($\lesssim 10$ nsec). After reducing the current density to 300 A/cm^2 on the same device [22] they achieved quasicontinuous lasing with a duration of 90 nsec. These observations cannot be explained by assuming that the upper working level is pumped by an ionization mechanism.

From this scheme for a plasma laser using photodissociating transitions it follows that the more strongly the recombination disequilibrium of the system is expressed the greater the fraction of pump energy that is converted to radiation (spontaneous and stimulated). This property is observed in analyses of experimental data. Thus, with quasistationary pumping of pure xenon the total luminous yield is 7% (6% + 1%) [15, 20], for helium–xenon it is $\gtrsim$10% [16], for argon it is $\gtrsim$ 10% [21], and in pulsed operation for xenon it is about 25% [10]. We recall that the parameter ξ changes in a similar way.

Thus, to construct plasma lasers using transitions of dissociating molecules it is necessary to satisfy the following relatively flexible conditions which, however, are unusual from the standpoint of traditional gas lasers.

Electron cooling must take place sufficiently rapidly. This reduces to a requirement of high gas density which, however, cannot be increased indefinitely in view of Eq. (10):

$$N_{thr} < N_1 < N_{cr}. \qquad (39)$$

The source energy expended in ionization and excitation of the gas must be sufficient to compensate for primarily the loss of radiation. An excessive concentration of input energy may cause overheating of the gas and termination of lasing:

$$W_{thr} < W < W_{cr}. \qquad (40)$$

For lasing in the afterglow the pumping-pulse fall time must be short and the duration of the entire pump pulse must not exceed the characteristic time for heating of the gas.

Inert Gas Hydride Molecules as the Working Component of an Active Medium

Elements of the Structure of Molecular Terms – the Most Critical for Lasing on Photodissociative Transitions

Table 1 lists the characteristics of the inert gases that will be important for the discussion to follow. Listed there are J, the ionization potential of the atom; E_{2M} and E_{2R}, the excitation energies of the lowest atomic levels; D, the dissociation energy of the state denoted by a subscript; r_0, the equilibrium internuclear distance; τ_{rad}, the radiative lifetime; $E \approx U_1(r_{02})$, the average energy of dissociation of the ground state of the molecule; $\hbar\omega_{vib}$, the elementary vibrational quantum; λ_0 the laser wavelength; $\Delta\omega$, the effective linewidth; and σ_{pd}, the effective cross section for the photodissociative transition.

The difficulty of lasing in the inert gases is due to causes closely related to the structure of the $R_2(2)$ molecule. First, the small photodissociation cross sections $\sigma_{pd} \approx 10^{-18}$ cm^2 compel use of fairly high values of W_{thr} and N_{thr}. Second, the dissociation energy E_d of the atoms, especially among the heavy inert gases, is not so small that we can neglect the danger of overheating the gas; it sharply limits W_{cr} and N_{cr}. It should also be noted that in the short-wavelength working radiation of the R_2 dimers, there is typically high absorption by impurities, by optical materials, and by the atmosphere. Thus the interest in evaluating other molecular transitions from class (b) is fully understandable. (See the Introduction.) We recall, for example, the attempts at a laser using mercury Hg_2 molecules [4]. Later [51] mercury vapor in argon was excited by a high-power electron beam; however, absorption was observed in the frequency range of the working transition. Without analyzing the reasons for the failure of this attempt, we point out that in view of conditions (14) and (39) photodissociation lasing media which much be prepared from high-boiling-point materials are not suitable.

The low temperature T_{cr} is a less favorable factor for dissociating-molecule schemes than the small photodissociation cross sections since the expression for the photodissociation coefficient contains the temperature T of the heavy particles in an exponent. Thus it is natural to consider molecules with a high dissociation energy for the ground state. This situation is typical of the compounds of the hard-to-polarize elements with small covalent radii, in particular hydrogen whose atoms seem to intrude into the electron shells of heavier atoms.

An important parameter of a molecular system with a photodissociating laser transition is the thermal stability of the upper working level (which is characterized by its dissociation energy D_2). In fact, the larger D_2 is, the higher the maximum attainable population of the given electronic state, the lower the probability of collisional deexcitation of the molecule, and the more thermodynamically favorable the formation of just this type of molecule becomes compared to other types (of the same or close-by degrees of electronic excitation) which might be formed from the same initial atomic components in the medium. If we assume that an excited molecule radiates from a group of low-lying vibrational levels then the following relations apply for the energy of the working photon and ΔE [see Eq. (14)]:

$$\hbar\omega_0 \approx E_2 - D_2 - E_d, \quad (41)$$

$$\Delta E \approx D_2 + E_d, \quad (42)$$

These relationships also indicate the appropriateness of choosing a molecule with high D_2. Up to now, nonempirical calculations of the potential curves of excited states have been done for only a very few dissociating molecules, and the situation is no better as regards their experimental determination. Thus a similarity technique is usually used for crude estimates of the parameters of an excited state. For example, it is assumed [33, 34] that for single-electron excitation if the state of one of the atoms which (state) is correlated to the lower excited state of the molecule is a Rydberg (roughly speaking $E_2 > J/2$) state, then the values of the molecular constants (such as D_2, r_{02}, ω_{vib}, and others) are close to the corresponding quantities for the ground state of the molecular ion obtained by removal of this electron. This approximation is more precise, as a rule, when the molecule (excited) and the molecular ion have the same orbital symmetry in their wave functions.

Characteristics of the Diatomic Hydrides of the Inert Gases

These considerations led us to examine the possibility of lasing on electronic transitions of the hydrides (HR) of the inert gases from a bound lower excited state to a dissociating ground state. This possibility for HHe and HNe was mentioned in the review [46]. Despite the fact that the existence of the (HR)* molecule was proposed a long time ago it was experimentally (mass spectroscopically) observed only later in the case of helium hydride [52], and in [53] an emission spectrum near $\lambda \sim 767$ nm was observed and identified as a system of bands of the transition between the neighboring lower electronically excited states, $^2\Pi \rightarrow {}^2\Sigma^+$, of the hydride (HAr) and deuteride (DAr) of argon. Quantum-mechanical calculations of the potential curves of several electronically excited states have also been done for HHe [54, 55] and HNe [56].

The ground state $'\Sigma^+$ of $(HR)^+$ ions correlates with the atomic states of $H^+ + R(1)$ for all the inert gases [57, 35] except xenon [58, 57] whose ionization potential is lower than that of the hydrogen atom ($J^H \approx 13.6$ eV). Similarly, the lower excited states $^2\Sigma^+$ of the HR(2) molecule correspond to combinations of the excited hydrogen atom H(2) with an R atom in the ground state (also, except for xenon). Considering the great similarity in the term structures of the $(HR)^+$ ions and the HR(2) molecules, as well as the similar character of the correlations between the molecular and atomic states, it is reasonable to turn to the affinities of the corresponding R atoms for protons† [59, 57] and to the parameters of hydrogen H(1) atoms scattering on R atoms [45] to estimate the parameters of the working states of the hydrides of helium, neon, argon, and krypton.

Table 2 lists the dissociation energies of the ground state $'\Sigma^+$ of $(HR)^+$ ions, their equilibrium internuclear distance r_{0+}, and the average energy of dissociation E_d of the ground state of the HR(1) molecule. ($E_d \approx U_1(r_{0+})$, where $U_1(r)$ is taken from [69] and is the empirical interaction potential of the H(1) and R(1) atoms.)

We estimate the radiative characteristics of the HR molecule as in [61]. The energy of the working photon and the corresponding wavelength of the photodissociative transition is found using Eq. (41) with $E_2^H \approx 10$ eV.

The Einstein coefficient for spontaneous decay

$$HR\,(2) \rightarrow H\,(1) + R\,(1) + \hbar\omega_0 \qquad (1a)$$

is clearly close to the coefficient A for decay of the resonance level of the hydrogen atom [52, 54], $A \approx 4 \cdot 10^8$ sec^{-1}. The effective linewidth $\Delta\omega$ of the photodissociative transition (except for the relative positions of the potential curves of the combination states) is determined by

TABLE 2

	He	Ne	Ar	Kr
$D_{(HR)^+}$, eV	1.84	2.28	3.5	4.5
r_{0+}, Å	0.77	0.99	1.31	1.47
E_d, eV	4.3	2.9	1.9	1.6

TABLE 3

	He	Ne	Ar	Kr
$\Delta\omega \cdot 10^{-15}$, sec^{-1}	6.5	4.3	2.9	2.4
$\hbar\omega_0$, eV	3.9	4.8	4.6	3.9
λ_0, nm	320	260	270	320
$\sigma_{pd} \cdot 10^{17}$, cm^2	1.5	1.5	2.5	4.1

† According to handbook data [59], a very large depth in the ion term is typical for multiatomic dissociating molecules containing hydrogen such as H_3, H_3O, and NH_4 which also may be of interest for laser purposes.

the force constant of the upper term, by the mechanisms for populating it, and by vibrational–rotational relaxation, all of which create the distribution of the HR(2) molecule over its internal degrees of freedom. In order of magnitude $\Delta\omega$ is usually comparable to E_d (cf Table 1); for crude estimates we set $\Delta\omega \approx E_d$. The resulting values of $\Delta\omega$, $\hbar\omega_0$, λ_0, and $\sigma_{pd} \approx \lambda_0^2 A/4\Delta\omega$ are given in Table 3. It is clear that the "center of gravity" of the laser spectrum of the hydrides HR with R lighter than xenon lie in the near ultraviolet. With transition (1a) having a wider bandwidth than (1), photodissociation of the inert gas hydrides is characterized by larger cross sections σ_{pd} (roughly an order of magnitude greater than the corresponding quantities for the dimers R_2). Thus the critical temperature for the hydride molecules HR is at least twice as high as for the dimers.

Some Peculiarities of Producing an Active Medium Containing HR(2) Molecules

In the preparation of media in which we may expect lasing on photodissociating transitions of diatomic hydrides of elements from the zeroth group, it is reasonable to start with a mixture of the appropriate inert gas and hydrogen, ionizing it with an external source. The presence of molecular hydrogen H_2 improves the heat capacity and thermal conductivity of the mixture and aids more rapid cooling of the plasma electrons. This ensures a strong recombination disequilibrium in the active medium and, therefore, a high rate of populating the upper working state. The relaxation of a dense plasma containing hydrogen mixed with an inert gas has in fact not yet been studied. Thus, on the basis of a comparison with the most important mechanisms for filling and emptying the upper working state in a pure inert gas plasma, we shall note only some peculiarities of the relaxation kinetics in this type of two-component plasma.

First of all, in such a medium there is a sharp increase in the number of possible plasma chemical reactions of charged and neutral particles. Charge exchange processes acquire an important role in the population kinetics of HR(2), as do exchange excitation and the formation of multiatomic clusters. As the plasma relaxes the first and second processes will aid in the accumulation of ions and excited neutrals which correlate with H^+ and H(2), respectively, since $J^H < J^R$. The existence of stable multiatomic ions (and molecules), despite the fact that their rates of formation may seem low (in particular because of their configuration factors), leads to branching out from the main working channels for relaxation of excited states. It is known that in a dense plasma containing hydrogen there are noticeable amounts of H_3^+ ($D \approx 4.4$ eV) and possibly H_5^+ ions. Ions of the type $(HR_2)^+$ may also have large dissociation energies. (According to a calculation in [62] the $(HH2_2)^+$ ion is the most stable of the particles of the type $(HHe_n)^+$.) The side channels for relaxation will not compete significantly with the main channel only if the formation of those molecular particles corresponding to the working channel is thermodynamically more favorable. From this point of view, of all the diatomic inert gas hydrides the greatest experimental interest should be in HKr. This does not, however, preclude the possibility of lasing on photodissociative transitions of multiatomic molecules such as $H_3(2)$, $HR_2(2)$, etc.

Excited molecules $(HR)^*$, as are the dimers R_2^*, are formed as a result of the recombination of molecular ions $(HR)^+$ with plasma electrons and association of R(1) and H^* atoms [although it is possible to point out more complicated reactions, for example, transfer of an excited atom in $H_2^* + R(1) \rightarrow (HR)^* + H(1)$]. In this case the predominance of dissociative recombination of the molecular ion is evidently unfavorable† to the production of a sufficient inverted population. If this is actually so, then, taking into account the different temperature

† If the products of this reaction are mainly the R^* and H(1) atoms.

dependence of the rate coefficients for three-body and dissociative recombination, it is possible in principle to find a relation between T_e and N_e which corresponds to the most efficient regime for populating the HR(2) state and with its aid to determine the optimum parameters of the active medium (particle density, pumping intensity, mixture composition).

Two principal mechanisms for emptying the upper working state are strong radiative decay and collisions with excited particles. If the relaxation is controlled by the molecular channel, and if during the relaxation it is mainly the lower vibrational levels of the HR(2) molecule which are filled, then processes such as (9a) are unlikely. On the other hand, inelastic collisions of HR(2) molecules with charged particles (ions and, especially, electrons) may cause substantial losses of HR(2) molecules. In particular, because of the relatively small difference in the energies of the combining states of the hydrides HR, the rate coefficient V_{12} for deactivation of HR(2) by electron impact must be roughly an order of magnitude greater than the value for the $R_2(2)$ dimer of the corresponding inert gas. The condition for dominance of radiative decay of the molecules over their collisional deexcitation by plasma electrons, $A > V_{12}N_e$, places an additional limitation on the order of magnitude of the free-electron density.

These features of the relaxation kinetics of an HR active medium and the resulting complications in achieving lasing are to some extent inherent in any (chemically) multicomponent scheme where the photodissociation of nonsymmetric (XY)* molecules is proposed as a laser transition.

The Possibility of Lasing on Vibrational Transitions of HR(2)

It has been mentioned several times previously that vibrational relaxation has a significant effect on the laser output from the electronic transition (1a). The stability of a photodissociative scheme relative to overheating of the gas and the magnitude of the cross section σ_{pd} are greatest if the recombination of the plasma leads mainly to filling of low-lying vibrational levels of the excited HR(2) molecules. At the same time, it is known that molecules with small reduced masses have a high energy in their elementary vibrational quanta $\hbar\omega_{vib}$, and, therefore, the rate of relaxation in vibrational–translational exchange must be low. Thus, in general the predominance of the atomic channel for populating the upper working level of the inert gas hydrides, particularly in the case of a deep term 2, would make amplification on the transition (1a) problematical. The situation improves because of the absence of a prohibition on radiative transitions between the vibrational levels of diatomic molecules with different nuclei.

Following [63] we shall consider the simplest scheme for vibrational relaxation of a molecule of the type HR(2). We shall assume that the loss of molecules is determined by their radiative decay, which is equally probable for any vibrational level n. For a low gas temperature $\theta = \hbar\omega_{vib}/T \gg 1$, the thermal populations in the levels, determined by the factor $\exp\{-\theta\}$, may be neglected. The system of equations for the populations $M_{2,n}$ in these levels takes the form

$$\frac{dM_{2,n}}{dt} = P_{n,n+1}M_{2,n+1} - (P_{n-1,n} + A)M_{2,n} + \eta_n, \tag{43}$$

where $P_{n,n+1}$ is the probability of a transition (per unit time) by the HR(2) molecule from the n + 1 level to the n-th level and η_n is the rate of the direct recombination excitation flux to the n-th level from higher electronic terms.

In the harmonic-oscillator approximation

$$P_{n,n+1} = \frac{n+1}{\tau_n}, \qquad \tau_n \equiv \tau$$

the stationary solution of the system (43) may be written as a finite sum of terms with different partial pumping:

$$M_{2,n} = \sum_{n'} M_n^{(n')}, \qquad M_n^{(n')} = \begin{cases} \eta_{n'}\tau \dfrac{(n')!\,\Gamma(A\tau + n)}{n!\,\Gamma(A\tau + n' + 1)} & \text{for } n \leqslant n', \\ 0 & \text{for } n > n'. \end{cases} \tag{44}$$

If the recombination pumping η_n ends up mainly on high levels n, then, according to Eq. (44), for $A\tau > 1$ the vibrational levels have a population inversion. Since A is usually 10^7–10^9 sec^{-1} and the time $\tau \gtrsim 10^{-6}$ sec (when stimulated processes may be neglected), in this model a population inversion over the vibrational levels of term 2 is a general property of class (b) molecules. (See the introduction to this chapter.)

Thus, in an active medium containing HR molecules lasing may be realized not only on the electronic transition (1a) but also on vibrational $(2, n) \to (2, n-1)$ transitions. This speeds up the recombination flux of excitation through the term HR(2) and results in a noticeable redistribution of the populations of its vibrational levels. Let us examine a quasistationary case in which lasing occurs simultaneously on electronic and vibrational transitions. Neglecting spontaneous and collisional transitions under these conditions, we write

$$\frac{dM_{2,n}}{dt} = B_{n+1}(M_{2,n+1} - M_{2,n}) - B_n(M_{2,n} - M_{2,n-1}) - B^{(n)}M_{2,n} + \eta_n = 0. \tag{45}$$

Here B_n and $B^{(n)}$ are the probabilities of the corresponding stimulated transitions, vibrational and electronic. Setting $B_n = b_n$, n, $b_n \equiv b$, $B^{(n)} \equiv B$ the system of equations (45) can conveniently be replaced by a single diffusion equation

$$\frac{d}{dn}\left(n\frac{dN}{dn}\right) = pN, \qquad p = \frac{B}{b}. \tag{45a}$$

For definiteness we assume that all the recombination flux goes to a high vibrational level n_0 of the working term HR(2), i.e., that $\eta_n = \eta\delta_{nn_0}$. This corresponds to the boundary condition $b[n(dN/dn)]_{n=n_0} = \eta$. The solution of this kind of problem is expressed in terms of Bessel functions as

$$M_{2,n} = \eta/\sqrt{Bbn_0}\,\frac{I_0(2\sqrt{pn})}{I_1(2\sqrt{pn})}. \tag{46}$$

From this it follows that for $p < 1$ it is mainly the low-lying vibrational levels which are populated and that for $p > 1$ the populations of all the vibrational levels of the HR(2) term lying below n_0 are of the same order of magnitude. By regulating the conditions for lasing on the vibrational levels it is possible to "shape" the population intensity (i.e., distribution) of the working term and to tune the frequency of the laser output from the electronic transition (1a) over a wide range.

An Attempt to Observe Electronically Excited Molecules Experimentally

Starting with the above ideas a preliminary attempt was made in [64] to study the luminosity of dense inert gases, hydrogen, and their mixtures in a transverse pulsed breakdown. The gas pressure in the discharge cell could be varied from several tenths to about 15 atm. The gas was broken down by a voltage of about 20–50 kV on a pulse-forming line after low-current pulsed preionization. For given electric field pulse formation conditions, the pulse had the shape of a train of short (about 5 nsec) spikes lasting about 80 nsec in all.

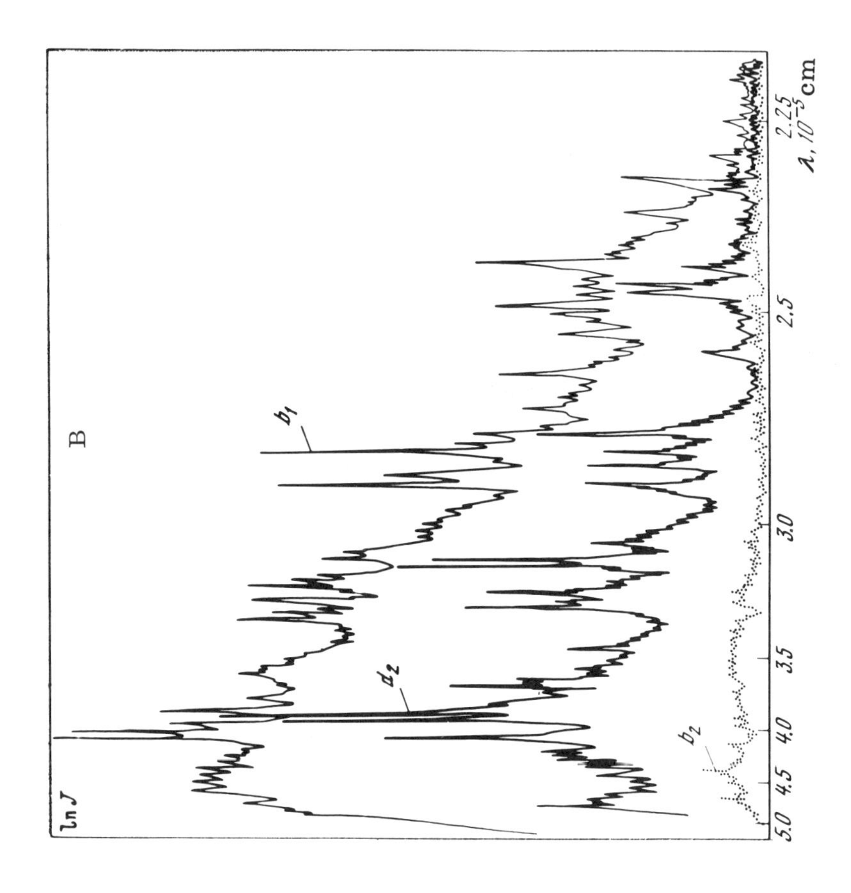
B
ln J
b_1
d_2
b_2
5.0 4.5 4.0 3.5 3.0 2.5 2.25
λ, 10^{-5} cm

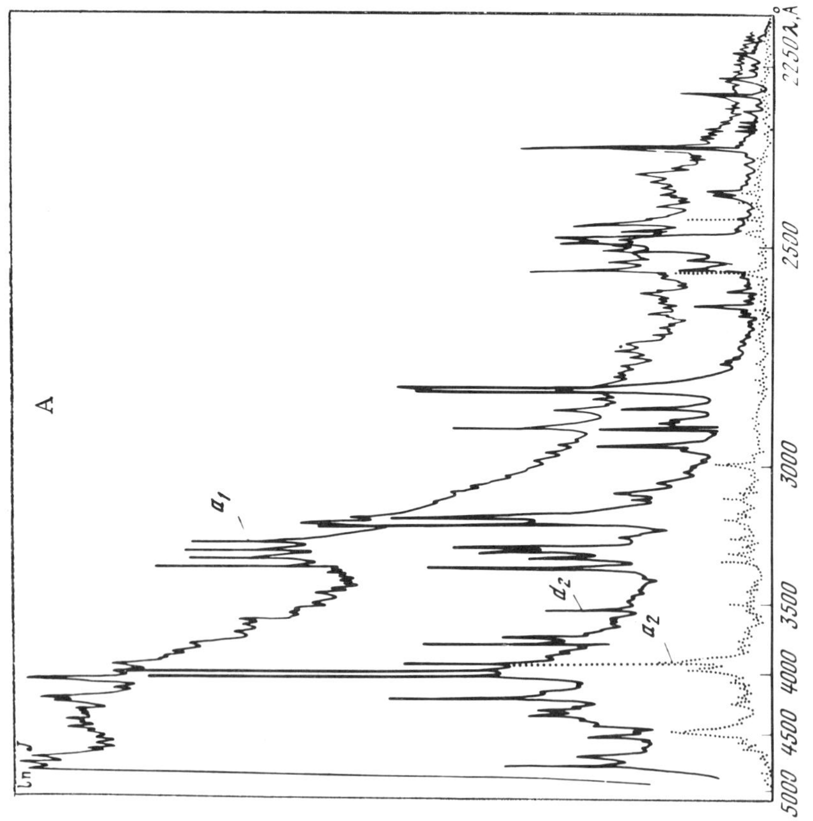
A
ln J
a_1
d_2
a_2
5000 4500 4000 3500 3000 2500 2250
λ, Å

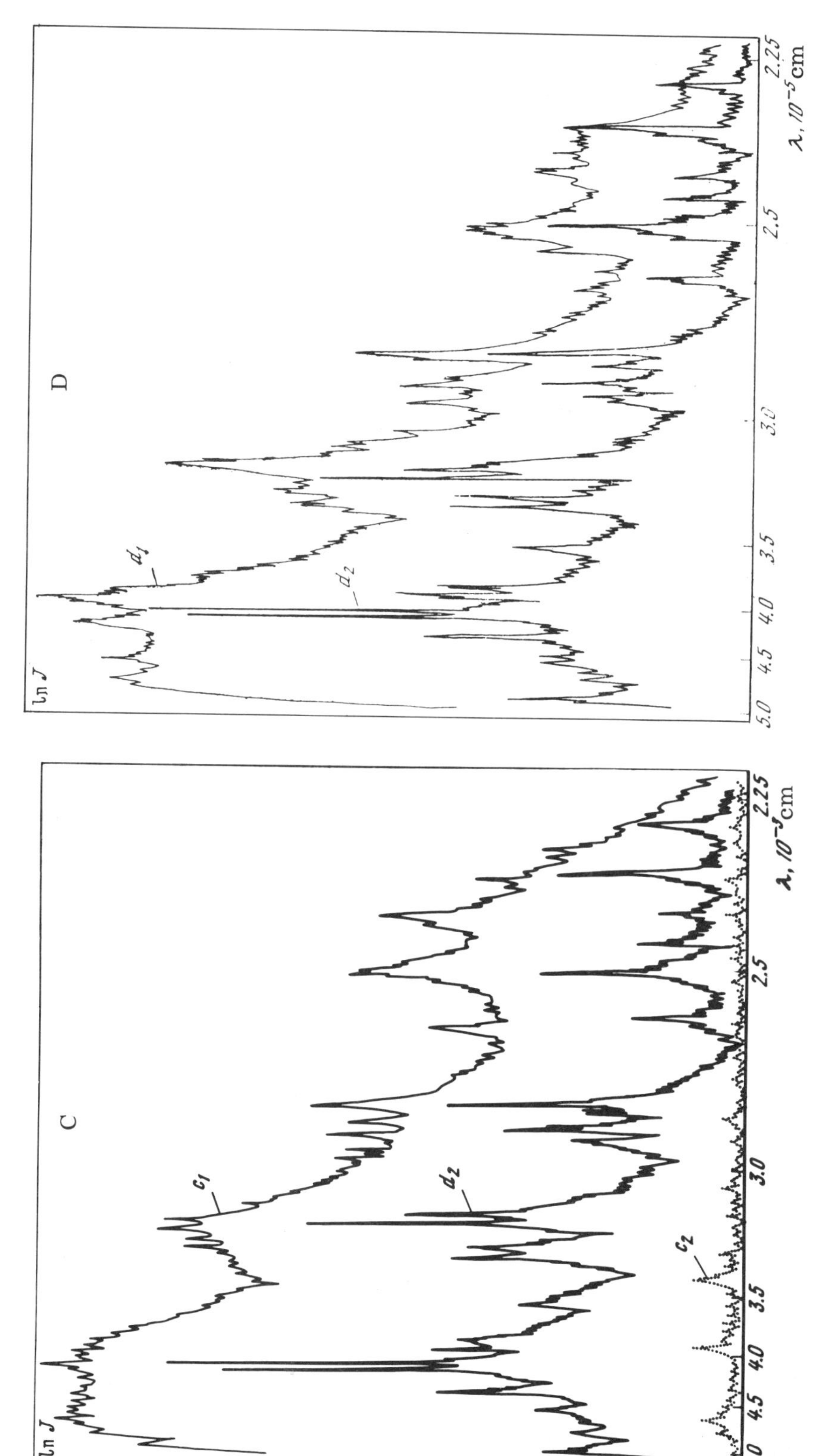

Fig. 4. Densitograms of the spectra of hydrogen, inert gases, and mixtures of these obtained under transverse pulsed breakdown conditions: A) a_2, 8 atm He; a_1, 4 atm H_2 and 8 atm He; B) b_2, 9 atm Ar; b_1, 2.5 atm H_2 and 10 atm Ar; C) c_1, 4 atm Kr and 6 atm H_2; c_2, 5 atm Kr; D) d_2, 5 atm H_2; d_1, 10 atm H_2.

According to the estimates of the amount of energy delivered to the gas, it remained cold after breakdown; evidently however, the character of the ionizing pulse did not permit the establishment of a strong recombination disequilibrium in the plasma that was created. The light from the discharge was recorded on an ISP-30 spectrograph over the wavelength range from 2000 to 5000 Å. Typical densitograms of the spectra obtained are shown in Fig. 4.

The experiments showed that as the pressure is increased from 1 to 12 atm the intensity of the light from pure (technical) hydrogen rose substantially (depending on the part of the spectrum by 2-5 orders of magnitude). The greatest rise is in the intervals λ 2350-2600, 3100-3400, and 3700-5000 Å. In the spectrum numerous narrow lines can be resolved against a very strong continuous background. The origin of these lines may be due, in the first place, to emission from impurities, or in the portion with $\lambda \geqslant 3600$ Å, to transitions between bound excited states of the H_2 molecule [35]. On the whole the spectrum is similar to the known continuum of hydrogen (observed at pressures of about 1-10 Torr) lying in the much shorter wavelength region $-\lambda_0 \sim 2000$ Å and corresponding to the $a^3\Sigma_g^+ \rightarrow b^3\Sigma_u^+$ transition of the diatomic molecule. In principle, at high pressures the wavelength of the intensity peak in the continuum spectrum of H_2 may shift to $\lambda_0 \sim 3200$ Å [65] (emission from the lower vibrational level of the $a^3\Sigma_g^+$ state). However, the trimodality of the intensity distribution in the recorded spectra leads us to suppose that emission from bound–free transitions of the H_3 molecule is making a certain contribution to their structure.

The luminous intensity of both the pure inert gases (He, Ar, Kr) and their mixtures with one another over this wavelength range is several orders of magnitude lower than of their mixtures with hydrogen, even when the hydrogen was present in impurity quantities (for example, when pure krypton was fed into a poorly evacuated cell). As the density of the mixture is increased, the intensity of its emission increases extremely rapidly while the spectrum becomes less spiky. The spectrum of pure hydrogen is the most spiky. It is interesting to compare these results with those of [66]. Thus the spectra of hydrogen and its mixtures with argon and xenon at pressures of about 1 atm were studied over the range $\lambda \sim 4000$-6000 Å. Under the conditions of that experiment [66] the discharge plasma was close to thermal. Two important features of the observed spectra were noted. First, the continuum spectrum of pure hydrogen differs somewhat from the calculated spectrum for the $a^3\Sigma_g^+ \rightarrow b^3\Sigma_u^+$ transitions. Second, the emission from the pure components has an additive contribution to the spectral-intensity distribution of the mixtures. This may be interpreted as meaning that, under the transverse breakdown conditions of [64] for $R + H_2$ mixtures, besides H_3 molecules some molecules of the type HR are formed (and as the density of R is increased their importance becomes still greater). Thus, to the extent that the preliminary nature of the experiments in [64] allow us to judge, the results are in agreement with our estimates above.

In [53] the observed diffuse structure in the bands of the assumed $^2\Pi(b) \rightarrow {}^2\Sigma^+$ transition of the HAr molecule, as opposed to the sharp structure for DAr, is explained by predissociation of the $HAr(2)^2\Sigma^+$ state due to intersection of the potential curve of this state with the ground-state term. Thus, replacement of hydrogen by deuterium may give more definite results. Later on, besides making more detailed studies of gaseous mixtures of hydrogen (or deuterium) with inert elements, it would be useful to study the luminescence of these media in the liquid phase which probably would permit the "freezing out" of some spurious processes. Finally, in order to reduce to a minimum the contribution of the H_3 spectrum which overlaps the HR(2) spectrum, it makes sense to add (instead of hydrogen) a compound that is rich in hydrogen but has a high ionization potential, such as methane (deuteromethane), to the gas.

The Monohalides of the Inert Gases

As is known, the inert gases form di-, tetra-, and hexahalides which are stable in the ground state. At the same time a high stability is typical of the ions $(RX)^+$ and excited molecules $(RX)^*$ of the monohalides, where X represents F, Cl, or Br [67-73, 59]. For example, the values of the dissociation energy of the $(RF)^+$ ions [67, 68] in order of ascending atomic number of R from helium to xenon are, respectively, 1.33; 1.65; >1.655; >1.58; 2.03 eV. Thus, radiative decay of dissociating $(RF)^*$ molecules must be regarded as a possible laser transition. The same conclusion was reached by Velazco and Setser [70] who observed intense luminosity of the $XeCl^*$ and $XeBr^*$ molecules near 305 and 280 nm, respectively. As an illustration, using data from a quantum-mechanical calculation [71] of the terms $(KrF)^{+1}\Sigma^+$ and $KrF(1)^2\Sigma^+$, let us evaluate the parameters of the photodissociating transition of the KrF(2) molecule in the same way as for the inert gas hydrides. We obtain $D \approx 1.94$ eV, $E_d \approx 1.95$ eV, $\hbar\omega_0 \approx 6.0$ eV, $\lambda_0 = 205$ nm, and assuming $A \approx 10^7$ sec^{-1}, $\sigma_{pd} = 4 \cdot 10^{-18}$ cm^2 (that is, no worse than for the dimer Xe_2^*). It should be noted that for the heavy inert gases the term $^2\Sigma^+$ of the RX molecule at small distances correlates with the ion configuration $R^+(^2P) + X^-(^1S)$ and not with the atomic configuration $R(^3P) + X(^2P)$. Thus, the wavelength λ_0' of the luminescence of XeF is greater than λ_0.

A photodissociative scheme for lasing on the monohalides of the inert gases has some features which favorably distinguish it from other dissociating molecule lasers. First of all, the combination of $R(1)^1S_0 + X(1)^2P_{3/2}$ atoms correlates with two repulsive molecular states, $RX(1) - {}^2\Sigma^+$ and $^2\Pi$. For $J^R < J^X$ the lowest electronically excited state of the inert gas monohalides which correlates with $R(2)^3P + X(1)^2P_{3/2}$ is the $RX(2)^2\Sigma^+$ state for which the radiative transitions $^2\Sigma^+ \rightarrow {}^2\Sigma^+$ and $^2\Sigma^+ \rightarrow {}^2\Pi$ are allowed by the selection rules. In principle it is possible to expect lasing on transitions from a common upper working level to any of the lower levels. In accordance with the change in the polarizability and the electron affinity of atoms in the series of halogens and inert gases, the dissociation energy D_2 of RX(2) molecules increases as the atomic number of X is reduced and as the atomic number of R is increased while the dissociation energy E_d falls as the atomic numbers of R and X are increased. Thus, by varying the composition of the active medium we may expect lasing on photodissociative transitions of RX(2) molecules in different parts of the ultraviolet range, from the near to the vacuum ultraviolet.

We emphasize once again that the case $J^X > J^R$ is of greatest interest since then the reactions

$$R\ (m) + X_2\ (1) \rightarrow RX\ (m) + X\ (1), \qquad m \geqslant 2 \tag{47}$$

serve as the main channel for populating the upper working state RX(2). Reactions of this type take place through the so-called harpoon mechanism [74] and are characterized by very large cross sections, $\sigma \sim 10^{-14}$ cm^2. For comparison we mention that the quenching constants for the metastable $^3P_2$ xenon atom and for the hydrogen and chlorine molecules differ by almost two orders of magnitude [75].

In recent experiments lasing has been achieved on molecular transitions of XeF, XeCl, KrF [76], and XeBr [77]. In [76] lasing was obtained during the trailing edge of an electron beam pulse ($W_{el} \sim 350$ keV, $j \sim 150$ A/cm^2, $\tau_{pulse} \sim 100$ nsec) at pressures above 2 atm, that is, under conditions corresponding to a recombination regime. The conversion efficiency (for KrF) of pump energy into radiation was about 0.4%. The comparatively low luminescence yield is probably due to the fact that the working components of the medium were strongly diluted with argon. The observed laser transitions lie in the near ultraviolet.

Lasing on dissociating RF molecules may find important application in the nuclear reactor-laser problem [78]. The working medium of the reactor contains the gaseous uranium compound UF_6 as the required fissile component together with added molecular fluorine F_2 to prevent chemical decomposition of the UF_6. The corrosiveness of this medium creates special difficulty in the choice of a laser active component. The possibility of using some inert gas for this purpose will depend on the extent to which the emission spectrum of the photodissociative transition of the RF(2) molecule lies in a transmission window of the absorption spectra of UF_6 and F_2. This indicates the timeliness of checking for lasing on the fluoride molecules of both the heavy and the light (He, Ne) inert gases.

Conclusion

The successes in obtaining lasing in inert gases [10-25] and their mixtures with the halides [76, 77] convincingly illustrate the advantages of the recombination principle for producing an active medium and demonstrate the feasibility of making plasma lasers using molecular transitions. The possibility of lasing on bound–free transitions and many details of the mechanisms involved in the photodissociative scheme are still not fully understood. At present the question of whether extremely short pulse lasers can be built using these transitions is unanswered. Media with complex chemical compositions have not been tested. Of considerable interest is the theoretical analysis of the feasibility of lasing on multiatomic dissociating molecules with stable electronically excited states.

Recently there was a report of intense lasing in a recombination regime [79, 80] on bound-bound molecular transitions. Such transitions are not less promising than photodissociative transitions in plasma lasers. Here we should expect in principle to obtain higher gains and that the limitation on the heavy-particle temperature would in many cases be less restrictive. The main difficulty in lasing on molecular transitions in class (c) is due to the problem of depopulating the lower working state, otherwise strong reabsorption makes these transitions self-limiting.

There is no doubt that the timeliness and wide practical applications of powerful coherent light sources are stimulating the development of new plasma laser schemes and more directed research on the processes occurring in dense low-temperature plasmas.

Literature Cited

1. L. I. Gudzenko and L. A. Shelepin, Zh. Éksp. Teor. Fiz., 45:1445 (1963).
2. L. I. Gudzenko, V. V. Evstigneev, Yu. I. Syts'ko, S. S. Filippov, and S. I. Yakovlenko, Preprint IPM, No. 63 (1971).
3. F. G. Houtermans, Helv. Phys. Acta, 33:933 (1960).
4. M. M. Litvak and R. J. Carbone, Solid State Res. Lincoln Lab. Mass. Inst. Technol., No. 4, p. 29 (1964).
5. R. J. Carbone and M. M. Litvak, Solid State Res. Lincoln Lab. Mass. Inst. Technol., No. 2, p. 21 (1965).
6. R. J. Carbone and M. M. Litvak, J. Appl. Phys., 39:2413 (1968).
7. A. J. Palmer, J. Appl. Phys., 41:438 (1970).
8. N. G. Basov, O. V. Bogdankevich, V. A. Danilychev, G. N. Kashnikov, O. M. Kerimov, and N. P. Lantsov, Kratk. Soobshch. Fiz., No. 7, p. 68 (1970).
9. N. G. Basov, V. A. Danilychev, Yu. M. Popov, and D. D. Khodkevich, Pis'ma Zh. Éksp. Teor. Fiz., 12:473 (1970).
10. H. A. Kochler, L. J. Ferderber, D. L. Redhead, and P. J. Ebert, Appl. Phys. Lett. 21:198 (1972).
11. H. A. Kochler, L. J. Ferderber, D. L. Redhead, and P. J. Ebert, Phys. Rev., A9:768 (1974).
12. P. J. Ebert, L. J. Ferderber, H. A. Kochler, R. W. Kuckuck, and D. L. Redhead, IEEE J, Quant. Electron., QE-10:736 (1974).
13. Wayne A. Johnson and J. B. Gerardo, Conference on Laser Engineering and Applications, New York (1973), Digest of Papers, p. 29.
14. Wayne A. Johnson and J. B. Gerardo, XIth International Conference on Phenomena in Ionized Gases, Prague (1973), Contributed Papers, p. 164.

15. J. B. Gerardo and Wayne A. Johnson, J. Appl. Phys., 44:4120 (1973).
16. Wayne A. Johnson and J. B. Gerardo, J. Appl. Phys., 45:867 (1974).
17. P. W. Hoff, J. C. Swingle, and C. K. Rhodes, Appl. Phys. Lett., 23:245 (1973).
18. P. W. Hoff, J. C. Swingle, and C. K. Rhodes, Opt. Commun., 8:128 (1973).
19. W. M. Hughes, J. Shannon, A. Kolb, E. Ault, and M. Bhaumik, Appl. Phys. Lett., 23:385 (1973).
20. E. R. Ault, M. L. Bhaumik, W. M. Hughes, R. J. Jensen, C. R. Robinson, A. C. Kolb, and J. Shannon, IEEE J. Quant. Electron., QE-9:1031 (1973).
21. W. M. Huges, J. Shannon, and R. Hunter, Appl. Phys. Lett., 24:488 (1974).
22. W. M. Huges, J. Shannon, and R. Hunter, Appl. Phys. Lett., 25:85 (1974).
23. S. C. Wallace, R. T. Hodgson, and R. W. Dreyfus, Appl. Phys. Lett., 23:672 (1973).
24. S. C. Wallace, and R. W. Dreyfus, Appl. Phys. Lett., 25:494 (1974).
25. M. Novaro and F. Lagarde, C. R. Acad. Sci., Ser. B., 277:671 (1973).
26. C. V. Heer, Phys. Lett., A31:160 (1970).
27. M. M. Mkrtchan and V. T. Platonenko, Pis'ma Zh. Éksp. Teor. Fiz., 17:28 (1973).
28. L. I. Gudzenko, and S. I. Yakovlenko, Dokl. Akad. Nauk SSSR, 207:1085 (1972).
29. V. L. Borovich and V. S. Zuev, Zh. Éksp. Teor. Fiz., 58:1794 (1970).
30. V. L. Borovich, V. S. Zuev, and D. B. Stavrovskii, Kvant. Élektron., 1:2048 (1974).
31. A. A. Belyaeva, R. B. Dushin, E. V. Nikiforov, Yu. B. Predtechenskii, and L. D. Shcherba, Dokl. Akad. Nauk SSSR, 198:1117 (1971).
32. S. E. Harris, A. H. Kung, E. A. Stappaerts, and J. F. Young, Appl. Phys. Lett., 23:232 (1973).
33. R. S. Mulliken, J. Chem. Phys., 52:5170 (1970).
34. R. S. Mulliken, Radiat. Res., 59:357 (1974).
35. Tables de constantes et données numériques, Vol. 17, Pergamon Press (1970).
36. D. C. Lorents, Radiat. Res., 59:438 (1974).
37. P. Moerman, R. Boucique, and P. Mortier, Phys. Lett., A49:179 (1974).
38. E. V. George and C. K. Rhodes, Appl. Phys. Lett., 23:139 (1973).
39. C. W. Werner, E. V. George, P. W. Hoff, and C. K. Rhodes, Appl. Phys. Lett., 25:235 (1974).
40. L. I. Gudzenko, I. S. Lakoba, and S. I. Yakovlenko, Zh. Éksp. Teor. Fiz., 67:2022 (1974).
41. S. I. Yakovlenko, Preprint IAÉ, No. 2174 (1972).
42. E. H. Fink and F. J. Commes, Chem. Phys. Lett., 30:267 (1975).
43. A. B. Callear and M. R. E. Hedges, Trans. Faraday Soc., 66:2921 (1970).
44. L. I. Gudzenko and S. I. Yakovlenko, Zh. Éksp. Teor. Fiz., 62:1686 (1972).
45. J. McDaniel, Collision Phenomena in Ionized Gases, Wiley (1964).
46. C. K. Rhodes, IEEE J. Quant. Electron., QE-10:153 (1974).
47. A. Gedanken, J. Jortner, B. Raz, and A. Szöke, J. Chem. Phys., 57:3456 (1972).
48. L. I. Gudzenko, L. A. Shelepin, and S. I. Yakovlenko, Usp. Fiz. Nauk, 114:457 (1974).
49. L. I. Gudzenko and S. I. Yakovlenko, Kratk. Soobshch. Fiz., No. 12, p. 13 (1975).
50. L. S. Frost and A. V. Phelps, Phys. Rev., A136:1538 (1964).
51. R. M. Hill, D. J. Eckstrom, D. C. Lorents, and H. H. Nakano, Appl. Phys. Lett., 23:373 (1973).
52. J. Gray and R. H. Tomlinson, Chem. Phys. Lett., 4:251 (1969).
53. C. J. W. Johns, J. Mol. Spectrosc., 36:488 (1970).
54. H. H. Michels and E. F. Harris, J. Chem. Phys., 39:1464 (1963).
55. S. A. Slocomb, W. H. Miller, and H. F. Shaefer, J. Chem. Phys., 55:926 (1971).
56. V. Bondybey, P. K. Pearson, and H. F. Shaefer, J. Chem. Phys., 57:1123 (1972).
57. H. P. Weise, H. U. Mittmann, A. Ding, and A. Z. Henglein, Naturforscher, 26a:1112, 1123 (1971).
58. C. Kubach and V. Sidis, J. Phys., B6:L289 (1973).
59. L. V. Gurvich et al., in: The Dissociation Energy of Chemical Bonds [in Russian], Nauka, Moscow (1974).

60. Yu. N. Belyaev, N. V. Kamyshov, and V. B. Leonas, Khim. Vys. Énerg., 4:260 (1970).
61. L. I. Gudzenko and I. S. Lakoba, Kratk. Soobshch. Fiz., No. 6, p. 3 (1975).
62. M. B. Mulleur, R. L. Matcha, and E. F. Hayes, J. Chem. Phys., 60:674 (1974).
63. L. I. Gudzenko, Yu. B. Konev, and V. S. Marchenko, Kratk. Soobshch Fiz., No. 9, p. 23 (1975).
64. L. I. Gudzenko, L. A. Kulevskii, I. S. Lakoba, and A. A. Medvedev, Kratk. Soobshch. Fiz., No. 1, p. 21 (1976).
65. N. D. Smith, Phys. Rev., A49:345 (1936).
66. G. Grandsire, Ann. Astrophys., 17:287 (1954).
67. J. F. Liebman and L. C. Allen, J. Am. Chem. Soc., 92:3539 (1970).
68. J. Berkowitz and W. A. Shupka, Chem. Phys. Lett., 7:447 (1970).
69. M. F. Golde and B. A. Truch, Chem. Phys. Lett., 29:486 (1974).
70. J. E. Velazco and D. W. Setser, J. Chem. Phys., 62:1990 (1975).
71. L. A. Kuznetsov, Yu. Ya. Kuzyakov, V. A. Shpanskii, and V. M. Khutoretskii, Vestn. Mosk. Gos. Univ., Ser. II, Khim., No. 3, p. 19 (1964).
72. B. Lie and H. F. Shaefer, J. Chem. Phys., 55:2369 (1971).
73. D. H. Liskow, H. F. Shaefer, P. S. Bagus, and B. Liu, J. Am. Chem. Soc., 95:4057 (1973).
74. D. R. Herschbach, Adv. Chem. Phys., 10:319 (1966).
75. J. E. Velazco and D. W. Setser, Chem. Phys. Lett., 25:197 (1974).
76. J. J. Ewing and C. A. Brau, Appl. Phys. Lett., 27:350 (1975).
77. S. K. Searle and G. A. Hart, Appl. Phys. Lett., 27:243 (1975).
78. L. I. Gudzenko, I. S. Slesarev, and S. I. Yakovlenko, Zh. Éksp. Teor. Fiz., 45:1934 (1975).
79. C. B. Collins, A. J. Cuningham, S. M. Curry, B. W. Johnson, and M. Stockton, Appl. Phys. Lett., 24:245 (1974); 24:477 (1974).
80. H. T. Powell, J. R. Murray, and C. K. Rhodes, Appl. Phys. Lett., 25:730 (1974).

CHAPTER IV

CHEMICAL DEPOPULATION OF THE GROUP STATE OF ATOMS IN PLASMA LASERS

The General Principles for Creating Plasma-Chemical Active Media

One of the most important current problems in applied physics is doubtless the construction of economical high-power light sources. At present the most promising method is lasing in a dense recombining plasma on transitions of atoms (molecules) from levels close to the continuum to ground state or low-lying levels. To obtain a population inversion on this type of transition there must exist rapid processes for the removal of "working" atoms (molecules) from the ground state. This is fairly difficult in a recombining plasma that efficiently amplifies light since the intensity of elementary processes leading to population of the lower energy levels is very great. Although the most effective populating processes involve light particles (electrons and photons), it appears that the most realistic relaxation processes for depopulating these states at the required rate are chemical reactions. The simplest chemical reactions of this type are the following:

$$T + ZY \rightarrow TZ + Y, \tag{1}$$

$$TZ \rightarrow T + Z, \tag{2}$$

where T may, in principle, denote either an atom or a molecule. Of these, reaction (1) is intended to create conditions for lasing on transitions from highly excited electronic states T^* to the ground state ($T^* \to T + hw$). We shall discuss this type of reaction here, and we may note the reactions of alkali metal atoms with the halogen molecules as an example. In the case of reaction (2), the decay of the ground state of an unstable or dissociating molecule TZ, it is reasonable to choose lasing on transitions of a large complex TZ from thermally stable excited terms into an unstable (or repulsive) ground-state term ($TZ^* \to TZ + hw$) rather than on transitions of the atom (or molecule) T. Lasing based on scheme (2) has already been realized in several types of lasers. A recombining plasma (at high pressure) of inert gases and their compounds serves as the active medium in these lasers.

Here we shall discuss the prospects for using a reaction of type (1) for creating a population inversion among the levels of T in the simplest variant, i.e., when T is an atom. It is necessary to begin this disscussion with an explanation of the mechanism by which these atoms appear in the active medium. Since it is necessary to have a high-density gas to obtain a high radiant energy output per unit volume of active medium, it is initially unclear where a sufficient (for effective lasing) density of T atoms would come from such a medium since they are lost in this fast chemical reaction. This problem is directly related to the nontrivial question of efficiently transferring a large amount of energy to a dense gas and ionizing it without overheating it. The experimenter will be most familiar with a gaseous-discharge scheme involving ionization by a longitudinal pulse or a steady-state electric field. A laser using this approach is unrealizable if it involves the processes being discussed here. At best it would result in disruption of the discharge into an arc accompanied by heating of the gas to high temperatures. A plasma laser in a dense gas may be pumped either by a transverse pulsed electric field or by fast "auxiliary" (relative to the relaxation processes in the gas) charged particles. At high energies beams of fast electrons or nuclear fission fragments (in nuclear reactor-lasers) will be the most promising means of delivering energy to a dense gas in high-efficiency pulsed or steady-state plasma laser devices.

We shall assume that the gas TZ is ionized by a hard ionizer which does not directly heat the heavy particles and free electrons. After ionization, the following events take place: The resulting free electrons are rapidly cooled in the dense, cold gas and recombine either with T^+ ions (from thermal dissociation of the TZ^+ ion) or by dissociative recombination with the TZ^+ ions themselves. In both cases there is a high probability that a T^* atom in a highly excited state will be formed. Then relaxation of the level populations of this atom takes place (both recombinational relaxation, in which there is a redistribution of the populations in all energy levels lying below this state, and chemical recombination (1) which removes an atom as part of a TZ molecule). We assume that reaction (1) is so fast that the T atoms cannot reach the gound state in large numbers as they undergo recombinational relaxation. In this situation the dense gas contains almost no atoms; those which are there (in stationary or pulsed regimes) stay a short time in relatively small amounts (compared to the total gas density) and are in excited states T^*.

Beginning with this relaxation mechanism scheme we shall formulate requirements for the chemical composition and parameters of the gaseous medium, fulfillment of which may result in effective population inversions over transitions to the ground state in T atoms.

I. In its ground state the T atom must enter reaction (1) with the compound YZ with a large cross section. At the same time, collisions with YZ molecules must not have a significant probability of deexciting T^* atoms in highly excited states.

II. The TZ molecule must be sufficiently stable in its ground electronic state. If this is not so, then besides the direct reaction (1) the inverse process, dissociation

$$TZ \to T + Z, \qquad (1a)$$

will occur at a substantial rate and lead to slowing down of the depopulating of the ground state or even to filling up of the lower working state of our planned active medium.

III. In order that the recombination flux move over the excited states to the ground state of the atom T, it is first necessary to choose a concentration of the component TW that essentially predominates in the TW + ZY mixture. Otherwise, recombination relaxation over the fragments of the molecular ion ZY^+ would lead, at least, to a reduction in the efficiency of the active medium.

IV. For this purpose the ionization potential of the T atom must be noticeably lower than the ionization potential of the W atom. Otherwise, not only can parasitic branching of the relaxation channels take place but a substantial ground-state population of T atoms may develop.

V. For this purpose the T and Z atoms must be chosen such that the molecular ion TZ^+ has a sufficiently shallow potential well in its ground-state electron term. Then the terms of highly excited electronic states of the neutral molecule TZ^* will also be thermally unstable.

It is easy to see that recombination of TZ^+ ions produced directly by the action of the ionizer on the gas may then take place in two ways and lead to almost the same result. In the first variant, recombination relaxation begins with thermal dissociation of this molecular ion ($TZ^+ \rightarrow T^+ + Z$) after which the T ion recombines in three-body collisions with two electrons or with an electron and a molecule ($T^+ + e + V \rightarrow T^* + V$). In the second variant, the recombination is dissociative ($TZ^+ + e \rightarrow T^* + Z$). In both cases the atom T is mostly in highly excited states.

VI. To ensure a high rate of removal of T atoms from the ground state in reaction (1) it is necessary that the gaseous mixture TW + ZY contain a sufficiently high concentration of ZY molecules. Since, as already noted, it is suitable to have the concentration of the TW component dominant, the chemical elements T and W must be chosen so the compound TW will have a low boiling temperature (at standard pressure).

From these requirements we can immediately limit the circle of possible chemical elements which might play the roles of T, W, Z, and Y. This naturally both lightens the search for the chemical composition of such a laser medium and makes a discussion of the prospects for solving this problem a bit more objective. We first consider condition I. A necessary condition for large cross sections of the reaction $T + ZY \rightarrow TZ + Y$ is that a strong bonding interaction between the incident particles begin at a distance of about 10 Å. In its "pure form" such a effective attraction between heavy particles at large distances perhaps can only be assured by the Coulomb interaction of oppositely charged particles. Such a situation might be realized for neutral particles entering reaction (1) if a preliminary charge-exchange interaction took place in which, while still at a large distance, an electron "jumped" from one reacting particle to another. Since condition IV implies that a low ionization potential for the T atom is desirable, we shall regard element T as a metal. In addition, condition IV determines the direction of the charge exchange ($T + ZY \rightarrow T^+ + ZY^-$). Thus, it is possible to formulate one further desired characteristic.

VII. The ZY molecule must have a sufficiently high electron affinity.

In fact the picture of the chemical reaction we have now sketched, beginning with charge exchange and then continuing with the Coulomb bonding of the reacting particles, has the character of a guiding statement. Reactions of this type take place in a single elementary act when the covalent and ionic terms corresponding to the molecular complexes T + ZY and $T^+ + ZY^-$ intersect at large distances of about 10 Å. Then the transition of the complex from the covalent part of the term to the ionic part takes place with a probability of close to unity. Then the T^+ and YZ^- ions approach one another rapidly, while the weaker $Y{-}Z^-$ bond disintegrates and the inverse transition to the covalent part of the term becomes impossible. Hence the following reaction takes place with a high probability:

$$T + YZ \rightarrow T^{+} + (Z^{-} - Y) \rightarrow TZ + Y.$$

Chemical reactions of this type are known. They have been studied both experimentally and theoretically and have the special name of "harpoon" reactions. The cross section for these reactions increases as the ionization potential of the T atom becomes smaller and as the electron affinity of the compound YZ becomes larger. This again indicates that the element T must be a metal. The YZ molecule must be a compound with a high electron affinity and be such that the negative ion $Z^{-}-Y$ can easily break up into Z^{-} and Y. It is natural to choose the halogen molecules in this case.

The metal atoms produced by recombination relax downward over the electronically excited states due to collisions with electrons and radiation. Still another mechanism for such relaxation is theoretically possible through the "harpoon" reaction. The ion term of the metal atom and a halogen molecule is usually made up of several of their covalent terms corresponding to the different electronic states of the metal atom. Thus, during the "harpoon" reaction after the transition to the ion term and the movement along it toward the intersection with the covalent term, a transition to this covalent term can occur which corresponds to a less excited state of the metal atom together with a flying apart (dissociation) of the interacting particles as

$$T^{*1} + YZ \rightarrow T^{+} + YZ^{-} \rightarrow T^{*2} + YZ + \Delta\varepsilon,$$

where T^{*1} and T^{*2} are the metal atom in the electronically excited states 1 and 2, respectively, and $\Delta\varepsilon$ is the energy release which may go into vibrations of YZ, into translational degrees of freedom, or into dissociating the bond $YZ \rightarrow Y + Z$.

In the review [1] several versions of the chemical composition of a plasma laser were considered as illustrations of the principle of chemical depopulation of the lower working level. Unfortunately it is rather difficult to satisfy all seven of the requirements given here. Of the various versions known to us the mixture $TlF + F_2$ [2] fits these requirements best. Here we shall deal specifically with that mixture. It should be said at once that the specific calculations done for it are highly preliminary in nature. In this case not only is there practically no information on the chemical reaction rates of the ions and electronically excited molecules and atoms but, moreover, the probabilities of a number of collisional transitions of the thallium atom which are important for the recombinational relaxation of a pure thallium plasma (as they determine the populations of the excited levels of the thallium atom) are only known with very low accuracy. However, the most important parameters of a plasma laser with chemical depopulation of the ground state of an atom are almost independent of these probabilities, at least under the conditions of greatest interest for applications.

The fitness of the mixture $TlF + F_2$ for this type of active medium is confirmed by the following list of its properties:

1. The comparatively low boiling temperature of TlF and F_2, ~ 850°C and ~ −190°C, respectively.
2. The molecular ion TlF^{+} is unstable in the ground state; its term has a depth $D_{TlF^+} \sim 0.2$ eV. It decays to form a Tl^{+} ion.
3. The TlF_2 molecule has a fairly stable ground state term with $D_{TlF} \sim 4.6$ eV.
4. The stability of the ground-state electronic term of the F_2 molecule is low at $D_{F_2} \sim 1.6$ eV.

5. The cross section of the chemical reaction $Tl + F_2 \rightarrow TlF + F$ is not known to us but the cross sections for the reactions of the thallium atom with other halogens (iodine and bromine) are very large at $\sigma \gtrsim 10^{-11}$ cm^2. There is no reason to suppose that the reaction with the fluorine atom has a smaller cross section.
6. The F_2 molecule has a very high electron affinity, $\gtrsim 3$ eV.
7. The ionization potential of the thallium atom is small at about 6 eV.
8. The mixture $TlF + F_2$ will allow extended steady-state utilization and does not require replacement of the chemical components.

Over the entire set of properties listed here this mixture composition is the most favorable of all the variants we considered. Only the mixture $TlCl + Cl_2$ comes close (but in a number of parameters it is still a bit inferior).

At this stage we shall limit ourselves again to radiative and collisional relaxation of Tl atoms including their loss from every level due to chemical reactions.

Normally in plasma chemical lasers a number of levels are inverted with respect to the ground state at once. Then the (most) highly excited states are to be preferred as an upper working level as they yield the maximum efficiency. The populations of these states are less subject to depopulation by chemical reactions (which is undesirable for the upper working level) and an inversion is often simpler to obtain for them than for lower-lying energy states. In this regard it is reasonable to consider the possibility of using laser transitions from more highly excited states than have been proposed previously, for example, the 7d–6p and 8s–6p transitions of the thallium atom.

In general the molecules (TlF^*) obtained as a result of the reactions of excited thallium atoms with fluorine are highly excited vibrationally. These molecules will also relax over the vibrational levels as well as over the electronic levels. A full analysis of this scheme also requires examination of this relaxation. It is possible to attempt lasing on vibrational–rotational transitions of TlF in order to increase the efficiency of the system. For the electronically excited states of TlF^* it is possible to assume, as shown in [3], an absolute inversion over the vibrational sublevels. The mechanism for this inversion resembles that for the inversion on electronic levels of an atom with chemical depopulation. In both cases there is a system of relaxing levels with simultaneous removal of particles from the system by a process whose rate exceeds the rate of internal relaxation. In the case of the vibrational level system, removal is by radiative and collisional transitions between the electronic levels of the molecule.

For the ground-state electronic term of the molecule there is no analogous process and internal relaxation will be faster due to v–v exchange with the vibrational reservoir of this branch. Thus, for the ground-state electronic term of TlF it is more realistic to count on a relative inversion over the vibrational–rotational levels [4]. Since the depth of the potential well of the ground-state term of TlF is fairly large ($D_{TlF} \sim 4.6$ eV), it is important to consider the possibility of lasing on vibrational transitions when attempting to increase the efficiency of the system.

Results of Specific Calculations

The calculation was made assuming that the characteristic times for changes in the free-electron density are much greater than the characteristic relaxation times for the excited levels (the stationary-sink approximation). The populations of thirteen levels of thallium (Fig. 1) were taken into detailed account. These levels were joined in six groups of states whose quantum numbers were denoted in brief by: 1) $-6P_{1/2}$; 2) $-6P_{3/2}$; 3) $-7S_{1/2}$; 4) $-7P_{1/2}$, $7P_{3/2}$, $6d^2D$; 5) $-8S_{1/2}$; 6) $-8p^2P$, $7d^2D$, $5f^2F$.

The population distribution within each of these groups of levels was assumed to follow a Boltzmann law with the free-electron temperature T_e. The cross sections for interactions

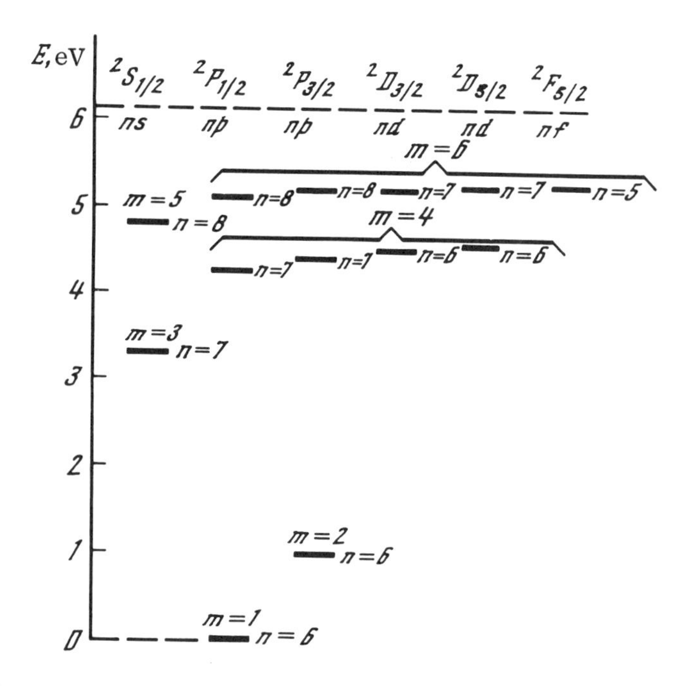

Fig. 1. Term diagram and brief notation for the thallium atom.

of Tl atoms with electrons were estimated using formulas derived in [5] from the Born approximation. The error in the estimates of the cross sections is not too important since, as already noted, the scheme depends weakly on the way they vary. For the (1-3, 1-2, 2-3) transitions we have used the results of more accurate calculations (the partial wave technique). The system of equations for the population distribution over the six equivalent levels in the quasi-stationary and stationary cases is as follows:

$$\frac{dN_m}{dt} = \sum_{\substack{m'=1 \\ m' \neq m}}^{6} K_{mm'}N_{m'} - (K_m + c_m)N_m = 0, \qquad m = 1, 2, \ldots, 5,$$

$$\frac{dN_6}{dt} = \sum_{m=1}^{5} K_{6m}N_m - (K_6 + c_6)N_6 + J = 0,$$

where $K_{mm'}$ is the probability of m' → m transitions, $V_{mm'} = \langle\sigma_{mm'}\, v\rangle$ is the rate constant for m' → m transitions in inelastic collisions with electrons, $A_{mm'}$ is the probability of a radiative m' → m transition, N_e is the free-electron density, $K_{mm'} = A_{mm'} + \langle\sigma_{mm'}v\rangle\, N_e$, $K_m = \sum\limits_{m'}^{6} K_{mm'}$ for $m' \neq m$, c_m is the rate of loss from the m-th level in the chemical reaction $Tl^*(m) + F_2 \rightarrow TlF^* + F$ and is given by $c_m = \langle\sigma^m_{reac}\, v\rangle N_{F_2}$, σ^m_{reac} is the cross section for the chemical reaction of thallium in the m-th state with fluorine, N_{F_2} is the density of F_2 molecules, and J is the recombination flux ($cm^{-3} \cdot sec^{-1}$) arriving at the sixth equivalent level.

Since the available information does not now allow us to choose a specific dependence of the chemical reaction rate c_m on the number of the level m, it is reasonable at this stage to obtain a solution under various assumptions about this dependence.

The results when the chemical reaction rate is assumed to be proportional to the statistical weight of the states are shown in Fig. 2. Clearly an inversion is achieved over the 7d−6p and 8s−6p transitions at much lower values of the chemical reaction rate than are inversions over the 6d−6p transition. This is because the chemical reaction has a greater effect on the population of the 6d level than it does on the 7d and 8s levels. It is reasonable to suppose that the

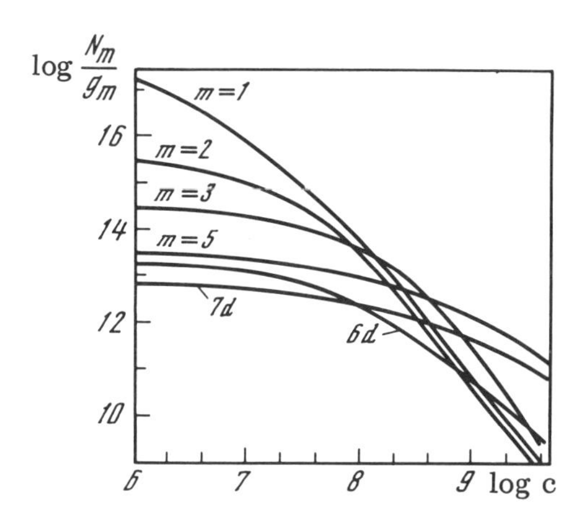

Fig. 2. The dependence for the population of the thallium levels on the rate constant c (sec^{-1}) for the loss of an atom in a chemical reaction during decay of a plasma with $N_e \simeq N_{Tl} = 10^{15}$ cm^{-3} and $T_e = 0.15$ eV. The rate coefficient for loss of excited Tl(m) atoms was given by the model $c_m = cg_m$, where g_m is the statistical weight of the state.

depths of the terms of the highly excited states of (TlF)* are also of the same order of smallness as the depth of the electronic ground state term of the $(TlF)^+$ ion. Then the equilibrium of the chemical reaction for the excited Tl* atoms which correlates with highly excited states of the molecule is shifted significantly toward dissociation. We shall thus consider a model in which only the levels 1, 2, and 3 enter into reactions with fluorine. The results of a calculation under this assumption are shown in Fig. 3. Here an inversion between the ground state and the 6d level develops earlier than one between 7d and the ground state, but the maximum inversion involves the 8s state, which is natural since this level is least subject to decay due to collisions with free electrons.

The unsaturated gain coefficients corresponding to the transitions, $\varkappa^0_{ab}$, were evaluated according to $\varkappa^0_{ab} = \sigma^{ph}_{ab}\Delta N_{ab}$; here $\sigma^{ph}_{ab} = (\lambda^2_{ab}/4)(A_{ab}/\Delta\omega)$ is the photoabsorption cross section, λ_{ab} is the wavelength of the b → a transition, ΔN_{ab} is the amount of inversion (given by $\Delta A_{ab} = N_a - (g_b/g_a) N_a$), N_b and N_a are the populations of the upper and lower laser levels, respectively, and $\Delta\omega$ is the linewidth of the working transition.

The transitions were assumed to be homogeneously broadened, the pressure was taken to be about 10 atm, and the gas temperature was taken to be about 1500°K. The gain coefficients for various transitions calculated with these assumptions are shown in Fig. 4 as functions of the free-electron density for $T_e = 0.15$ eV. From the graph it is clear that for each transition there is an optimal (for gain) free-electron density for given values of the rate of loss in chem-

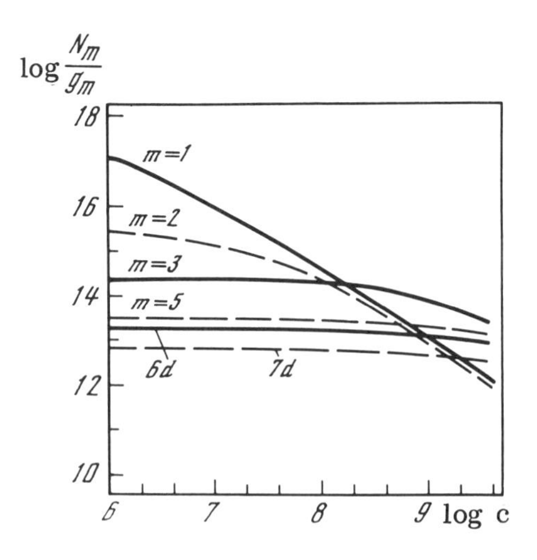

Fig. 3. Dependence of the level populations of thallium on the rate coefficient for loss in a chemical reaction assuming $c_m = 0$ for m > 2. The plasma parameters are the same at $N_e \simeq N_{Tl} = 10^8$ cm^{-3}, $T_e = 0.15$ eV.

Generally speaking the loss of "working" atoms from excited states as well as from the ground state due to the chemical reaction eases the requirements on the rate of this reaction, so that an inverted population may be obtained at smaller concentrations of the "nonworking" chemical reagent. But when it is possible to choose reagents it must be recalled that loss of excited atoms in excited states due to chemical reactions is primarily harmful in terms of the basic problem of converting the energy of ionizing the medium into laser radiation. This process, which opens up a parasitic channel for loss of atoms from the upper working state, thereby reduces the gain coefficient, lowers the efficiency of the laser, and speeds up heating of the active medium. This is conveniently illustrated by a discussion of the degree of utilization of the electron recombination flux as laser radiation. It might be expected that for a sufficiently high rate of stimulated transitions almost all the recombination flux arriving at an upper working level close to the continuum would go to the ground state of the atom in the form of laser radiation. But this, of course, depends on the ratio of the rates of loss due to the chemcal reaction from the ground and excited states of the atom. Thus, for example, in a chemical reaction scheme in which the coefficients c_m are proportional to the statistical weight of the Tl(m) state ($c_m = cg_m$) with $N_e \simeq 10^{14}$ cm^{-3}, $T_e \simeq 0.15$ eV, and a rate coefficient of the chemical reaction of $c \sim 10^8$ sec^{-1}, only a tenth of the entire recombination flux can enter the lasing channel. We note that these values of the laser parameters correspond to a maximum laser energy output of about 20 W/cm^3. In the other limiting scheme of chemical depopulation, evidently better than the chemical reaction discussed here, only the loss of thallium atoms from low-lying energy states due to chemical reactions is included. In this scheme the laser radiation may completely take up the recombination flux in its energy range, but the concentration of fluorine required to attain an inversion and, more important, to sustain such intense lasing, must be much larger ($c \gtrsim 3 \cdot 10^9$ sec^{-1}).

Factors Which Complicate This Relaxation Model

The preceding simplified model of the relaxation of a supercooled $TlF_2 + F_2$ plasma mixture yields some optimistic conclusions. A more detailed analysis requires knowledge of a number of interaction constants of excited atoms and molecules. These constants are unknown to us but a further examination (by analogy, in view of our ignorance) compels us to insist less resolutely on the favorable properties of this kind of plasma. We thus hope to turn the attention of experimenters to a number of difficulties which have not been included here.

I. Beginning with his measurements of the lifetime of the excited state Tl($7S^{1/2}$) in TlI vapors, Brus [6] concluded that the cross section which determines the quenching of these states of the TlI molecule is large, $\sigma_{TlI} \sim 2 \cdot 10^{-14}$ cm^2. He interprets this in terms of the harpoon formation of the complex $Tl_2^+I^-$ which then decays in a time of the order of the rotation period. If this supposition is true and allows generalization to analogous complexes (in particular, to $Tl_2^+F^-$) then the lifetime of the excited states of the thallium atom is very small, e.g., $\tau(Tl^*) \lesssim 5 \cdot 10^{-10}$ sec for $N_{TlF} \sim 3 \cdot 10^{19}$ cm^{-3}. In this case, to obtain a population inversion on the Tl(m) → Tl(1) transition the condition $N_{TlF} \lesssim N_{F_2}$ must be satisfied, and this impairs the lasing characteristics of the medium.

II. The properties of the active medium are also impaired by attachment of free electrons to molecules,

$$e + TlF \rightarrow Tl\,(6P^{1/2}) + F^-, \qquad (2)$$

$$e + F_2 \rightarrow F^- + F. \qquad (2a)$$

These processes are harmful primarily because they create a branch in the recombination flux of the electrons which goes past the upper working level and immediately to the lower level.

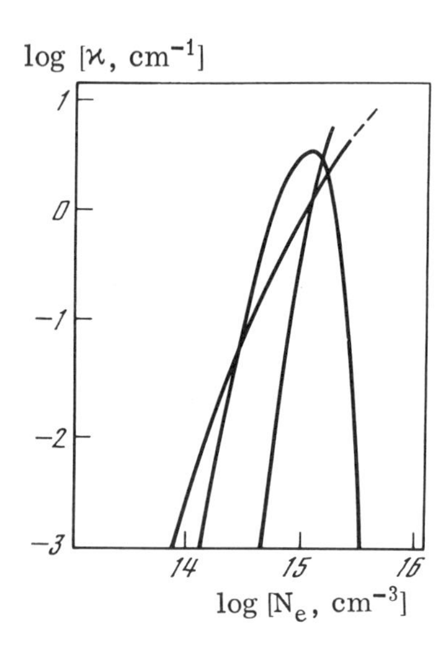

Fig. 4. The dependence of the gain coefficient on the electron density for several transitions into the ground state with $c = 10^9$ sec^{-1}, T = 0.15 eV.

ical reactions (or, equivalently, of the corresponding F_2 concentration). This is easy to understand. If there is not a sufficiently strong recombination flux it is impossible to plan on a high population in the upper working level, while for very strong recombination fluxes the chemical reaction cannot succeed in depopulating the lower working level, so the gain coefficient is also reduced.

The unsaturated gain coefficient on the $7d-6p^2P_{3/2}$ transition for a chemical reaction rate $c = 10^9$ sec^{-1} (or a fluorine concentration of about 1 atm) with $N_e \sim 10^{15}$ cm^{-3} and $T_e \sim 0.15$ eV reaches 3 cm^{-1}. The threshold rate of the chemical reaction at which this level is inverted relative to the ground state of the atom may easily be reduced by reducing the recombination flux. This is apparent from the graph of Fig. 5 which shows the dependence of these "threshold" values c_{cr} of the rate of loss through chemical reactions on the free-electron density N_e. For both models of the dependence of the chemical reaction rate on the electronic excitation mentioned above, an inversion over the $8s-6P_{1/2}$ transition develops at a reaction rate of about 10^8 sec^{-1} ($N_e \sim 0.1$ atm) for an electron density and temperature of $N_e \sim 10^{14}$–10^{15} cm^{-3} and $T_e \sim 0.15$ eV. This rate is an order of magnitude less than that obtained previously from estimates for the 6d level. We note that then the gain coefficient does not fall.

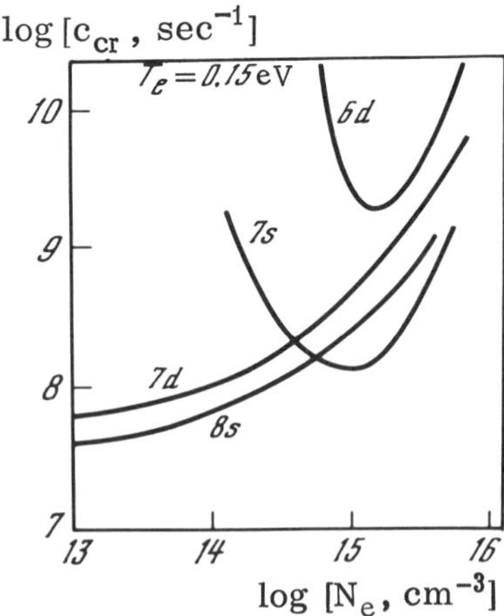

Fig. 5. The dependence of the critical rate coefficient for loss of thallium in a chemical reaction on the electron density.

In addition, both processes (2) and (2a) nonproductively reduce the density of free electrons. If processes (2) and (2a) predominate over three-body recombination $Tl^+ + e + Z \rightarrow Tl^* + Z$ then the recombination relaxation picture deteriorates.

Under what conditions do these processes least threaten our active medium?

I'. If the upper working level is a Tl^* state lying at or above $Tl(8S_{1/2})$ then its deactivation due to formation of the complex Tl_2F is less effective than the large-cross-section σ_e inelastic transitions due to collisions with free electrons (if their density N_e is sufficiently high). We use the notation $\alpha_0 \equiv N_e/N_{TlF}$. Then $\alpha_0 \gg \langle\sigma'_{TlF} v_{Tl}\rangle/\langle\sigma_e u_e\rangle$ must hold. Setting $\sigma'_{TlF} \sim 10^{-14}$ cm^2, $v_{Tl} \sim 10^4$ cm · sec^{-1}, $\sigma_e \sim 10^{-3}$ cm^2, and $v_e \sim 10^7$ cm · sec^{-1}, we obtain

$$N_e \gtrsim 10^{-3}\ N_{TlF}. \tag{3}$$

For a $TlF + F_2$ mixture with $N_{TlF} : N_{F_2} \simeq 1:1$ this requirement is difficult to satisfy, but if it is diluted substantially with an inert gas the situation improves. If we take $Ar + TlF + F_2$ with density ratios $N_{Ar} : N_{TlF} : N_{F_2} \simeq 98:1:1$, then for a mixture with a density $N \sim 3 \cdot 10^{20}$ cm^{-3}, we must have $N_e \gtrsim 3 \cdot 10^{15}$ cm^{-3}, which is quite possible.

II'. The harm caused by dissociative attachment (2) and (2a) is reduced as the rate of three-body recombination becomes greater. We note that recombination involving a molecule as the third particle takes place somewhat more rapidly than is customarily assumed. The attachment of an electron to a thallium ion in this case may be related to a loss of energy due to vibrational–rotational excitation of TlF:

$$Tl^+ + e + TlF\ (u, j) \rightarrow Tl^* + TlF\ (u', j'), \quad u' \geqslant u, \quad j' \geqslant j. \tag{4}$$

Here u is the vibrational quantum number and j is the rotational quantum number. Dissociative attachment will be suppressed if

$$N_e\ (k_{TlF} + k_e) \gg \langle\sigma_{at} v_e\rangle N_{TlF},$$

where k_{TlF} (cm^3 · sec^{-1}) $\sim 10^{-27} N_{TlF}$ (cm^{-3}), $k_e \simeq 10^{-23} N_e$ for $T_e = 0.15$ eV.

Literature Cited

1. L. I. Gudzenko, L. A. Shelepin, and S. I. Yakovlenko, Usp.Fiz. Nauk, 114:457 (1974).
2. L. I. Goodzenko, S. I. Yakovlenko, and V. V. Yevstignejev, Phys. Lett., 53A:105 (1975).
3. L. I. Gudzenko, Yu. B. Konev, and V. S. Marchenko, Katk. Soobshch. Fiz., No. 9 (1975).
4. L. I. Goodzenko and B. F. Gordietz, Phys. Lett., 41A:59 (1972).
5. L. A. Vainshtein, I. I. Sobel'man, and E. A. Yukov, Cross Sections for Excitation of Atoms and Ions by Electrons [in Russian], Nauka, Moscow (1973).
6. L. E. Brus, J. Chem. Phys., 52:1716 (1970).

CHAPTER V

NUCLEAR REACTOR-LASERS

The Idea of a Nuclear Reactor-Laser [1]

There is no need to prove that the most promising energy source is presently the nuclear reactor. As the natural supplies of traditional fuels are exhausted nuclear power stations are rapidly acquiring a decisive role in the energy economy of a number of countries. These power plants, which employ the newest equipment, the best materials, and the most advanced technical solutions for their manifold components, certainly are among the most important achievements of engineering and physics. But (however paradoxical it may seem to the majority of physicists not directly involved in the development of reactors) it is necessary to begin this chapter with the statement that in all nuclear power stations a basically archaic, fundamentally backward method is used to obtain electrical energy. This is not only our personal opinion; the most prominent physicists and atomic scientists continually regret this fact. This view of nuclear reactors is not new, having arisen even before the development of the first such installations, and was reflected in the very names given, apparently by E. Fermi, to them – "atomic furnace" or "atomic boiler." This somewhat abusive "steam engine" tint to the names for this remarkable modern device is due to the fact that in nuclear reactors the fission energy released is not initially converted directly into electrical energy but is first converted to heat. Yet the awareness that heat is the lowest grade of energy is already associated with S. Carnot. Heat can be converted only partially into electricity, in accordance with the comparatively low efficiency of a heat machine. Thus, even in the best nuclear power stations the efficiency (of conversion of the nuclear fission energy released in the reactor into electricity) only approaches a third. This chapter deals with a means for removing the energy from nuclear reactors which would radically improve their characteristics. From the following it will be clear that the bases of this new method are closely related to recent progress in several "nonnuclear" areas of physics; hence, the continued unproductive regret about the thermal utilization of nuclear energy is not so surprising.

In order to avoid the thermal path, it is necessary to prohibit the thermalization of the reacting medium or, in other words, not to allow the high-grade kinetic energy of the nuclear fragments released in the fission process to "equipartition" itself over the degrees of freedom of these fragments. Thus energy must be removed from the nuclear reactor while it is still in a substantially nonequilibrium stage, i.e., over times less than the relaxation time of the medium. Naturally, to solve this problem successfully, on the one hand, the processes leading to thermal equilibrium in the reactor must be slowed down as much as possible and, on the other, the fastest methods for removing energy from its volume must be used. Compared with ordinary nuclear power reactors it is possible to increase the relaxation time by several orders of magnitude. To do this it is necessary to go from traditional reactors with solid "fuel" (uranium, plutonium) to so-called gaseous reactors in which the fissible material is a chemical compound of the same uranium (plutonium) that is a gas under the working conditions. The fastest reasonable means of removing the nuclear fission energy from the reactor must obviously begin with its conversion into electromagnetic radiation. This radiation must be capable of being extracted from the reactor without great losses. In such a case it is natural to study the possibility of converting most of the energy of the fragments into the characteristic radiation of atoms and molecules. But to rely on the spontaneous emission of the medium filling the reactor, which goes in all directions and in wide frequency bands and is reabsorbed by the medium itself, is not suitable for this problem. Here we can only speak of using stimulated (induced) transitions provided the medium efficiently amplifies the corresponding radiation. Amplification of light (lasing) assumes the existence of a population inversion among the energy levels of the medium due to a thermodynamic disequilibrium produced by the nuclear fission products. Thus, we realize the value of analyzing the possibility of converting the energy of the fission fragments in a gaseous-phase nuclear reactor into laser radiation. Twenty or thirty years ago posing the problem in this way would have seened strange, at best, to the overwhelming majority of physicists. Even now it may seem premature to some scientists. This caution is easily understood since many problems of different kinds arise at once.

First of all the question of the reality of gas reactors in general is appropriate. It would hardly do if, when a reactor were filled with a gaseous uranium (or plutonium) compound, a critical mass were unachievable due to the additional (compared to pure fissile materials) losses of neutrons which always occur in such compounds. It may be that such devices would also be extremely impractical, for example, very heavy and large, extremely expensive, poorly controlled, etc. All of these questions, of course, have to be answered in detail. However, a detailed discussion would take a lot of time and, as we shall try to explain here, belongs more to the stage of choosing and evaluating the best design. In order not to answer each such question individually here, we note that research gas reactors (using uranium hexafluoride) with generally acceptable parameters do exist. The weight of this gas (in a reactor) with sufficient enrichment is fairly small; 20-30 kg of gas is sufficient for a critical mass at roughly 90% U^{235}. The pressure of this gas might not exceed atmospheric and the inside diameter of a spherical reactor might be small, of the order of 1 m. The core material of such reactors is uranium hexafluoride UF_6, a gaseous compound which is fairly widespread in reactor technology since it is used as an intermediate product in the separation of uranium isotopes. We shall not list here the numerous advantages of gas reactors over traditional ones. It is important to mention that such power plants would have some specific drawbacks. Not the least important of these would be the need to redesign the components to ensure safe operation of the reactors. In fact, in our case the fissible material is a gas which is not only radioactive but extremely poisonous. Gas reactors have not yet reached the same category as other nuclear power plants largely because of this deficiency, especially as a clear stimulus for the expenditure of material and time on the development and introduction of such devices has not perhaps existed until recently.

Of no less fundamental importance to us is a number of problems associated with the process of fission energy conversion into laser radiation. The possibility of effective energy conversion is determined by several starting assumptions.

1. The bulk of the nuclear fission energy is concentrated in the kinetic energy of the fission fragments which are multiply charged ions (of chemical elements with atomic weights of roughly half that of U or Pu).

2. The bulk of the energy of the fast multiply ionized particles goes into ionization as they move in the medium.

3. The ionization of a cold dense gas by fast charged particles leads, after a fairly short time, to a recombination disequilibrium in the resulting plasma. This kind of disequilibrium is characterized by an enhanced density of free electrons, recombination relaxation of charged particles, and extremely high populations in the electronically excited states of atoms (molecules and ions), which may exceed the equilibrium values by ten or more orders of magnitude.

4. A recombining dense plasma with appropriately chosen composition efficiently amplifies the characteristic radiation of its atoms (molecules, ions). A number of advantages of such active media over the more traditional ones make it possible to expect "plasma lasers" with high energies [2].

Transformation of the kinetic energy of the fragments into stimulated emission requires a sufficient plasma disequilibrium, and a corresponding rate of recombination which also means a nuclear fission rate exceeding some threshold value (i.e., $G > G_{thr}$). On the other hand, since in any case a large part of the fission energy is lost as heat, while for excessive gas temperatures $T > T_{cr}$ the population inversion over the atomic (molecular, ionic) levels is cut off, yet another limit must be placed on the rate of fissions which, if it is exceeded, leads to termination of lasing due to overheating of the gas ($T > T_{cr}$ for $G > G_{cr}$). Thus, the reactor must be designed so the limiting fission rate per unit volume of the gaseous medium determined by its parameters obeys (with a reliable margin) the condition

$$G_{cr} > G_{thr}. \tag{1}$$

The basic design of this type of nuclear reactor is determined by the distinctiveness of one of its linear dimensions as well as by the fact that the fuel is gaseous. Efficient laser removal of the energy is possible only from a reactor long enough that light is amplified sufficiently in its laser-active medium. Several quantities traditionally included in reactor designs will also affect the devices being discussed here in specific ways. Of first importance is the fact that the mean free paths of neutrons in the medium are comparatively long while the mean free paths of the fission fragments are rather short. In a gas of density ($N \sim 3 \cdot 10^{20}$–10^{21} cm^{-3}) suitable for this problem the mean free path of the fast fragments (ions with a charge roughly twenty) is of order $l_{mfp} \lesssim 10^{-1}$ cm. The diffusion of heavy charged particles may be neglected in the fairly cold dense plasmas analyzed below, so the recombination zone of the medium practically coincides with the zone in which it is ionized by the fragments. Thus, lasing may occur only in those regions where nuclear fissions take place; that is, the nuclear- and laser-active zones must coincide. Thus, when discussing efficient laser removal of energy from a nuclear reactor, we must speak of a single power-producing device, the reactor-laser (RL), rather than of some sort of laser located inside a nuclear reactor [3].

We have already to speak of a whole series of modifications of reactor-lasers and, thus, of their classification according to various features, one of which is primarily nuclear, the others of which are mainly optically based. The mode of operation of the nuclear reactor is first used to distinguish steady-state thermally pulsed reactor lasers. In reactors of the first type the required thermal regime ($T < T_{cr}$) is ensured by continuous heat release from the active zone. The second type of reactor assumes the use, first of all, of the inherent heat capacity of the system, but here (in the case of power applications with light pulses repeated as often as possible) it is also impossible to proceed without forced removal of heat from the active zone. Thus, the design of any such system must involve dividing the device into a fairly large number of energy-producing elements which are bathed in a coolant and have at least one comparatively small dimension in a direction perpendicular to the direction along which light is amplified in the active medium. The overall neutron field, which links the entire RL into a single device, is fed by (and uses) the gaseous fuel in each of these elements. The transverse dimensions of the elements must also not be too small, otherwise the fission fragment energy lost at their walls will reduce the efficiency and the neutron energy losses will increase unacceptably.

In terms of their light characteristics, pulsed reactor-lasers may in principle be optically quasistationary or optically pulsed. In addition, reactor lasers form two classes (both stationary and pulsed) in terms of their radiant self-containment: (1) with mirrors for creating laser feedback (in such lasers the unsaturated amplification of light as it passes through the active medium must exceed the cavity losses), and (2) avoiding the use of vulnerable (damaged by radiation, the fission fragments, and the corrosive medium) mirrors but requiring the input of fairly powerful radiation from an external laser at the same frequency. A compromise version is also possible – lasing without mirrors in a "superradiant" regime. By analogy, in the "nuclear" classification it is of interest to examine the properties which are crucial to reactor-lasers and their subcritical variants (booster and hybrid) in terms of the neutrons.

We now proceed to a more specific discussion of several of these general questions.

The Disequilibrium in the Active Medium of a Reactor-Laser

We first consider the deviation from thermal equilibrium of the gaseous medium filling the energy (producing) elements.† At first this medium goes strongly out of equilibrium with

† As opposed to the case of ordinary nuclear reactors, the traditional name "fuel element" does not reflect the way energy is removed from a reactor-laser.

respect to the velocity distribution of the heavy particles since the fission energy is converted mainly to kinetic energy of multiply charged ions, the fission products. As a rule, immediately after fission a pair of fragments fly apart with energies of about 100 MeV each. These particles fly into the comparatively cold gas in the energy elements where the overwhelming bulk of the molecules have a small energy corresponding to a temperature $T \lesssim 0.3$ eV. Thus, some of the particles, having a mean density $\rho_{frag} \simeq M_U G \tau_{frag}$ (where M_U is the mass of a uranium nucleus and τ_{frag} is the characteristic time over which the fragments lose their energy) which is very small, have an energy (per particle) about 10^9 times greater than the mean energy per particle of the gas. Of course the spatial and temporal coexistence of two such groups of heavy particles characterizes an extremely nonequilibrium situation with low entropy. Such a situation makes it possible to convert the bulk of the fission energy released into work.

The next stage of energy conversion involves ionization of the cold, dense gas in the energy elements by the fission fragments. During this external ionization a plasma with a small degree of ionization $\alpha \equiv N_e/N$ is formed (where N_e is the free-electron density and N is the gas density). This value of α is, however, much greater than the thermodynamic equilibrium value of the degree of ionization corresponding to this gas density and to the free-electron temperature T_e obtained under these conditions. Thus (we emphasize again) the energy elements are filled with a superionized plasma. This may be said in another way: At the free-electron density produced in this plasma by the entire set of processes accompanying nuclear fissions, the mean energy (temperature) of the free electrons is below equilibrium. The concepts of "superionization" of the plasma and "supercooling" of its free electrons are equivalent. They characterize the recombination disequilibrium of the plasma. In such a homogeneous plasma the relaxation to equilibrium is evidently determined by volume recombination of free electrons with ions.

As we have already noted, the direct interaction of the fast multiply charged ions with the gas in the energy elements reduces almost completely to ionization of the molecules. Elastic impacts of these ions which heat the gas and inelastic collisions which excite the molecules to discrete levels have very much lower probabilities than the ionizing collisions. The efficient relaxation interaction of the free electrons formed in this way with the cold dense gas does not allow the fast ions to pump collective plasma oscillations. This interaction causes the disequilibrium in the distribution of the electrons in the continuum to make the population of a number of bound states of the gas molecules differ strongly from a Boltzmann distribution. Part of the discrete energy levels are overpopulated and part are underpopulated. Because of this the characteristic radiation of the gas molecules in the energy elements can be used to remove the energy from a reactor-laser. Lasing may occur on a whole set of different transitions (electronic, vibrational, vibrational–rotational) at once, rather than on a single transition. However, for concreteness we shall mainly consider lasing on molecular (atomic) transitions between electronic terms. Relaxation of a supercooled plasma forms a recombination flux of electrons over the energy levels from the continuous spectrum of the molecules to their ground states. This mainly leads to overpopulation of the higher-lying states which are closest to the continuum and to underpopulation of the lower-lying states next to the ground state. For a reasonable choice of conditions on the transitions a population inversion develops mainly between these groups of levels and may be used for lasing.

The recombination flux due to ionization of the gas by the fission fragments is clearly the pump for the "laser component" of the nuclear reactor. This pumping may be either pulsed or steady state. In the pulsed variety there can (in principle) be such a strong burst in the number of fissions that the resulting growth in the free-electron density corresponds to a substantial disequilibrium in the free-electron distribution, but this time of the ionization type. An inversion in the populations of the discrete levels can also be associated with this type of disequilibrium in N_e and T_e. It is just this type of disequilibrium (with a few overheated electrons present and ionization taking place) which characterizes the active medium of a different type of gas laser which is more customary in quantum electronics than plasma lasers (whose active medium is a recombining plasma). However, it is important to keep in mind that gas laser schemes cannot work under reactor-laser conditions. It has already been noted that to obtain a critical mass of uranium in the entire set of energy elements while keeping the whole reactor to a reasonable size it is necessary to have a sufficiently dense gas filling. When the density is increased it becomes more difficult to overheat the free electrons at the low gas temperature required to produce the ionization disequilibrium. The fission pulse must be made still shorter, but the gas overheats even faster. The gas technique practically loses all its effectiveness somewhere around a gas density of $N \sim 10^{18}$ cm^{-3}. A density at least two or three orders of magnitude greater than this is needed for productive operation of a reactor-laser. After termination of the fission pulse recombination relaxation begins and during that time the disequilibrium may already be fairly effectively realized with the help of lasing and removal of energy in this manner. However, under the conditions discussed here, the relaxation of the molecular and atomic populations takes place rather rapidly over times $\tau \lesssim 10^{-8}$–10^{-7} sec, so even with the shortest technically realizable fission pulse lasing should occur before the end of the pulse. Thus, in analyzing the operation of a reactor-laser it is necessary to consider mainly the quasistationary (or stationary) regime.

We begin with the remark made above that for a quasistationary regime a recombinational disequilibrium is always established in the gaseous filling of the energy elements. Let us explain it, limiting ourselves to spatially homogeneous stationary conditions. Let z denote the average number of electrons generated in the gas by each "cascade" of ionizations caused by a single fission of a uranium nucleus. To estimate this number we may write $z \simeq E_{fis}/E_{pair} \simeq 2 \cdot 10^8$ eV/20 eV = 10^7. Here $E_{fis} \simeq 200$ MeV is the initial kinetic energy of the disintegration fragments and $E_{pair} \sim 10$–30 eV is the average energy expended in a single ionization (appearance in the plasma of a new molecular ion–electron pair) in the accompanying ionization cascade. The rate of change in the free-electron density is determined by the fissions of nuclei, by volume recombination of charged particles, and by ionization of the gas by its own electrons. Assuming that the degree of ionization of the medium is extremely small ($\alpha \lesssim 10^{-5}$) we shall assume for simplicity that the main recombination process is three-body collisions involving a neutral molecule. Assuming, in addition, that the electron and ion densities are the same in view of the small degree of ionization, we have

$$dN_e/dt = zG - K_e N_e^2 + S N_e N. \tag{2}$$

The time variation in the free-electron temperature T_e is determined by the power density of the reactor-laser and the efficiency with which the gas removes the energy of the electrons; that is,

$$\frac{3}{2}\frac{d(N_e T_e)}{dt} = W - \langle \sigma v_e \rangle N_e N, \tag{3}$$

where $W = GE_{fis}$ is the power released by fissions per unit volume of the medium, $\sigma = \sigma' + \sigma''$ is the effective cross section for exchange of energy between electrons and molecules, $\sigma' = (2m/M)\sigma_{el}(T - T)$ is the cross section for elastic energy transfer, m and M are the electron and ion masses, σ_{el} is the momentum transfer cross section, $v_e = (2T_e/m)^{1/2}$ is the electron thermal speed, and σ'' is the total characteristic of the probabilities of electronic and vibrational–rotational transitions of molecules during collisions with free electrons. In a dense gas with a large concentration of molecules $\sigma'' > \sigma'$ as a rule. Heating the gas beyond the critical temperature T_{cr} (which, depending on the gas composition and the choice of working levels,†

† In quantum electronics it is customary to refer to the energy levels of the active medium between which stimulated transitions determine the gain of the medium as the working levels.

usually lies within the limits $T_{cr} \sim 0.1$-0.2 eV) violates the condition for effective lasing. Assuming that $T < T_{cr}$ it is possible, as a rule, to neglect the equilibrium emission of the gas in the first approximation. However, the interaction of the molecules with the supercooled (because of this) free electrons leads to a noticeable disequilibrium in the population of the excited states. As a result of this interaction there appears, in particular, a population inversion of the molecules over the working transitions. Taking this into account the equation for the rate of change of the gas temperature T (more precisely, the temperature of the heavy particles) can be written in the form

$$\frac{3}{2} N \frac{dT}{dt} = [\langle \sigma' v_e \rangle + (1 - r) \langle \sigma'' v_e \rangle] N_e N - Q, \tag{4}$$

where the dimensionless parameter r effectively describes the fraction of the excitation energy of the molecules due to electron impacts which goes into emission from the gas and Q is the average rate of heat transfer to the walls of the energy elements from the gas. In a more detailed description of the relaxation of the active medium it is necessary to include the sublevel population kinetics. Then, in place of the effective parameters σ'' and r in Eqs. (2)-(4) there is a much larger group of quantities, both parameters and variables. We shall not dwell on that description here as it has to be done anew for each specific chemical composition of the medium. We shall also not yet include the inhomogeneity of the conditions across the perpendicular cross section or along the length of the energy elements.

In Eqs. (2)-(4) it is most common to set $dN_e/dt = 0$, $dT_e/dt = 0$ assuming that the time dependence of T_e and N_e may be related only to the variation T(t) in the gas temperature. This "quasistationary analysis" when the gas temperature might not be constant is justified since at low degrees of ionization when $\alpha \ll T/E_{pair}$ (and it only makes sense to consider such a plasma here) the characteristic time τ_T for a change in the gas temperature greatly exceeds the relaxation times of the electron density and temperature, τ_{N_e} and τ_{T_e}. Taking $T \leq T_{cr}$, where $T_{cr} = 0.1$-0.2 eV, and assuming that the energy to form a molecular ion–free electron pair in the plasma is $E_{pair} \sim 10$-30 eV, we see that quasistationarity is already assured at $\alpha \sim 10^{-4}$. When the quasistationary solution of Eq. (2) is evaluated with the actual gas and electron temperatures put in it, it clearly yields a nonequilibrium value $N_e = N_e^0(1 + \eta)$ since the term zG determines how much the energy flux brought by ionization into the gas (as if from outside) differs from zero. The other reason for the disequilibrium (the difference in the temperatures of the electrons and heavy particles) is less important in determining the value of N_e. Let us estimate η, the relative deviation (caused by fission fragments in the energy elements) in the free-electron density from its thermodynamic equilibrium value N_e^0. The value of N_e^0 itself is found from Eq. (2) by setting $G = 0$ in it; N_e^0 corresponds to the equilibrium value at the actual values of the gas density N and electron temperature T_e in this plasma†:

$$K_e N_e^0 = SN. \tag{2a}$$

According to Eq. (2) we have also

$$\eta = \frac{1}{2} (\sqrt{1 + \beta} - 1), \qquad \beta \equiv 4 \frac{zGK_e}{S^2 N^2}. \tag{5}$$

Since the parameter β is positive, in this problem $\eta > 0$ always; that is, the plasma is not in recombinational equilibrium (i.e., is superionized or, equivalently, supercooled). The value of the dimensionless parameter β then characterizes the degree of recombinational disequilibrium.

† The value of the gas temperature must be set equal to the electron temperature ($T_e = T$) when determining N_e^0, but this has almost no effect on the parameters S, K_e, and z.

Equation (5) demonstrates the simple relationship between this parameter and the relative deviation from equilibrium of the electron density when it is maintained in a quasistationary state by a hard source. In particular,

$$\eta \simeq \beta/4 \quad \text{for} \quad \beta \ll 1 \quad \text{and} \quad \eta \simeq \sqrt{\beta/4} \quad \text{for} \quad \beta \gg 1. \tag{5a}$$

$\beta \gg 1$ is the condition for intense volume recombination of the plasma. If it holds, then

$$N_e \simeq N_e^0 \eta = \sqrt{\frac{zG}{K_e}}, \tag{5b}$$

that is, the term SNN_e, which determines the rate of ionization of the gas by its own electrons, may be neglected in Eq. (2). It is clear that this applies not only to the electron density, but also to the populations determined by N_e. The physical conditions for this are formulated directly in Eq. (5) for β: For sufficiently high power densities of the external ionization source and low free-electron temperatures† it is permissible to assume that these plasma electrons‡ do not ionize the molecules but only recombine, thereby compensating for the "external" ionization of the gas.

The Composition and Parameters of the Active Medium [4]

We now consider the general features of a population inversion among the discrete levels of atoms (molecules, ions) in an intensely recombining plasma. We begin with the fact that inelastic collisions of heavy particles with electrons have the most important effect on the excited level populations in a superionized plasma with a sufficiently high free-electron density ($N \gtrsim 10^{14}$ cm^{-3}). Collisional transitions between energetically close states (of the particles) whose energy differences do not exceed the electron temperature have the greatest probabilities. Because of the effectiveness of such transitions local "blocks" appear in the population distribution of the discrete levels. Within each of these blocks the populations follow a Boltzmann law with a unique total population for each block and a common effective temperature, equal to the free-electron temperature T_e of the plasma, for all the blocks. When this is taken into account the analysis of the population relaxation is simplified. In the first stage of the calculation closely lying energy levels are combined into single states (blocks) with a total population $N_{(M)} = \sum_{m \in M} N_m$ in each block and statistical weights $g_{(M)} = \sum_{m \in M} g_m$ (Fig. 1). The populations of a system formed in this way are of a much lower order (of difficulty) to deal with: It is simpler than the atom being analyzed here and the energy levels in the system are located relatively far away from one another. In the second stage of the calculation the populations of the individual levels within a block are found simply by using the Boltzmann formula.

Of special interest is the band of close-lying electronic energy levels adjacent to the continuum (free) electrons. The populations of these levels relax rapidly, having been filled by the recombination flux directly from the electron continuum. Because of the closeness of these levels the rate of the flux is determined by transitions due to free-electron impacts. As the energy of the atomic state is reduced the rate of collisional relaxation falls, but in general the probability of its spontaneous radiative deexcitation rises. In addition, under certain conditions

†According to the above [see Eq. (3)] the low (for a significant free-electron density) temperature $T_e \lesssim 1$ eV is maintained for sufficient gas densities N by cooling of electrons by the heavy particles of the gas.

‡The external source for ionizing the gas might be something other than nuclear fragments in a broad class of similar problems, for example, electrons introduced by an "external" beam.

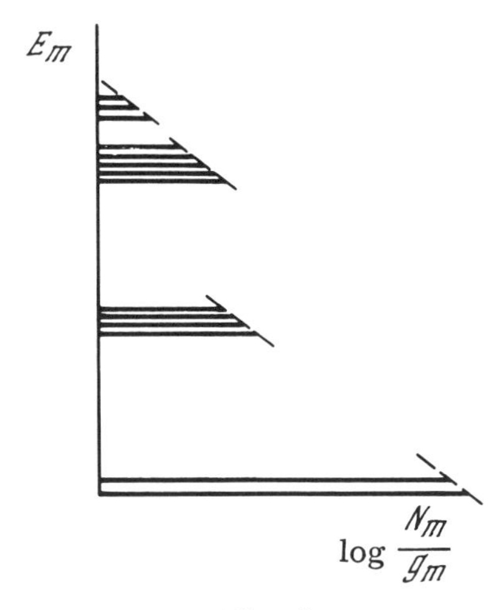

Fig. 1

collisions of the heavy particles among themselves have an effect on these excited states, often leading to emptying (a sharp reduction in the populations). In such a situation a population inversion over some pair or other of levels develops and light can be amplified at the frequency of the transition between them.

Let lasing take place on the X(b) → X(a) transition of the atom (molecule, ion) X. This transition is known as the working transition. We shall use the indices a and b to denote characteristics of its lower and upper "working" levels, which have energies E_a and E_b, where $E_a < E_b$. Because of an inversion in the populations of these levels ($N_a/g_a < N_b/g_b$) radiation with frequency $\omega_{ab} = 1/\hbar(E_b - E_a)$ is amplified by stimulated radiative transitions X (b) + $\hbar\omega_{ab}$ → X (a) + $2\hbar\omega_{ab}$ between these levels. At small intensities J of this radiation its passage through the laser medium has almost no effect on the populations. The radiation itself increases in intensity in proportion to the distance it has propagated within the medium; that is,

$$dJ = J\varkappa_0 dz, \qquad \varkappa_0 = \sigma_{ab}\left(N_b - \frac{g_b}{g_a}\right)N_a = \sigma_{ab}N_b(1 - \sigma_{ab}), \tag{6}$$

where σ_{ab} is the photoabsorption cross section for the working transition, $\delta_{ab} \equiv g_b N_a/g_a N_b$ is the inversion coefficient, and $\varkappa_0$ is the unsaturated gain coefficient. As the intensity increases the population of the upper working level falls and the population of the lower working level rises. The gain falls and saturation occurs. For nearly complete saturation $J \gg J_s$ the medium is almost transparent; i.e.,

$$N_b - \frac{g_b}{g_a}N_a \ll N_b, \qquad \varkappa \ll \varkappa_0.$$

Then the amplification of light in the medium is proportional to the path length traversed and no longer depends on the field intensity; that is, dJ = Cdz. Since we are interested in the preparation of a laser-active medium, we shall not now include the effect of the amplified light on the populations, and of the radiative transitions we shall consider only the spontaneous transitions. Let us examine the simplest laser scheme, the "open two-level model." We assume that there are no intermediate levels between the working levels, that the energy difference $E_b - E_a$ is not small, and that in the electron recombination flux flowing from the continuum along the discrete levels, it is possible to neglect transitions which bypass state b (dashed arrow in Fig. 2). Then the equations for the populations of the working levels may be

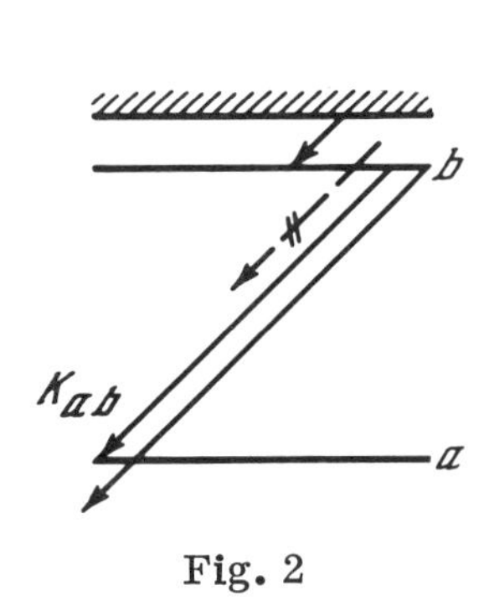

Fig. 2

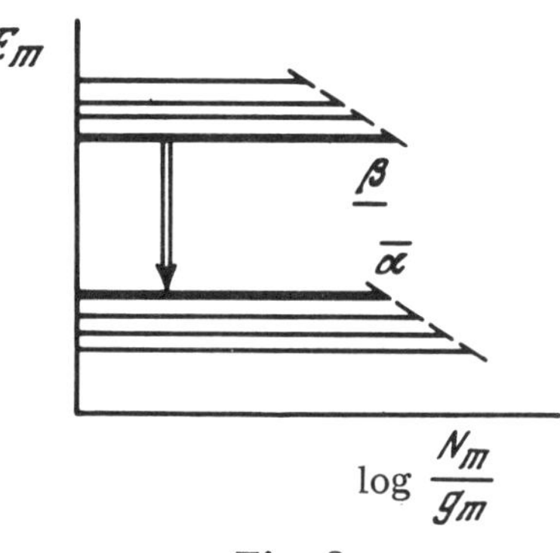

Fig. 3

written in the form

$$\frac{dN_a}{dt} = K_{ab}N_b - K_aN_a, \qquad \frac{dN_b}{dt} = \frac{N_+}{\tau_+} - K_bN_b. \tag{7}$$

Here K_a, K_b, and K_{ab} are kinetic coefficients and N_+ and τ_+ are the density and characteristic lifetime of the X^+ ions. The "stationary-sink" conditions are valid in a wide range of problems; that is, the rates of change of the populations of the working levels are small compared to the partial rates at which they are both filled ($K_{ab}N_b$ and N_+/τ_+) and emptied (K_aN_a and K_bN_b). Thus in estimating the populations of the working levels we may set $dN_a/dt = dN_b/dt = 0$ in Eq. (7). Then the condition for an inversion, which must be satisfied by the kinetic coefficients, takes the form

$$\delta_{ab} = \frac{g_bK_{ab}}{g_aK_a} < 1. \tag{8}$$

From Eq. (7) we also obtain a simple formula for the population of the upper working level,

$$N_b = 1/K_b\, N_+/\tau_+. \tag{9}$$

If instead of two energetically isolated states a and b of the atom X there are blocks of close-lying levels ($\alpha_1, \alpha_2, \ldots, \alpha_p$) and ($\beta_1, \beta_2, \ldots, \beta_q$) with $E_\beta - E_\alpha \geq |E_{\alpha'} - E_{\alpha''}|, |E_{\beta'} - E_{\beta''}|$, then equations of the form of Eq. (7) determine the kinetics of the total populations of these blocks. Usually in such a case it is appropriate to choose as the upper working level the most populated state $\underline{\beta}$, that is, the lowest level of block (β), and as the lower working level, the least populated state $\bar{\alpha}$, that is, the uppermost level of block (α). The condition for a population inversion on the $\underline{\beta} \rightarrow \bar{\alpha}$ transition may be much more lax than that of Eq. (8) for the total populations of blocks (α) and (β). This is illustrated by Fig. 3.

The decay of a supercooled dense plasma may begin with various recombination processes: three-body ($X^+ + e + Z \rightarrow X^* + Z$), dissociative ($XY^+ + e \rightarrow X^* + Y$), ion-ion ($X^+ + Y^- \rightarrow X^* + Y$ or $X^+ + Y^- + Z \rightarrow XY^* + Z$), charge exchange ($X^{++} + Y \rightarrow (X^+)^* + Y^+$), and so on, but all of them lead with overwhelming probability to the filling of highly excited electronic states of the atom (molecule, ion) obtained in the initial state of relaxation. It may thus be said that recombination of the plasma is a universal mechanism for "pumping" the upper working level. The more intense the recombination (that is, the greater the ratio N_+/τ_+) is, the higher the population N_b of this level [according to Eq. (9)] will be. The gain coefficient $\varkappa$ increases in proportion to N_b and therefore to the ratio N_+/τ_+. It is only necessary that condition (8) for a population inversion be satisfied. This condition is assured if the rate K_a of depopulating the lower working level is high enough. The mechanisms for depopulation must be examined more individually as they are based on quite different principles. The possibility and efficiency of their operation depend both on the general conditions (on N, N_e, T, and T_e) and on the chemical composition of the plasma.

The simplest and most general depopulating mechanism is the spontaneous radiative decay of the lower working level,

$$X(a) \rightarrow X(0) + h\omega_{0a}. \tag{10}$$

In a brief discussion we cannot deal separately with depopulation of an excited state due to collisions with free plasma electrons. We have already noted that the result of such a process is the formation of blocks of close-lying states in the population distribution of the discrete levels and that within these blocks local Boltzmann distributions are rapidly set up. Depopulation of excited levels during collisions between heavy particles most often occurs either through ionization of an atom or molecule of an easily ionized impurity by an excited particle in Penning ionization,

$$X^* + Y \rightarrow X + Y^+ + e, \tag{10a}$$

or through loss of particles in a chemical reaction such as

$$X + Y + Z \rightarrow XY + Z, \quad X + YZ \rightarrow XY + Z. \tag{10b}$$

It is significant that chemical reactions may remove particles (atoms, molecules, ions) both from excited and from the ground electronic states. We can point out two other mechanisms which may depopulate both the electronically excited ground states. In the case of an ion it may recombine into a neutral particle (or into an ion of lower charge),

$$X^+ + e \rightarrow X^*,$$

and in the case of a thermally unstable or (even) dissociating molecule XY the emptying of such a term takes place through the disintegration of the molecule

$$XY \rightarrow X + Y, \tag{10c}$$

For nuclear fission energy to be effectively removed from a nuclear reactor in the form of directed light it is necessary that a number of conditions occur, of which we note two: (1) the range of energy of the working transitions of the medium which go into radiation must be close to the ionization potential of the "working" atoms (molecules), and (2) the intensity of the radiation at these transitions within the energy elements must be sufficiently large that spontaneous emission may be neglected compared to induced emission.

The latter condition in fact means that the energy elements must each individually or all together provide enough amplification of the light. The threshold gain coefficient which would compensate for the losses of working radiation in the medium and in the cavity is given by the value at which $\exp(\varkappa_{thr} L) = 1 + \xi$, where L is the length of the active medium and ξ is the fraction of the luminous intensity lost in the cavity in one pass. If $\xi \ll 1$, then

$$\varkappa_{thr} = 1/L\,(\ln(1 + \xi) \simeq \xi/L. \tag{11}$$

Here $\varkappa_{thr}$ may serve only as a sort of orientation since at the self-excitation threshold the intensity of a laser is too small. As J increases, the gain coefficient, as noted above, falls; thus, for effective operation of a reactor-laser it is ncessary to pump the upper working state much more strongly, at least one or two orders more than the threshold value. Since we have not yet discussed the specific relaxation mechanisms and chemical composition of the active medium, we shall assume a priori that effective emptying of the lower working level allows us to neglect its population and to assume $\delta_{ab} = 0$. The threshold value W_{thr} of the power density of the reactor-laser in a quasistationary regime will now be found using Eq. (2) with $SNN_e \simeq 0$. Having obtained

$$N_+/\tau_+ = K_e N_e^2 = zG,$$

we insert this together with the condition $\delta_{ab} = 0$ into Eq. (6). Then for the number of fissions it follows that

$$G_{thr} = \frac{K_b \varkappa_{thr}}{\sigma_{ab} z} = \frac{K_b \xi}{\sigma_{ab} z L}. \tag{12}$$

The cross section σ_{ab} for absorption on the working transition may be written in the form $\sigma_{ab} \simeq (\lambda_{ab}^2/4)(A_{ab}/\Delta\omega_{ab})$, where $\Delta\omega_{ab}$ and λ_{ab} are the frequency width and wavelength of this spontaneous transition b → a and A_{ab} is its probability. The linewidth under these conditions as a rule is due to Doppler broadening, so we may set $\Delta\omega_{ab} = 10 v_T/\lambda$, where $v_T = (2M/T)^{1/2}$ is the thermal speed of the heavy (lasing) particles. From this we obtain

$$G_{thr} = \frac{1}{Lz} 40\xi \frac{v_T}{\lambda_{ab}^3} \frac{K_b}{A_{ab}}. \tag{12a}$$

The total probability K_b of decay of the upper working level of the active medium is made up of the probability of a spontaneous radiative transition to the lower working level as well as to all other energy levels of the working particle X lying below b, together with the probabilities of collisional transitions from level b to all states of this particle during collisions with electrons and heavy particles, and finally, with the probability of loss of X(b) in chemical reactions. If the densities of both the free electrons and the impurities (relative to the working particles X) are comparatively small, then the probability of decay of the upper working level is minimal and is determined solely by its spontaneous emission. Often in this case it is possible to include only the working radiative transition since the other spontaneous transitions from the upper level are strongly attenuated due to reabsorption in the dense gas. Denoting the corresponding minimal threshold rate of nuclear reactions by G_{min}, we write $K_b \simeq A_{ab}$, and also

$$G_{min} = 40\xi \frac{v_T}{zL\lambda_{ab}^3}, \qquad W_{min} = G_{min} E_{fis}. \tag{13}$$

From these formulas it is clear that the efficiency of the luminous channel for removal of energy from a reactor-laser (otherwise the same as the lasing condition for any laser) is determined by the intensity J of the total pumping through a cross section perpendicular to the direction of the radiant flux leaving the optical cavity. In our case $J \equiv WL$, where W as before is the power density of the reactor, the pump power released in 1 cm^3 of the amplifying medium. For the minimal threshold intensity we have, according to Eq. (13),

$$J_{min} = W_{min} L \simeq 40\xi \frac{v_T}{\lambda_{ab}^3} E_{fis}. \tag{14}$$

In order to estimate the resulting orders of magnitude of these quantities, we now specify the actual numerical values of the parameters of an active medium. For example, let $\xi = 0.1$, $z = 1 \cdot 10^7$, $E_{pair} = 20$ eV, $\lambda_{ab} = 4000$ Å, and $v_T \simeq 10^5$ cm/sec. Then the three lengths $L^{(1)} = 2$m, $L^{(2)} = 10$ m, and $L^{(3)} = 50$ m for the active medium of an energy element correspond to the following minimal values of the fission rate and power density:

$$G_{min}^{(1)} = 4 \cdot 10^9 \text{ fissions/cm}^3 \cdot \text{sec}, \qquad W_{min}^{(1)} = 0.1 \text{ W/cm}^3,$$
$$G_{min}^{(2)} = 8 \cdot 10^8 \text{ fissions/cm}^3 \cdot \text{sec}, \qquad W_{min}^{(2)} = 0.02 \text{ W/cm}^3,$$
$$G_{min}^{(3)} = 1.6 \cdot 10^8 \text{ fissions/cm}^3 \cdot \text{sec}, \qquad W_{min}^{(3)} = 0.004 \text{ W/cm}^3.$$

All three lengths correspond in this case to the same threshold fission intensity $J_{min} = 20\,W/cm^2$. From this it is clear that these very modest values (for a nuclear reactor) are crude underestimates. Thus, under reactor-laser conditions the rate of collisional and chemical depopulation of the upper working level would (as a rule) exceed the rate of spontaneous radiative decay of this level. Thus, the actual threshold pumping intensity J_{thr} must exceed J_{min} by ten or more times which, however, still does not make the intensity J_{thr} too high. For more rigorous calculations it is necessary to consider the specific composition of the gaseous mixture and the thermal conditions in the energy elements.

Removal of an appreciable portion of the energy of the fission fragments in the form of light can ease the thermal operating regime of a nuclear reactor. In estimating (with a margin) the critical temperature characteristics of a reactor-laser we shall assume for simplicity that the entire energy of fission of the uranium nuclei goes into heat in the energy elements. The need to examine the thermal regime of a reactor-laser arises for three reasons. First, the free electrons must be cold enough since only when they are substantially supercooled is it possible to achieve intense recombination pumping of the upper working level X(b). The cold thermostat which cools the free electrons is the heavy gas particles in this case. Second, a rise in the gas (and electron) temperature above some value characteristic of each given chemical composition (of the active medium) leads to disruption of the mechanisms for relaxation of the plasma which ensure emptying of the lower level. We now illustrate this with several examples. This is clear in the case of the collision of heavy particles X with electrons leading to the establishment of local Boltzmann distributions in blocks of close-lying states. As the electron temperature T_e is increased the populations within the blocks are leveled out, thus depriving this mechanism of its effectiveness. An increase in the gas temperature T results in a slowing down of the rate of loss of particles in chemical reactions of the type (10b) due to shifting of the equilibrium constants toward the reverse reactions. In the case of the dissociative reaction (10c), as the temperature T is increased there is a rise in the number of fragments X and Y which have enough energy so that, along with the working photodissociation reaction $(XY)^* \rightarrow X + Y + \hbar\omega$, photoassociation with absorption of radiation at the working frequency could take place with a large probability [2]. The third reason for a limitation in the temperature of the energy elements is technological and is traditional for nuclear reactors.

For a sufficiently high gas density (assuming that $N \gtrsim 3 \cdot 10^{20}$ cm^{-3}) and a low degree of ionization the electron temperature exceeds the gas temperature by a negligible amount. Thus, for effective lasing it is necessary to ensure a sufficiently low gas temperature in the energy elements. Without detailing the peculiarities of different chemical compositions of the active medium, we shall begin with crude estimates for $T < 3000°K$. Let us estimate the corresponding limit in the stationary power density of the reactor when the gas in the energy elements is cooled solely by molecular heat transfer to the walls of these energy elements. The walls are efficiently cooled by a coolant which maintains a stationary gas temperature T in the center of the energy elements. It is assumed that a cylindrical energy element of comparatively great length is bathed in a flow of forced-circulating coolant which rapidly removes the heat. Taking Eq. (4) into account, we write

$$Q = \frac{N}{\tau_{cool}}(T - T_0), \qquad \tau_{cool} \approx \frac{R^2}{4\chi}, \tag{15}$$

where τ_{cool} is the characteristic time for cooling the gaseous active medium situated in a cylindrical container (the energy element), χ is the thermal diffusivity of this medium, R is the radius of the vessel, and T_0 is the temperature of its walls. Here we shall limit ourselves to crude estimates and not include the appreciable inhomogeneity in the distribution of the nuclear fuel within the energy element that is determined by the large radial temperature gradient. (The nonuniform density of uranium leads to a radial dependence for the number of fissions and, thus, for the heat release within the energy element. The same role is played by the dependence on the gas temperature of the interaction of uranium nuclei with neutrons.) We now set $\chi = v_T/\sigma_{el}N$, where σ_{el} is the elastic scattering cross section of the atoms (molecules) of the gas fill inside the energy elements, and [using Eqs. (4) and (15)] write the stationary temperature in the form

$$T = T_0 + \frac{W}{N}\tau_{cool} = T_0 + W\frac{R^2\sigma_{el}}{4v_T}. \tag{16}$$

Using the last equation, we now evaluate W_{cr}, the maximum allowable power density at which a stationary temperature regime in the reactor-laser is possible (for effective lasing):

$$W_{cr} = \frac{4(T_{cr} - T_0)v_T}{\sigma_{el}R^2}. \tag{16a}$$

Setting $T_{cr} - T_0 = 0.2$ eV, $\sigma_{el} = 10^{-15}$ cm^2, R = 1.5 cm, and $v_T = 10^5$ cm/sec in this equation, we obtain

$$W_{cr} \approx 4 \cdot 10^{19}\ \mathrm{eV/cm^3 \cdot sec} \approx 5\ \mathrm{W/cm^3}. \tag{17}$$

This corresponds to a maximum permissible rate of nuclear fissions, $G_{cr} \sim 2 \cdot 10^{11}$ fissions/cm^3 · sec. In a gaseous core thermal reactor with a density $N(U^{235}) \sim 3 \cdot 10^{19}$ cm^{-3}, this value of G_{cr} corresponds to a neutron flux of order 10^{13}–10^{14} neutrons/cm^2 · sec. It should be noted that this (17) value of the critical power density is given with a substantial margin. Inclusion of the dependence of the radial distribution of the heat release on the gas temperature results in a substantial rise in the allowable power density. There is still a more important margin in (17) associated with the neglect of the possibility of more rapid heat removal from the central regions of the energy element by natural convection or forced radial circulation of the fuel.

What should be the chemical composition of the active medium of a nuclear reactor-laser? We begin with the fact that the composition of the gas in the energy elements must include uranium hexafluoride primarily because, of all the possible known uranium compounds, it has the lowest boiling point, $T_b \simeq 70$°C. In analyzing the prospects for using it in lasers it is also important that UF_6 is transparent in the visible. In order to avoid complicating the problem due to separation of UF_4 and UF_2 from UF_6 and their settling on the walls of the energy elements, an adequate amount of fluorine impurity must be added. If it were clear that the mixture UF_4 and UF_2 assured laser action with highly efficient conversion of the energy spent in ionizing it into light energy removed from the energy element, it would be appropriate to settle primarily on this choice for the gas to fill the energy elements. However, the properties of uranium hexafluoride have hardly been studied so far in this regard. Thus, it is necessary here to choose some examples of chemical compositions for the gas in which, besides UF_6 and F_2, there is included a component which is laser active. Among the possible components, which of course must be chemically passive relative to UF_6 and F_2, we may include, for example, the inert gases and such fluorides as have boiling points which are not too high.

The feasibility of lasing in media which are suitable both in chemical composition and in the values of the plasma parameters for effective lasing in energy elements has unfortunately not yet been analyzed in any detail. Nevertheless, it is now still possible to form a general impression of the expected characteristics on the basis of a number of efforts made on various recombining plasmas. We must first recall the numerical estimates of the characteristics of stationary lasing in an energy element filled with a mixture of helium with uranium hexafluoride and fluorine. There it was planned to use the emission of helium atoms in the $He(n = 3) \rightarrow He(n = 2)$ transitions or of helium molecules in transitions between correlating (with given states) electronic terms of the He_2 molecule. These transitions span only a small part of the ionization potential of helium, so the efficiency of conversion of the ionization energy of this plasma into directed light is extremely small. This sort of active medium cannot be regarded as one of the filling materials for a power reactor-laser, but is of some interest for illustrating certain basic features of lasing caused by nuclear fragments. This medium is already now comparatively accessible for theoretical examination. It is perhaps simpler to make model experimental tests for it than for other chemical compositions. In this case the mechanism for producing the required populations in the working levels is as follows. Pumping of the upper level is determined by recombination of helium ions. Depopulation of the lower level is mainly due to deexcitation in Penning collisions (primarily with UF_6 molecules),

$$He\ (2) + Z \rightarrow He\ (1) + Z^+ + e,$$

or

$$He_2\ (2) + Z \rightarrow He\ (1) + He\ (1) + Z^+ + e.$$

Some estimates [4] show that lasing might be obtained from such an active medium in a laser of length $L \simeq 30$ m with densities $N(He) \simeq 3 \cdot 10^{20}$ cm^{-3} and $N(UF_6) \simeq 3 \cdot 10^{19}$ cm^{-3}, 90% enrichment of the uranium, energy elements of radius $R \sim 1$ cm, and an overall cross-sectional area of the reactor of about 1 m^2. More precisely, these calculations demonstrate that with such a choice of parameters the condition $G_{cr} > 2G_{thr}$ holds. This treatment also indicates that it would be desirable with this chemical composition to reduce the density of the UF_6 component by about an order of magnitude. This reduction in $N(UF_6)$ has several consequences. We begin with the fact that the rate K_b of depopulating the upper level He(3) by ionizing collisions is considerably reduced – this is exactly the main reason for reducing $N(UF_6)$. The rate K_a of depopulating the lower level He(2) is also reduced somewhat, of course, but this does not lead to termination of the inversion over the working transition or to a noticeable drop in the unsaturated gain coefficient. Slowing down the emptying of the upper working level, therefore, makes it possible to lower the recombination flux (and thus the free-electron density and the fission rate) without reducing the gain and to heat the amplifying medium less. But to ensure neutron criticality of the reactor it is necessary in this case to make a corresponding increase in the perpendicular dimensions of the reactor-laser. (That is, in the simplest scheme for assembling a reactor-laser from parallel straight elements, their number must be substantially increased. See Fig. 4.)

It should be noted that in those relatively few plasma laser active media compositions for which there are laboratory studies as well as relaxation calculations, the theory fits experiment fairly well. This is true first of all of the results of some calculations of the lasing char-

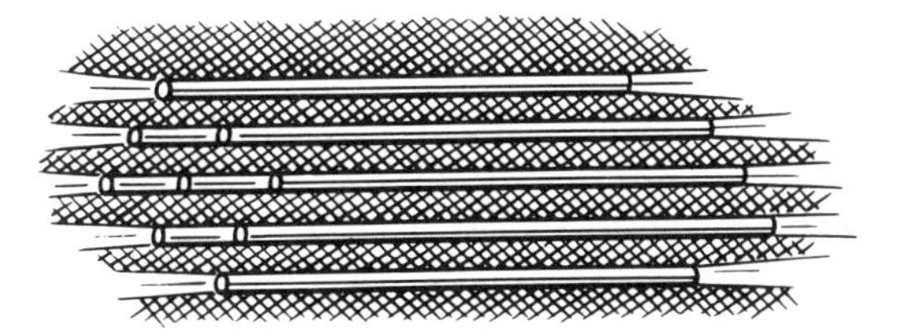

Fig. 4

acteristics of a decaying plasma on transitions of alkaline earth elements and the successful firing of pulsed lasers using this type of plasma [5]. The data available to us indicate reasonable agreement between theoretical [6] and experimental [7] work on lasing using transitions of dimer molecules of inert gases from bound electronically excited states into a repulsive ground-state term. In this work the decaying plama was obtained by injecting a beam of fast electrons into a dense inert gas of high chemical purity. Unfortunately, the experience on these two types of plasma lasers does not transfer directly to the problem of running a reactor-laser. Lasing can never be achieved on ionic transitions or with chemically pure inert gases in the medium which fills the energy elements.

An Example of an Active Medium for a Steady-State Reactor Laser [8]

The relaxation processes in an intensely recombining plasma of well-chosen composition favor the amplification of high-energy luminous fluxes. According to some fairly general considerations and calculations it is precisely under supercooled plasma conditions that it is possible in principle to expect conversion of ionization energy into directed radiant energy with high efficiencies $\eta \sim 0.7$-0.8. In order to explain this we recall that the upper working level, which in this kind of plasma is fed from the electron continuum, may be located rather closely to the ionization limit. The lower working level in a relaxing plasma, however, may lie near the minimum energy of the ground state of the working particle (atom, molecule). The most convincing example of this is the successful generation of photodissociation radiation of disintegrating molecules on a transition whose lower working "level" is a relatively low-lying range of energies in the ground (repulsive) state electronic term. We note that in this case emptying of the lower working state is apparently by means of the fastest of the chemical reactions, as the fragments of the molecule fly apart in its repulsive term. Steady-state lasing on such transitions of dissociating molecules is not possible primarily because of the smallness of the cross section σ_{ab} (photoabsorption at the working transition) which is due to the unusually wide bandwidth of the corresponding luminescence line, which takes up $\Delta\omega \sim 10^{14}$ sec^{-1}. Because of this bandwidth the gain coefficients required for triggering lasing in the energy elements (and even more, to achieve efficient removal of the energy by means of laser light) can be achieved only at extremely high populations of the upper working level. Such values of N_b, now obtained experimentally by ionizing a dense gas with a high-current relativistic electron beam, exceed the values which traditionally occur in laser technology by several orders of magnitude. Any method for ionizing a gas in this sort of regime will require intensive injection of energy and involve a correspondingly rapid heating up of the gas. In the beam experiments mentioned here, with lasing on photodissociative transitions of ground-state dissociating molecules,

$$XY(2) \rightarrow X(1) + Y(1) + \hbar\omega_{12}, \quad (18)$$

the gas is heated to beyond the critical temperature T_{cr} over a time $\Delta t \sim 10^{-7}$-10^{-6} sec. For $T > T_{cr}$ there are already so many fast atoms in the gas for which photodissociation,

$$X\,(1) + Y\,(1) + \hbar\omega_{12} \rightarrow XY\,(2),$$

is fully possible that the medium no longer amplifies but absorbs light at these frequencies.

From this we may conclude generally that strong radiative transitions with fairly narrow luminescence lines are preferable as working transitions under stationary reactor-laser conditions. It is natural to attempt to solve the problem more radically, concentrating on the analysis of the possibilities for lasing on atomic transitions of the mixture inside the energy elements. As the lower working level we shall choose the ground or a low-lying state of the atom

X, while for the upper level we try to use levels lying near the continuum. This is important for highly efficient conversion of energy into light, but the oscillator strength of an atom for such transitions and the photoabsorption cross section decrease with increasing principal quantum number. At first it seems, in view of the excessively large number of equivalent variants, that the analysis would be difficult without specifying the choice of element X. However, this is not so. The number of acceptable elements X is unfortunately very low. In this analysis it is necessary to include a whole range of factors, an important one of which is the transparency of all components of the mixture to the working radiation. It might be expected that this side of the problem would not lead to substantial limitations since transition of an X atom from a high-lying state to the ground state (or one adjacent to it) could generally be effectively split up. Thus, if for a sufficient population in level b and an inversion over the $b \rightarrow a$ transition its frequency ω_{ab} is too high for some component of the mixture, but there exists an intermediate level c such that $E_b > E_c > E_a$, then, instead of trying to use lasing at ω_{ab}, one should go to longer-wavelength lasing at frequencies ω_{ac} and ω_{cb}. Multiquantum as well as two-quantum subdivision of the working transition may be used. We now turn to another important question. How, in a dense, relatively cold gas, can we obtain a sufficient concentration of the required type of atoms? It would seem that only the inert gases (of all the chemical elements) could exist as atoms in a gaseous mixture containing uranium hexafluoride and fluorine, but even these associate to form molecules at substantial densities of their excited states. Therefore, as possibly the only solution to this problem in view of these remarks, we must discuss all plasma processes during which atoms originate from molecular compounds as they are ionized (for a while) into excited states. Depending on the extent of recombinational relaxation the atoms are lost in chemical reactions to form the initial compounds. If the rate of capture (combining) of working X atoms in their ground state is sufficiently high, then it is possible to plan on a population inversion over transitions of these or (energetically) adjacent states. To ensure the necessary capture rate we must find an appropriate large-cross-section reaction of element X and, in addition, have a large concentration of the component Y which reacts with X. The overwhelming portion of the ionization energy of the plasma must go into formation of excited X^* atoms. Thus, the density N(XZ) of the component which yields the X^* component must be at least an order of magnitude greater than the concentrations of the other components. Beginning with the condition $N(UF_6) \gtrsim 10^{19}$ cm^{-3}, we can already see that concentrations $N(XZ) \gtrsim 10^{20}$ cm^{-3} are required. From this it immediately follows that the compound XZ must be sufficiently volatile, otherwise this concentration cannot be obtained at a moderate temperature $T < T_{cr}$. These considerations, together with the previous requirements on the constituents of the energy elements, severely restrict the possible choices. At present the most suitable candidate atom for amplification of light in a steady-state reactor-laser seems to us to be thallium and its compound (one of the main components of the mixture in an energy element, in terms of concentration) thallium fluoride (cf. the following and [9]).

Hence we shall assume that the mixture contains three components with the following densities:

$$N\,(\mathrm{TlF}) \sim 3\cdot 10^{20}\ \mathrm{cm^{-3}},\quad N\,(\mathrm{UF_6}) \sim 3\cdot 10^{19}\ \mathrm{cm^{-3}},\quad N\,(\mathrm{F_2}) \sim 10^{19}\ \mathrm{cm^{-3}}.$$

In this mixture relaxation begins with the fast "cascade" electrons (produced during passage of uranium fission fragments through the medium) causing the creation of molecular ions and excited thallium fluoride molecules. The molecular ion $(TlF)^+$ has very low thermal stability, with a ground-state electron term of depth only two thousand degrees, i.e., $D(TlF)^+ \simeq 0.2$ eV. The same is true of the higher electronically excited states of the neutral molecule $(TlF)^*_{el}$ as well. Thermal decay of the $(TlF)^+$ ion and three-body recombination of $(TlF)^+$ or Tl^+, as well as dissociative recombination of $(TlF)^*_{el}$ and mixing of the recombinational relaxation channels of Tl^* and $(TlF)^*_{el}$, and any related processes must (in view of this) lead to thallium atoms becoming predominant among the electronically excited particles in the gas which fills the energy elements.

The cross section of the reaction

$$\mathrm{Tl}\ (m) + \mathrm{F_2} \rightarrow \mathrm{Tl\ F}\ (m) + \mathrm{F} \qquad (19)$$

of an unexcited thallium atom (m = 1) with a fluorine molecule is one of the largest cross sections for a binary chemical reaction with $\sigma_{chem} \sim 10^{-14}\ cm^2$. The large depth of the ground-state term of thallium fluoride, D(TlF) $\simeq$ 4.6 eV, makes the probability of the reverse of reaction (19), which would produce unexcited thallium atoms, negligible. It is otherwise in the case of electronically excited molecules, at least those whose energy is close to the ionization energy of the molecule, J(TlF) $\simeq$ 6 eV. The terms of these molecular states are at most several tenths of an electron volt and, at the gas temperature needed for a concentration N(TlF) $\sim 3 \cdot 10\ cm^{-3}$, the equally fast recovery of excited atoms Tl(m) from the molecules [the inverse of reaction (19)] makes the chemical depopulation of such thallium levels with m > 1 ineffective, even with the comparatively high probabilities of collisional transitions. It is significant that at a fluorine concentration $N(F_2) \sim 10^{19}\ cm^{-3}$ the rate of depopulation of the ground state Tl(1) of thallium by chemical reaction (19) is already enough both for a population inversion and for efficient lasing on several transitions of the thallium atom to the $6P^2_{7/2}$ state, especially from the 8S and 7D levels.

This scheme for obtaining excited thallium atoms as the molecular gas is ionized by fission fragments and removing them when they are in the ground state is not yet complete since the initial and final stages are not the same. That is, the thallium cycle is not closed. Ionization takes place mainly from low-lying vibrational levels of the ground-state term of the molecule, and, after relaxation of the thallium atom (including radiation at the working transitions) and loss of Tl(1) in chemical reaction (19), a molecule $(TlF)^*_{vib}$ is formed in the ground-state electronic term but in excited vibrational–rotational levels of that term (Fig. 5). Since the depth of the potential well of the ground-state term is not small compared to the ionization potential, the question of how the resulting molecule relaxes becomes important. If all the energy D(TlF) of the ground-state term goes into heat, then the conversion of ionization energy into radiation does not exceed 50% even in the best relaxation situation, which is in fact hardly realizable.

The one-way flow Tl $\rightarrow (TlF)^*_{vib}$ due to binary chemical reactions (19) or three-body collisions of the type Tl(1) + F(1) + TlF $\rightarrow (TlF)^*_{coll}$ + TlF leads to the formation of molecules at high vibrational levels. Here we can speak of a supercooled, nonequilibrium dissociated gas.

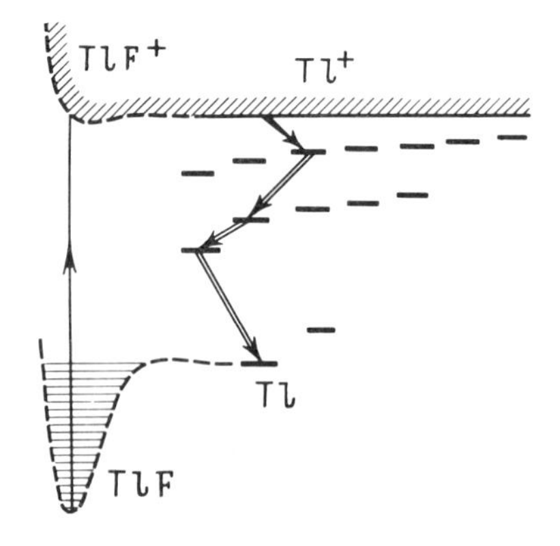

Fig. 5

Relaxation over the vibrational levels of a deep term in this case may produce a population inversion in an entire chain of vibrational–rotational transitions. This is because of the high density of energy which fills the vibrational degree of freedom as thallium is associated into a molecule while the rotational energy of the molecule rapidly equilibrates with the translational degrees of freedom of the cold gas. Furthermore, as shown in [10], under similar conditions an absolute inversion (without a shift in the rotational state) is produced over a substantial fraction of the vibrational transitions. Thus, in order to understand how high the conversion efficiency (of fission fragment energy into radiation extracted from the energy elements) may be in this three-component gaseous mixture $TlF + UF_6 + F_2$, we must analyze the efficiency of lasing on vibrational–rotational transitions of the thallium fluoride molecule as well as on electronic transitions of the thallium atom.

Pulsed Reactor-Lasers [11]

Clearly, under otherwise similar conditions, greatly increasing the rate G of nuclear fissions simplifies the problem of starting up a nuclear reactor, makes it possible to obtain more intense emission, improves lasing in the energy elements, makes it possible to greatly expand the range of possible active media, and permits construction of a relatively short reactor-laser. (In general, it eases the technical arrangement, improves the basic characteristics of reactor-lasers, and makes them more diverse.) This explains the importance of examining the operating features of a reactor-laser in forced pulsed regimes where a very high power density over short periods of time, $\tau_{pulse} \sim 10^{-6}$–$10^{-4}$ sec, is typical. In designing such devices for scientific research purposes it is possible to plan for single-pulse operation. For power applications we must consider periodic sequences of this kind of fission pulse between which the nuclear reactor-laser cools down by means of some sort of organized heat transfer. Thus, the pause intervals may be times of order $\tau_0 \sim 10^{-4}$–1 sec. We shall limit ourselves here to order-of-magnitude estimates of the nuclear-reactor and optical parameters of the simplest self-modulated, thermally pulsed, fast (neutron) reactor-laser. Neglecting the effect of delayed neutrons in such devices and ignoring the spatial inhomogeneities in the problem (and therefore, the characteristic time for redistribution of the neutrons over the active volume), we write an equation for the power density in such a reactor laser:

$$\frac{dW}{dt}(t) = \frac{\rho(t)}{\tau} W(t), \tag{20}$$

where $\rho(t) = (k-1)/k$ is the reactivity of the device, k is the multiplication factor for the neutrons over a single "generation," and τ is the mean lifetime of the neutrons in this reactor. In this self-modulated reactor-laser scheme the nuclear fission pulse is formed by means of a rapid transfer of the system from a (neutron) subcritical state to a supercritical state; that is, the function $\rho(t)$ changes sign at the turn-on time t = 0, or

$$\rho(t) < 0 \quad \text{for} \quad t < 0; \qquad \rho(0) = \rho_0, \quad \rho_0 > 0.$$

After this the reactivity falls with increasing temperature T of the gaseous active medium, and the rate of fissions rises ever more slowly, finally falling and causing the reactor to cease operating.

As in the steady-state problem, we shall use the concept of a critical temperature T_{cr}, which if exceeded leads to termination of fission energy removal as light from the reactor-laser. Let W^p_{cr} be the critical power density corresponding to this temperature. We shall show that under thermally pulsed reactor-laser conditions this quantity may exceed the previously given $W_{cr} = W^{st}_{cr}$ for steady-state conditions by many times. Denoting the fraction of the energy released (from the turn-on time t = 0) in nuclear reactions that goes into heating the medium by q, for the general case we write

$$\frac{dT}{dt}(t) + \frac{1}{\tau_{cool}}[T(t) - T_0] = \frac{q}{N} W(t); \qquad T_0 = T(0), \tag{21}$$

where τ_{cool} is the characteristic cooling time of the gaseous medium and T_0 is the constant wall temperature. In a steady-state regime, when for $t > 0$, $W(t) \equiv W_0$, we obtain

$$T(t) = T_0 + \frac{q}{N} W_0 \tau_{cool} \left[1 - \exp\left(-\frac{t}{\tau_{cool}}\right)\right].$$

W_{cr}^{st} is defined for an established constant temperature $T(\infty) = T_{cr}$ as

$$W_{cr}^{st} = \frac{N}{q\tau_{cool}}(T_{cr} - T_0). \tag{22}$$

In the pulsed reactor-laser problem of interest to us the changes in $\rho(t)$ and $W(t)$ control the operating regime. Thus, under fairly general conditions the differential equations (20) and (21) must be solved simultaneously, employing the relationship between reactivity and temperature. We have not yet had the opportunity to specify the contributions of various mechanisms which depend on the design of the reactor-laser to the decrease in reactivity when the temperature is increased. Under reactor-laser conditions this decrease is mainly determined by the reduction in both the effective density of the gaseous fissile material and the neutron interaction cross section of U^{235} as $T = T(t)$ rises. We shall limit ourselves to a linear dependence, $\rho(t) = \rho_0 - \gamma[T(t) - T_0]$. We shall assume that the characteristic cooling time of the medium is fairly long, or, more precisely, that $\tau_{cool} \gg \tau$. Then Eq. (21) implies that the temperature rises in proportion to the fission energy released in the medium, i.e.,

$$T(t) \simeq T_0 + \frac{q}{N} \int_0^t W(\xi)\, d\xi.$$

Under these conditions the relationship between the reactivity and the power density takes the simple form

$$\rho(t) = \rho_0 - \Gamma \int_0^t W(\xi)\, d\xi, \qquad \Gamma = \frac{q}{N}\gamma,$$

or, equivalently,

$$\frac{d\rho}{dt}(t) = -\Gamma W(t), \qquad \rho(0) = \rho_0. \tag{23}$$

The solution of the nonlinear system of two first-order differential equations (20) and (23) with the initial conditions $\rho(0) = \rho_0$ and $W(0) = W_0$ may be written in the form

$$W(t) = 4W_m \exp\left(\frac{t_m - t}{\tau_0}\right)\left[1 + \exp\left(\frac{t_m - t}{\tau_0}\right)\right]^2. \tag{24}$$

Here, as can be easily seen, $W_m = \rho_0/2\Gamma\tau$ is the peak power density, t_m is the time at which it occurs, and $\tau_0 = \tau/\rho_0$ has the meaning of a general (in this model of a pulsed reactor-laser) relaxation time of the nuclear reactor system. According to Eqs. (20) and (23) the character of the pulse (24) is determined by the fact that from the time the laser is turned on, $t = 0$, the reactivity decreases monotonically, thereby slowing down the growth (for $\rho > 0$) in the power density. At time $t = t_m$ the reactivity, which is falling, crosses zero and afterwards $W(t)$ also

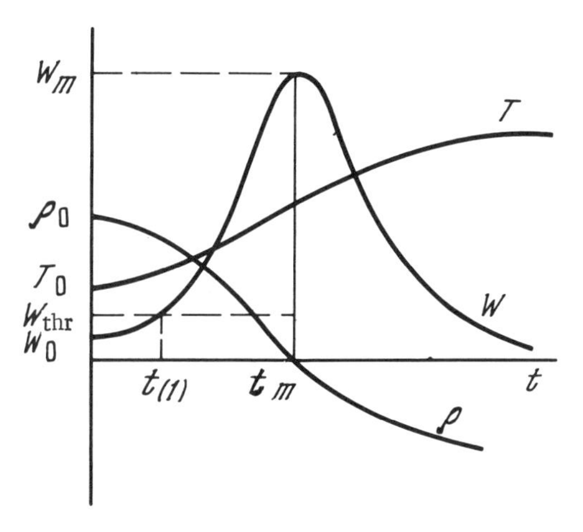

Fig. 6

falls. Thus, a pulse of nuclear fissions is produced in this reactor-laser scheme by means of temperature self-modulation of the reactivity. The light pulse is generally somewhat shorter.

Figure 6 shows the time dependences of the nuclear reactor characteristics in the self-modulating scheme. The characteristic cooling time of the free electrons in the plasma inside the reactor laser is fairly small. At the molecular gas densities $N \gtrsim 10^{20}$ cm^{-3} of interest to us, it is in the nanosecond range and may be neglected in most cases. The relationship between the recombination time τ_{Ne} of the plasma and the nuclear reactor relaxation time τ_0 may be different. We suppose at first that $\tau_{Ne} \ll \tau_0$. Then the beginning $t_{(1)}$ of lasing coincides with the moment the power density in the reactor reaches the threshold value $W(t_{(1)}) = W_{thr}$. Of course, the onset of lasing requires that $W_m > W_{thr}$. A reactor with $\tau_{Ne} \gg \tau_0$ should be designed so that at time t_m (when maximum power density is attained) the temperature does not exceed T_{cr}, the critical value. The time $t_{(2)}$ at which lasing ceases is determined under these conditions by the smaller of the two limiting values $t_{(2)} = \min(t'_{(2)}, t''_{(2)})$ and $T(t'_{(2)}) = T_{cr}$, where $W(t''_2) = W_{thr}$. According to Eq. (5.5) the times of maximum power density and initiation of lasing differ from the relaxation time by the logarithmic factors

$$t_m \simeq \tau_0 \ln\left(4 \frac{W_m}{W_0}\right), \qquad t_{(1)} \simeq \tau_0 \ln\left(\frac{W_{thr}}{W_0}\right).$$

For rational use of the fission energy the device parameters should naturally be chosen so that $T_{(2)} \gtrsim t_m$. The quantity q reflects the efficiency with which energy is removed from the reactor by lasing, which of course depends on the temperature T. Nevertheless, in our estimates we shall neglect the T dependence of q (which is strongest just before the critical temperature is reached). For $t_{(2)} \gtrsim t_m$ we have in this case that

$$\left[1 + \exp\left(\frac{t_m - t_{(2)}}{\tau_0}\right)\right]^{-1} = \left[1 + \exp\left(\frac{t_m}{\tau_0}\right)\right]^{-1} + \frac{(T_{cr} - T_0)\,N}{4W_m q\tau_0}.$$

From this it follows that the peak power of the light is greatest for $t_{(2)} = t_m$, and the duration of the laser pulse is estimated by $\Delta' = t_m - t_{(1)} \simeq \tau_0 \ln(4W_m/W_{thr})$. If also $W_m \gg W_0$, then

$$(W_m)_{\max} \simeq \frac{N(T_{cr} - T_0)}{2q\tau_0}, \tag{25}$$

which corresponds to the simple relation between the parameters of this pulsed reactor-laser scheme,

$$\rho_0 = \gamma\,(T_{cr} - T). \tag{26}$$

Assuming $W_{cr}^{p} = (W_m)_{max}$, from Eqs. (22) and (26) we obtain

$$W_{cr}^{p} \simeq \frac{\tau_{cool}}{2\tau_0} W_{cr}^{st} \tag{27}$$

We recall that in the steady-state reactor-laser scheme discussed above the energy elements are in the shape of long tubes with small radii R and heat removal is due to thermal conductivity of the medium toward the walls. In the case of ordinary molecular thermal conductivity the cooling time of the gas is $\tau_{cool} \simeq \sigma_{el}/R^2N/(4v_T)$. Setting the elastic collision cross section equal to $\sigma_{el} \approx 10^{-15}$ cm^2, taking the thermal speed to be $v_T \approx 10^5$ cm/sec, and taking $R \approx 1$ cm and $N \simeq 3 \cdot 10^{20}$ cm^{-3}, we obtain $\tau_{cool} \simeq 1$ sec. By using natural or forced convection in the energy elements this characteristic time may be reduced by two or three orders of magnitude. The relaxation time of the reactor depends strongly on the initial reactivity. Evidently, we have to begin with values $\rho_0 \approx 10^{-2}$–10^{-1}. For $\tau \simeq 5 \cdot 10^{-6}$ sec, we then obtain $\tau_0 \approx 5 \cdot 10^{-5}$–$5 \cdot 10^{-4}$ sec. Thus, according to Eq. (28) the critical power density in a pulsed reactor may exceed that estimated for a steady-state reactor-laser without convection by three or four orders of magnitude.

If we envision optimal conversion of fission energy into light rather than maximum radiant intensity in the pulsed regime, then it is necessary to go to reactor-laser parameters such that $T(t_{(2)}) \simeq T_{cr}$. Then

$$t_{(2)} \simeq \tau_0 \ln [(16W_m^2)/W_0W_{thr}],$$

so that the duration of the output pulse increases by a factor of two over the previous estimate to

$$\Delta'' \simeq t_{(2)} - t_{(1)} \simeq 2\tau_0 \ln (4W_m/W_{thr}) \approx 2\Delta'.$$

The critical power density of the reactor in this regime is less than W_{cr}^{p},

$$W''_{cr} \approx \frac{N(T_{cr} - T_0)}{4q\tau_0} - \frac{W_{thr}}{4} \approx \frac{1}{2}\left(W_{cr}^{p} - \frac{W_{thr}}{2}\right).$$

Realization of the other limiting relationship between the characteristic times of a reactor-laser, $\tau_{Ne} \gg \tau_0$, seems more remote to us now, but still possible. Under these conditions the actual value of the maximum power density of the reactor has no fundamental role in the process of removing the energy by means of light. All of the luminous characteristics are determined solely by the total fission energy over the entire pulse, $E_p \simeq \int_0^\infty W(t)\,dt$. In analyzing the operating conditions of a pulsed reactor-laser in this limit, there is perhaps no sense in appealing to a self-modulating scheme and using Eq. (24) for the change in the power density with time. The parameters for such a reactor-laser must be chosen starting with the unacceptability of overheating of the medium beyond a critical temperature T_{cr} characteristic of the laser problem. Then we need only rely on the heat capacity C of the gas, ignoring its thermal conductivity during the pulse, i.e., $C(T_{cr} - T_0) = E_{max}^{p}$. It is then necessary to consider the possibility of using both thermal and optical pulsed operation of the reactor-laser, that is, Q-switching of the optical cavity (and, perhaps, mode locking). This kind of solution makes it possible to use a reactor-laser to produce light pulses with extremely high energies and lasting $\tau_{light} \sim 10^{-8}$–$10^{-9}$ sec.

The results of these estimates indicate that even with the simplest self-modulated thermally pulsed reactor-laser, laser parameters that are extremely high compared both with existing and with long-term predicted laser systems will be achieved. Photodissociative transitions of dissociating molecules could also be used in pulsed reactor-lasers. (Such active media, as is known, overheat rapidly.) There already exists an appreciable reserve for further extension of the pulsed reactor-laser regime. It involves going to two-cascade, self-modulating schemes as well as to subcritical systems. In the latter case the reactivity is always negative and the nuclear reactor converts a neutron flux generated outside itself by an electron beam (a "booster" system) or by an as-yet-not-built pulsed thermonuclear machine (a "hybrid" system). In this regard it is now appropriate to note the important possibility of using pulsed reactor-lasers in a pulsed controlled thermonuclear device and in this way obtaining a periodically pulsed unitary energy system. In the latter scheme a pulse of high-energy thermonuclear neutrons passes through a subcritical nuclear reactor laser made without enriching the uranium in the isotope U^{235}. The light pulse produced by uranium fission is then partially directed toward producing the next thermonuclear pulse.

Further Work on This Problem

It seems to us that the preferred directions for verifying and realizing the basic ideas of reactor-lasers are indicated by the outlines described here. The priority of such work at this time is determined, of course, not by the obvious need of mankind of research on methods of rational energy use, but by the practicality of the proposed schemes for solving this problem.

Assuming the feasibility of the schemes discussed here, we now list (without intending an ordering in importance or time) the main measures for realizing these schemes as practical devices.

1. It is necessary to verify the laser activity of several fillers for the energy elements. The most reasonable experimental approach now is to analyze the optical characteristics of a dense plasma obtained in the afterglow of a sharply curtailed transverse field pulse. As soon as possible it is desirable to begin the study of the properties of uranium hexafluoride mixed with fluorine and of thallium fluoride mixed with fluorine.

2. An analysis of the feasibility of lasing in dense gases of this composition ionized by an electron beam, under both steady-state and pulsed conditions, is needed. It is necessary to proceed to the study of the properties of media with combined laser-active and nuclear-active components, by placing cuvettes of them in the neutron field of the required intensity (in a standard nuclear reactor).

3. An analysis of the simplest gaseous core reactors of the type described here is needed.

4. The search for media to fill the energy elements and amplify radiation under supercooled electron conditions must be continued. Besides laboratory study of the optical properties of UF_6 and TlF in mixtures with fluorine it is necessary to make chemical and thermodynamic investigations of the conditions for prolonged maintenance of the chemical composition of the required gaseous mixtures.

5. Work is needed on the theoretical analysis and laboratory realization of various regimes for removing heat from the gas in the energy elements.

6. Attention must be directed to the search for solutions to the chemical engineering problems involved in the protracted coexistence of the walls of the energy elements and their extremely active contents. Here we must point out the problem of maintaining (in a working state) the end mirrors and optical windows for removal of high-energy light from the reactor.

7. The configuration of the energy elements must be studied. As possible competing schemes we may mention (more traditional for lasers) straight tubes with mirrors on their ends, a large number of which (about 10^4) ensure criticality, and (less traditional) spiral tubes of much greater length.

8. It is necessary to carry out a series of neutron and reactor studies of possible active media, to analyze the behavior of the neutrons in long tubes and the stability conditions for them to have a uniform distribution along the length of the tubes, and to examine the density

distribution of the gaseous fissile material within the energy elements. The propagation of fast neutron pulses in a nonstationary reactor-laser must be studied. Methods for rapidly turning on the reactivity in a high-energy pulsed gas-core reactor must be considered.

9. Physics departments must begin the preparation of specialists in this area. Information about nonstandard techniques for creating the dense supercooled plasma needed for this research and experience in the theoretical analysis of such media are as of now absent in the programs of universities and engineering schools.

Literature Cited

1. L. I. Gudzenko and S. I. Yakovlenko, Kratk. Soobshch. Fiz., No. 2, p. 14 (1974).
2. L. I. Gudzenko, L. A. Shelepin, and S. I. Yakovlenko, Usp. Fiz. Nauk, 114:457 (1974).
3. K. Thom and R. Schneider, AIAA, Vol. 10, No. 4 (1972).
4. L. I. Gudzenko, I. S. Slesarev, and S. I. Yakovlenko, Zh. Tekh. Fiz., 75:1934 (1975).
5. E. L. Latush and M. F. Sém, Zh. Éksp. Teor. Fiz., 64:2017 (1973).
6. L. I. Gudzenko and S. I. Yakovlenko, Dokl. Akad. Nauk SSSR, 277:1085 (1973).
7. L. I. Gudzenko, I. S. Lakoba, and S. I. Yakovlenko, Zh. Éksp. Teor. Fiz., 67:2022 (1974).
8. C. K. Rhodes, IEEE J. Quant. Electron., QE-10:153 (1974).
9. L. I. Gudzenko and S. I. Yakovlenko, Kratk. Soobshch. Fiz., No. 12, p. 13 (1975).
10. L. I. Gudzenko and V. S. Marchenko, Trudy FIAN, 90:90 (1976) [Chapter IV of present volume].
11. L. I. Goodzenko and B. F. Gordietz, Phys. Lett., 41A:59 (1972).
12. L. I. Gudzenko, I. S. Slesarev, and S. I. Yakovlenko, Kratk. Soobshch. Fiz., No. 2, p. 34 (1976).

Part II

INVERSE PROBLEMS IN THE THEORY OF OSCILLATIONS

CHAPTER VI

A SCHEME FOR MEASURING KINETIC COEFFICIENTS

Statement of the Problem

Due to the discovery of astronomical sources of narrow-band radiation in recent years, there has been a rapid rise in the number of publications analyzing the various possible mechanisms for efficient amplification of light under natural conditions. A number of models of astrophysical objects have been proposed which (in the authors' opinion) yield maser effects. The most systematic way of verifying the correctness of interpretations of specific sources and the reasonableness of the hypotheses advanced has been direct correlation analysis of the observed emission from outer space. The statistical analysis of a signal recorded for a long-enough time by a sensitive, low-noise apparatus will yield a classification of the radiation sources according to the type of kinetic equation. This can be understood from the discussion below. Here we shall explain why these equations must describe the population kinetics in the emitting medium. Such classifications could make it possible to accept one or another "maser" hypothesis, could suggest the actual mechanism when these hypotheses are rejected, and, finally, could be of intrinsic importance for astrophysics.

However, correlation analysis of kinetic equations is appropriate for a number of laboratory physical problems as well as for truly uncontrollable objects. Among such problems there is no doubt that the measurement of various kinetic coefficients is now outstanding. We shall deal with this problem here. It seems that it is unnecessary to explain the need for developing technically realizable methods of measuring the probabilities of collisional atomic (molecular, ionic) transitions in a low-temperature plasma. The form of the kinetic equations for the time variation of the averaged populations N_i of the quantum levels in the observed volume of plasma is usually known. For large populations, whose fluctuations may be neglected, the simplest linear variant of the kinetic equations is

$$\frac{d}{dt} N_i(t) = \sum_{\substack{j=1,\\ j\neq i}}^{J} K_{ij} N_i(t) - K_{ii} N_i(t) + H_i, \qquad i = 1, J. \tag{1}$$

Here $N_i = \frac{1}{\theta} \int\limits_{t-\theta}^{t} \mathrm{N}_i(\tau)\, d\tau$; N_i are the actual ("instantaneous") populations, θ is the smoothing interval, which we shall assume to be small compared to the characteristic times of the dynamic system (1), K_{ij} are the kinetic coefficients with $K_{ii} = \sum\limits_{j=0,\, j\neq i}^{J} K_{ji}$, H_i is the intensity with which level (i) is filled from levels which are not included in the sequence $\mathbf{i = 1, J}$ (i.e., i = 1, 2, ...,J), and the symbol $\mathbf{i, j = J_1, J_2}$ denotes i, j = J_1, J_1 + 1, ..., J_2.

The parameters K_{ij} of Eq. (1) are uniquely related to the transition probabilities including the unknown radiationless transitions. As will be shown below (see also [1]), taking the fluctuations due to the finite number of particles in the observed volume into account leads, for steady-state conditions, to the addition of stationary (in a broad sense) random functions of time, $\Phi_i(t)$, to Eq. (1):

$$\frac{d}{dt} \mathrm{N}_i(t) = \sum_{j=1,\, j\neq i}^{J} K_{ij} N_j(t) - K_{ii} N_i(t) + H_i + q_i \Phi_i(t), \qquad i, j = 1, J; \tag{2}$$

where

$$q_i = \sqrt{2K_{ii} \langle N_i \rangle / \theta}, \qquad \langle \Phi_i \rangle \equiv 0, \quad \sigma^2 \Phi \equiv 1,$$
$$\langle \Phi_i(t)\, \psi_i(t+\tau) \rangle = 0 \qquad \text{for} \quad \tau \geqslant \theta. \tag{3}$$

The set J of autocorrelation functions of the populations,

$$R_i(\tau) \equiv \langle (N_i(t) - \langle N_i \rangle)(N_i(t+\tau) - \langle N_i \rangle) \rangle, \quad i = 1, J,$$

includes complete information on the kinetic matrix K_{ij}. Since there is a relationship between the autocorrelation functions for the populations and for the radiation fields, it is pertinent to consider the technical feasibility of measuring the radiationless transition coefficients from spectral lines.

Correlation studies of the radiation field of atoms, ions, and molecules has long been a topic of the corresponding branches of physical spectroscopy which study the spectral intensities of the emission lines of transitions between quantum levels of these particles, that is, the Fourier transformations of the correlation functions of the field. Despite the fact that the characteristics of spectral line shapes (and, primarily, the half width of a line) are related to the probabilities of transitions between the corresponding levels, it is useless to approach the problems being discussed here by the traditional methods of spectral analysis. In any general case it is impossible to find radiationless transition probabilities from spectrum lines. This is indicated by the negative results accumulated in an immense number of reports on spectroscopy. The basic reason for the failure is the fact that the line shape is determined primarily not by the kinetic matrix $\|K_{ij}\|$ but by all sorts of broadening mechanisms (Doppler effect, fluctuations of microscopic fields, etc.) which have no direct bearing on the population kinetics. It is impossible to escape this "background"; thus, most often the results of experiments yield a form of parameter that is in some way related to the kinetic matrix, i.e., the spectral line width.

Turning from the spectral intensity to other characteristics of a spontaneous emission field does not change the situation because under very general conditions the radiation field of a large number of particles which form a stationary Gaussian process is determined fully by its correlation matrix and, therefore, the "spectrum" already contains complete information about it.

As has already been noted there is no doubt that there are fairly general methods for measuring the kinetic coefficients K_{ij}. From the kinetic equations (2) it is apparent that these coefficients are directly related to the populations $N_i(t)$, which in turn may be determined experimentally in some way. At the same time, Eqs. (2) show that the average values of the stationary populations $\langle N_i \rangle$ yield too little information on the kinetic coefficients. This also follows from the fact that in the system of J equations

$$\frac{d}{dt}\langle N_i \rangle = \sum_{j \neq i,\, j=1} K_{ij}\langle N_j \rangle - K_{ii}\langle N_i \rangle + H_i = 0, \tag{4}$$

there are still J unknown quantities H_i besides the $J(J+1)/2$ unknown kinetic coefficients K_{ij}. Thus, to find the coefficients K_{ij} it is necessary to analyze the time variation of the populations.

Cross-correlation analyses of the populations as opposed to spectral studies of the radiation field make it possible to determine all the coefficients in the kinetic matrix. Before describing the techniques, we make one further remark. From Eq. (1) it is clear that if, for example, it were possible to control experimentally the various initial values of the average populations, then, with sufficiently accurate detection of the succeeding time variation of the populations due to relaxation, it would be possible to determine the kinetic coefficients by the inverse problem technique. Another logically simple conceptual way of finding the matrix $\|K_{ij}\|$ is to detect and numerically process the responses of the medium (in the form of a vector of the level populations) to fast interactions, for example, periodic modulation of the populations which are also controlled by the experimenter. However, a purely dynamic approach can hardly find practical application in any general case because under actual conditions it is impossible either to create a given initial nonstationary distribution of average populations or to interact with the level populations in a measured way.

Of course the impossibility of reliably determining the matrix $\|K_{ij}\|$ from spectroscopic analyses, from comparisons of the average stationary populations, and, in fact, even by the dynamic approach, is an indication of the difficulty of finding this matrix. But the results of much physical research depend radically on the nature of the population kinetics. This both indicates the importance of the problem and signifies that effective methods for measuring the kinetic coefficients may be found. First of all, we must choose a signal that directly reflects the population kinetics.

Choosing a Signal

Determining the dynamic coefficients of system (2) (i.e., the kinetic matrix $\|K_{ij}\|$) from the correlation functions of its signal is a typical inverse problem of the theory of oscillations [2-4]. In accordance with this it would be best, for simplicity of the later mathematical treatment, to choose the instantaneous populations of the levels of interest as the signal. However, the populations $N_i(t)$ are not directly detected by the apparatus. The populations appear in a photodetector measurement of the number $m_{ki}(t)$ of photons spontaneously emitted by the object over a small time interval $(t - \theta, t)$ in radiative $i \rightarrow k$ transitions of atoms from level (i) to some lower-lying energy level (k), with $E_k < E_i$.†

Let the receiver-detector contain a number of channels (k, i) which permit evaluation of the number of photons spontaneously emitted by the atoms of the object in $i \rightarrow k$ transitions. Each channel consists of three devices: (1) an input filter which blocks all frequencies outside the band $(\omega_{ki} - \Delta\omega_{ki}, \omega_{ki} + \Delta\omega_{ki})$, (2) a radiation detector at the outlet of the channel ("photomultiplier") which works in a photon-counting regime and responds "instantaneously" to an "accepted" photon with a short standard electrical current pulse, and (3) a correcting filter which is in front of this detector and eliminates the dependence of the entire channel sensitivity on the frequency of the light within the band $(\omega_{ik} - \Delta\omega_{ik}, \omega_{ik} + \Delta\omega_{ik})$. We shall assume that no other radiation from the object enters the bandwidth of the input filter except photons spontaneously emitted in the $i \rightarrow k$ transition and that the entire spectral line of this transition is contained within this band. The average value of the smoothed current at the photomultiplier output of each channel (j, i) is proportional to the average photon count rate and thus to the frequency at which photons spontaneously emitted by the medium on the $i \rightarrow j$ transition appear. We emphasize that the components of the vector signal $J_{ki}(t)$ detected by this apparatus are not made up of the total spontaneous emission fields (or of unique functions of them such as the emission intensities), but are the number of photons emitted over a time θ. The average values of the components of the detected vector are proportional to the smoothed values of the populations $N_i(t)$ of the corresponding quantum levels of the object.

A steady state of the object does not always correspond to stationary average populations of its levels. Thus, for example, a laser may be in a self-modulating regime corresponding to motion in a limiting cycle in phase space $\{N_i\}$. In this case, if we neglect small fluctuations the populations seem to be periodic functions of time. For simplicity we shall assume in advance that the state of the object is in the neighborhood of a stable stationary point of the kinetic equation. Then the determination of the coefficients of Eq. (1) from the correlation matrix for J_{ki} reduces to a known procedure [2-4]. True, to justify it in this case it is still necessary to demonstrate short-term correlation of the population fluctuations, and in order to estimate the accuracy of the K_{ij} coefficients obtained, it is necessary to know the dispersion of these fluctuations [i.e., Eq. (3) must be made more specific]. In addition, the parameters of the additive noise which arises along the path from the instantaneous populations to the detected sig-

† A variant, which is based on detection and analysis of absorption (instead of emission) by the medium and is fundamentally also acceptable, will not be considered here.

nal J_{ki} must be known and the effect of smoothing the signals in the detector must be taken into account.

In deriving Eqs. (2) and estimating the additive noise it is appropriate at first to remove oneself from the specific conditions characteristic only of collisions of atoms, molecules, or ions among themselves or with free electrons and examine a sufficiently general situation, which might, for example, include chemical or nuclear reactions as well. Such a treatment is mathematically similar to schemes which arise in queuing theory. In this regard it is natural to turn to the stochastic Poisson flux [1].

The Stochastic Poisson Flux

By a flux of events in the theory of probability we mean a sequence of random events which take place in successive moments (or time intervals). Poisson fluxes play a particular role in queuing theory. The simplest, or stationary, Poisson flux is characterized by a single number, the flux density λ, while a nonstationary Poisson flux is determined by a regular function of time, $\lambda(t)$, the instantaneous flux density [5, 6]. A generalization of the concept of Poisson fluxes which we need in our problem is the stochastic Poisson flux introduced in [1]. The stochastic Poisson flux associated with a random vector process $\mathbf{Z}(t)$ is defined as follows.

A. For any nonoverlapping time intervals the number of events falling in one interval is independent of the number falling in the others (absence of aftereffects).

B. The probability of two or more events falling in a sufficiently small time interval is negligibly small compared to the probability of a single event (ordering).

C. Almost all trajectories $\mathbf{Z}$ have a "conditional instantaneous density"

$$\lambda(t, \mathbf{Z}(t)) = \lim_{\Delta t \to 0} \mathscr{E}\left[\frac{l(t-\Delta t, t)}{\Delta t} \Big/ \mathbf{Z}(t)\right] < \infty, \tag{5}$$

where $l(t_k, t_m)$ is the number of events occurring in the time interval $t_k \leqslant t \leqslant t_m$, and $\mathscr{E}[\alpha/\mathbf{Z}(t)]$ is the mathematical expectation of the random quantity α under the condition that at time t the random vector $\mathbf{Z}$ takes the value $\mathbf{Z}(t)$.

Stationary, nonstationary, and stochastic Poisson fluxes have properties A and B in common but differ in property C. For almost all sampling trajectories $\mathbf{Z}$ a stochastic Poisson flux forms a nonstationary Poisson process corresponding to the sample. For the problem being discussed here, λ does not depend explicitly on t; that is, $\lambda = \lambda(\mathbf{Z}(t))$. The conditional one-dimensional distribution of the number of events on an interval of fixed length θ obeys the Poisson formula

$$P_m[t-\theta, t/\mathbf{Z}(t)] = P_m(t-\theta, t/\Lambda) = \frac{\Lambda^m}{m!}\exp(-\Lambda); \quad \Lambda \equiv \int_{t-\theta}^{t} \lambda(\mathbf{Z}(\tau))\,d\tau, \tag{6}$$

where $P_m[t_p/Z(\tau)]$ is the probability that m events will occur over the time interval $t_p < t \leq t_q$ when the realization $Z(\tau)$ is fixed over this interval. The probabilities P_m depend on the behavior of $Z(\tau)$ over the interval $(t-\theta, t)$ only through Λ. To emphasize this we shall use the notation $P_m(t-\theta, t/\Lambda)$.

As an example of a one-dimensional Poisson flux we consider a preliminary schematic description of the operation of a photon counter. We denote the instantaneous frequency at which photons are incident on the counter by x(t). Let q be the constant probability of detection of a photon, i.e., of the response of the counter (to a photon) by a standard current pulse. We shall assume that the frequency of incident photons is not so large that the inertia of the photocathode of the counter or nonlinearity in its broadband electrical circuit will have any effect. We shall also neglect the dark current of the counter for now. Then the acts of detecting photons may be described by a stochastic Poisson flux with a conditional instantaneous density $qx(t)$ and

$$\Lambda(t) \equiv \int_{t-\theta}^{t} \lambda(\tau)\,d\tau = q\int_{t-\theta}^{t} x(\tau)\,d\tau.$$

We shall be interested in the characteristics of the number $u(t)$ of photons detected by the counter over a time interval $(t - \theta, t)$. Using the Poisson formula (6), we obtain

$$\mathscr{E}(u/\Lambda) = \sum_{\nu=0}^{\infty} \nu P_\nu(t-\theta, t/\Lambda) = \bar{e}^{\Lambda}\sum_{\nu=0}^{\infty} \nu\frac{\Lambda^\nu}{\nu!} = \Lambda,$$

$$\mathscr{E}(u^2/\Lambda) = \sum_{\nu=0}^{\infty} \nu^2 P_\nu(t-\theta, t/\Lambda) = \bar{e}^{\Lambda}\sum_{\nu=0}^{\infty} \nu^2\frac{\Lambda^\nu}{\nu!} = \Lambda + \Lambda^2,$$

$$\mathscr{E}(u^2/\Lambda) - [\mathscr{E}(u/\Lambda)]^2 = \Lambda.$$

Denoting the conditional fluctuations in the number of counts from the detector by $\omega(t) \equiv u(t) - \Lambda(t)$, we find

$$\mathscr{E}(\omega/\Lambda) \equiv 0, \quad \mathscr{E}(\omega^2/\Lambda) = \Lambda,$$
$$\mathscr{E}[\omega(t)\omega(t+\tau)/\Lambda(t), \; \Lambda(t+\tau)] = 0 \quad \text{for} \quad \tau \geqslant \theta. \tag{7}$$

The last equation is a consequence of property A. Averaging Eq. (7) over the distribution Λ, we go from the conditional mathematical expectations to the unconditional ones:

$$\langle\omega\rangle = 0, \quad \sigma^2\omega = \langle\omega^2\rangle - \langle\omega\rangle^2 = \langle\Lambda(t)\rangle,$$
$$\langle\omega(t)\omega(t+\tau)\rangle = 0 \quad \text{for} \quad \tau \geqslant \theta.$$

Thus, the number $u(t)$ of photon counts over the instantaneous time interval $(t - \theta, t)$ may be written in the fairly simple form

$$u(t) = \Lambda(t) + \omega(t) = \Lambda(t) + \sqrt{\langle\Lambda(t)\rangle}\varphi(t), \tag{8}$$

where $\varphi(t)$ are the normalized short-term correlated fluctuations with $\langle\varphi(t)\rangle \equiv 0$, $\sigma^2\varphi(t) \equiv 1$, $\langle\varphi(t)\varphi(t+\tau)\rangle = 0$ for $|\tau| \geq \theta$.

Radiative Transitions from the Detected States

We are studying a collection of gas particles of a single type within a fixed volume W which change their state at random times. The purpose of the analysis is to evaluate the characteristics of the time changes in a vector consisting of a relatively small number (J) of random, positive integer functions $\{N_i(t)\}$, where $i = 1, J$. Each of the component functions $N_i(t)$ (the instantaneous population of the i-th quantum state of the object) of this vector is equal to the number of particles inside W in a state with internal energy E_i at time t. For brevity the J (in some way) distinguishable states of the gas particles in this study will be referred to as detectable. The spontaneous transitions from just these states are recorded in the experiment. Changes in the populations of the detectable states are related to random processes of three types:

(a) $i \rightarrow j$ transitions between two detectable states $(j, i = 1, J)$;

(b) $0 \rightarrow i$ transitions from various undetectable states to a detectable level ($i = \overline{1, J}$);
(c) $i \rightarrow 0$ transitions from a detectable state ($i = \overline{1, J}$) into some undetectable ones.

The region W being analyzed is assumed to be homogeneous and the general macroscopic characteristics of the medium (such as the particle density, temperature, etc.) are assumed to be constant for now. The linear dimensions of this region are not too small, i.e., they are such that changes in the detected populations $N_i(t)$ due to the arrival of particles from outside or the loss of particles from inside volume W are negligibly small compared to the changes in $N_i(t)$ due to transitions occurring within W. In addition, it is assumed that the number of particles in W is large so the limitation placed on the change in the populations by (the law of) conservation of the particles may be ignored. We shall proceed from the following assumptions:

I. The transitions between levels take place instantaneously (a practical, sufficient condition is that the duration of a transition $\tau_0 \ll \theta \sim 10^{-8}$ sec) and the probability that the state of a particle be unchanged over a time τ is an exponential function of τ.
II. Under the experimental conditions being treated here transitions between the levels, which change the populations $N_i(t)$ of the detectable states, take place in every particle independently of the state of the other particles.
III. In the fixed time intervals the number of $(i) \rightarrow (k)$ and $(i') \rightarrow (k')$ transitions, where $i \neq i'$ or $k \neq k'$, are independent of one another.
IV. The transitions $(i) \rightarrow (k)$ between two states, where $i, k = \overline{0, J}$ and $i \neq k$, form stochastic Poisson fluxes, which are formed by the vector random process $\mathbf{Z}(t) \equiv \{N_i(t), i = \overline{1, J}\}$. The instantaneous densities of the fluxes λ_{ji}, which (fluxes) describe the partial rates of change of the populations for given values of $N_k(t)$, also depend on the populations of the undetectable states. The time variation of the latter and their variation within the confines of the observed volume W are ignored here; thus, having stationary conditions in mind, we shall write $\lambda_{ji} = \lambda_{ji}(\{N_k(t)\}; k = \overline{1, J})$.

The principal limitations placed on experiments by condition II are the following. In order to exclude the effect of stimulated radiative transitions and the effects of coherence of the states, the observed volume W of gas must be practically transparent and must be beyond the range of strong external resonant fields. The populations of the detectable (and a number of other) excited states of the X atoms (molecules, ions) must be small so that (for simplicity) for a given accuracy the effect of reactions of the type $X(i) + X(j) \leftrightarrow X(k) + X(m)$, which reduce or increase the populations of two detectable states at once, may be neglected. Since the basic purpose of this chapter is to discuss the general principles of the correlation scheme for evaluating the kinetic coefficients, it is not yet appropriate to consider complicated variants of the theory which permit analysis of measurements in a dense medium and in intense fields.

We shall consider the number $m_{ji}(t)$ of transitions from state (i) to state (j) over a time $(t - \theta, t)$. The fixed quantity θ has the significance of a characteristic averaging time in the receiver-detector which is the same for all channels (j, i). We shall assume conditions I and II to be fulfilled. Then from conditions III and IV, we obtain

$$\mathscr{E}(m_{ji}/\Lambda_{ji}) = \Lambda_{ji}, \quad \mathscr{E}(m_{ji}^2/\Lambda_{ji}) - [\mathscr{E}(m_{ji}/\Lambda_{ji})]^2 = \Lambda_{ji}, \tag{9}$$

$$\mathscr{E}(m_{ji}(t)m_{ji}(t+\tau)/\Lambda_{ji}(\tau),\ \Lambda_{ji}(t+\tau)) = \Lambda_{ji}(t)\Lambda_{ji}(t+\tau) \text{ for } \tau \geq \theta, \tag{10}$$

$$\mathscr{E}(m_{ji}(t)m_{j'i'}(t')/\Lambda_{ji}(t),\ \Lambda_{j'i'}(t')) = \Lambda_{ji}(t)\Lambda_{j'i'}(t') \text{ for } i \neq i' \text{ or } j \neq j', \tag{11}$$

where

$$\Lambda_{ji}(t) \equiv \int_{t-\theta}^{t} \lambda_{ji}(\{N_k(\tau)\}, \quad k = 1, \mathrm{J})\, d\tau. \tag{12}$$

We now briefly explain the method of proving these equations. Equation (9) is derived with the help of the Poisson formula (6) in a way similar to that used for the simplified model function u(t) of the photon counts. The absence of aftereffects in the stochastic Poisson fluxes makes it possible to write the conditional mathematical expectation of the product of time-shifted processes with the same subscripts in the form

$$\begin{aligned} &\mathscr{E}\,[m_{ji}(t)m_{ji}(t+\tau)/\Lambda_{ji}(t),\ \Lambda_{ji}(t+\tau)] = \\ &= \mathscr{E}\,[m_{ji}(t)/\Lambda_{ji}(t)]\,\mathscr{E}\,[m_{ji}(t+\tau)/\Lambda_{ji}(t+\tau)] = \Lambda_{ji}(t)\Lambda_{ji}(t+\tau) \quad \text{for} \quad \tau \geqslant \theta. \end{aligned}$$

Equation (10) is proven in the same way. Finally, taking the independence of transitions in different particles (condition III) into account, we obtain an expression for the product of processes with at least one different subscript,

$$\begin{aligned} &\mathscr{E}\,[m_{ji}(t)m_{j'i'}(t')/\Lambda_{ji}(t),\ \Lambda_{j'i'}(t')] = \mathscr{E}\,[m_{ji}(t)/\Lambda_{ji}(t)] \times \\ &\times \mathscr{E}\,[m_{j'i'}(t')/\Lambda_{j'i'}(t')] = \Lambda_{ji}(t)\Lambda_{j'i'}(t') \quad \text{for} \quad i \neq i' \quad \text{or} \quad j = j', \end{aligned}$$

which proves Eq. (11). Equations (9)-(12) completely describe the second-order statistical properties of the numbers $m_{ji}(t)$ of transitions from state i to state j. With these formulas we may consider the fluctuations in the kinetic equations.

The Role of Fluctuations in the Kinetic Equations

For the transition to the population fluctuations we note that the rates of change in the smoothed populations, $dN_i(t)/dt$, are related to $m_{ij}(t)$ by the simple equations

$$\frac{d}{dt} N_i(t) = \frac{1}{\theta} \sum_{j=0,\, j\neq i}^{J} [m_{ij}(t) - m_{ji}(t)]. \tag{13}$$

Introducing the smoothed values of the fluxes to the detectable states,

$$l_i(t) = \frac{1}{\theta} \sum_{j=0,\, j\neq i}^{J} [\Lambda_{ij}(t) - \Lambda_{ji}(t)], \qquad i = 1, \mathrm{J} \tag{14}$$

and the short notation for the vector function $\Lambda(t) = \{\Lambda_{ij}(t),\ i, j = 0, \mathrm{J}\}$, we now consider the simplest characteristics of the rate distribution $\left\{\frac{dN_i}{dt}(t),\ i = 1, \mathrm{J}\right\}$. Using Eqs. (13) and (14) and the properties (9)-(11) of the m_{ij}, we obtain

$$\mathscr{E}\left[\frac{dN_i}{dt}(t)/\Lambda(t)\right] = l_i(t), \quad i = 1, \mathrm{J}; \tag{15}$$

$$\mathscr{E}\left[\left(\frac{dN_i}{dt}\right)^2 \Big/ \Lambda(t)\right] = [l_i(t)]^2 + \frac{1}{\theta^2} \sum_{j=0,\, j\neq i}^{J} [\Lambda_{ji}(t) + \Lambda_{ij}[t)], \qquad i = 1, \mathrm{J}; \tag{16}$$

$$\mathscr{E}\left[\frac{dN_i}{dt}(t)\, \frac{dN_j}{dt}(t)/\Lambda(t)\right] = l_i(t)\, l_j(t) - \frac{1}{\theta^2} [\Lambda_{ij}(t) + \Lambda_{ji}(t)], \tag{17}$$

where $j \neq i$, $i, j = 1, J$, and

$$\mathscr{E}\left[\frac{dN_i}{dt}(t)\frac{dN_i}{dt}(t+\tau)/\Lambda(t), \Lambda(t+\tau)\right] = l_i(t)\, l_j(t+\tau) \quad \text{for} \quad \tau \geqslant \theta, \; i, j = 1, J. \tag{18}$$

For example,

$$\mathscr{E}\left[\left(\frac{dN_i}{dt}(t)\right)^2 \Big/ \Lambda(t)\right] = \frac{1}{\theta^2}\mathscr{E}\left[\left(\sum_{j=0,\, j\neq i}^{J}(m_{ij}-m_{ji})^2/\Lambda\right]\right. =$$

$$= \frac{1}{\theta^2}\mathscr{E}\left\{\sum_{j=0,\, j\neq i}^{J}(m_{ij}^2+m_{ji}^2) + 2\sum_{j=0,\, j\neq i}^{J}\sum_{=j+1,\, j'\neq i}^{J}(m_{ij}\,m_{ij'}+m_{ji}m_{j'i}) -\right.$$

$$\left.- 2\sum_{j=0,\, j\neq i}^{J} m_{ij}\sum_{j=0,\, j\neq i}^{J} m_{ij}\Lambda\right\} = \frac{1}{\theta^2}\left\{\sum_{j=0,\, j\neq i}^{J}(\Lambda_{ij}+\Lambda_{ji}+\Lambda_{ij}^2+\Lambda_{ji}^2) +\right.$$

$$\left.+ 2\sum_{j=0,\, j\neq i}^{J}\sum_{j'=j+1,\, j'\neq i}^{J}(\Lambda_{ij}\Lambda_{ij'}+\Lambda_{ji}\Lambda_{j'i}) - 2\sum_{j=0,\, j\neq i}^{J}\Lambda_{ij}\sum_{j=0,\, j\neq i}^{J}\Lambda_{ij}\right\} =$$

$$= \frac{1}{\theta^2}\left\{\sum_{j=0,\, j\neq 0}^{J}(\Lambda_{ij}+\Lambda_{ji}) + \left(\sum_{j=0,\, j\neq i}^{J}(\Lambda_{ij}-\Lambda_{ji})\right)^2\right\} = \frac{1}{\theta^2}\sum_{j=0,\, j\neq i}^{J}(\Lambda_{ij}+\Lambda_{ji}) + [l_i(t)]^2.$$

Equations (15), (17), and (18) are derived analogously.

We shall factor out the conditional fluctuations in the rates of change of the smoothed populations and set [in view of Eq. (15)]

$$\frac{dN_i}{dt}(t) = l_i(t) + h_i(t), \qquad i = 1, J.$$

Using the properties of dN_i/dt we find from Eqs. (15)-(18) that

$$\mathscr{E}(h_i(t)/\Lambda(t)) = 0, \qquad i = 1, J;$$

$$\mathscr{E}(h_i^2(t)/\Lambda(t)) = \frac{1}{\theta^2}\sum_{j=0,\, j\neq i}^{J}[\Lambda_{ij}(t)+\Lambda_{ji}(t)], \qquad i, j = 1, J;$$

$$\mathscr{E}(h_i(t)\,h_j(t)/\Lambda(t) = -\frac{1}{\theta^2}[\Lambda_{ij}(t)+\Lambda_{ji}(t)], \qquad j \neq i, \quad i\,j = 1, J;$$

$$\mathscr{E}(h_i(t)\,h_j(t+\tau)/\Lambda(t), \; \Lambda(t+\tau)) = 0 \quad \text{for} \quad \tau = \theta, \quad ij = 1, J.$$

Transforming now from the conditional to the unconditional mathematical expectations, we find for the rate fluctuations that

$$\begin{aligned} &\langle h_i(t)\rangle = 0, \qquad \langle[h_i(t)]\rangle = \frac{1}{\theta^2}\sum_{j=0,\, j=i}^{J}[\langle\Lambda_{ji}(t)\rangle + \langle\Lambda_{ij}(t)\rangle]; \\ &\langle h_i(t)\,h_j(t)\rangle = \frac{1}{\theta^2}[\langle\Lambda_{ij}(t)\rangle + \Lambda_{ji}(t)], \qquad i \neq j; \\ &\langle h_i(t)\,h_j(t+\tau\rangle \quad \text{for} \quad |\tau| \geqslant 0, \quad i, j = 1, J. \end{aligned} \tag{19}$$

Returning to the standard notation, we finally write the time rates of change of the smoothed populations in the form

$$\begin{aligned} &\frac{dN_i}{dt} = l_i(t) + q_i\Phi_i(t), \qquad i = 1, J; \\ &q_i = \frac{1}{\theta}\left[\sum_{j=0,\, j\neq i}^{J}(\langle\Lambda_{ji}\rangle + \langle\Lambda_{ij}\rangle)\right]^{1/2}. \end{aligned} \tag{20}$$

Here $\Phi_i(t)$ are the stationary short-term correlated fluctuations, for which

$$\langle \Phi_i(t)\rangle \equiv 0, \quad \sigma^2\Phi_i(t) \equiv 1, \quad \langle \Phi_i(t)\Phi_j(t+\tau)\rangle = 0 \text{ for } |\tau| \geqslant \theta; \tag{21}$$

$$\langle \Phi_i(t)\Phi_i(t+\tau)\rangle = \begin{cases} 0 & \text{for } |\tau| \geqslant \theta, \\ 1 - \frac{|\tau|}{\theta} & \text{for } |\tau| < \theta; \end{cases} \tag{22}$$

$$\langle \Phi_i(t)\Phi_j(t)\rangle = -\frac{\langle \Lambda_{ij}(t)\rangle + \langle \Lambda_{ji}(t)\rangle}{q_i q_j \theta^2}. \tag{23}$$

In deriving Eq. (22) it was noted that in principle θ may be taken to be very small compared with the remaining characteristic times of the problem. This means that the actual unsmoothed noise $\Phi_i(t)$ may be regarded as δ-correlated. Thus, the noise smoothed over a time θ must have a "triangular" autocorrelation function (22). For a sufficiently large number of particles the results of Eqs. (20)-(23) are not related to the form of the kinetic equations, so their range of applicability is much wider than the problems being examined here. We shall assume under our experimental conditions that the kinetic equations (1) are valid for a large number of particles; that is, the conditional instantaneous densities are linear functions of the populations of the detectable states, i.e.,

$$\lambda_{ij}(t) = K_{ij}N_j(t), \quad \lambda_{i0}(t) = K_{0i}(t)N_i(t), \quad i, j = 1, J. \tag{24}$$

Using the condition for stationarity,

$$\sum_{j=1,\, j\neq i}^{J} K_{ij}\langle N_j\rangle + H_i = K_{ii}\langle N_i\rangle,$$

we write the equations for the fluctuating deviations $\eta_i(t)$ of the detected populations $N_i(t)$ from their average values $\langle N_i(t)\rangle$ in the form

$$\frac{d\eta_i}{dt}(t) + K_{ii}\eta_i(t) = \sum_{j=1,\, j\neq i}^{J} K_{ij}\eta_j(t) + q_i\Phi_i(t);$$
$$q_i = \sqrt{\frac{2}{\theta}K_{ii}\langle N_i\rangle}, \quad \langle \Phi_i(t)\Phi_j(t)\rangle = -\frac{1}{2}\frac{K_{ij}\langle N_j\rangle + K_{ji}\langle N_i\rangle}{\sqrt{K_{ii}K_{jj}\langle N_i\rangle\langle N_j\rangle}}, \tag{25}$$

where the properties of the functions $\Phi_i(t)$ are given by Eqs. (21) and (22).

For the appropriate experimental conditions this system of equations is far from degenerate and the cross correlation coefficients of the fluctuation components η_i and η_j are small. In this case the dispersion $\sigma^2\eta_i$ cannot be much less than $\langle N_i\rangle$. Since the matrix $\|K_{ij}\|$ is not known in advance, we shall proceed from the minimum value of the dispersion and assume that

$$vN_j \equiv \frac{\sigma N_j}{\langle N_j\rangle} \equiv \frac{\sigma\eta_i}{\langle N_j\rangle} \simeq \frac{1}{\sqrt{\langle N_j\rangle}}, \tag{26}$$

that is, we shall proceed from the least favorable case. Until a scheme for finding the dynamic parameters K_{ij} from the correlation functions is known, we must recall that the apparatus does not directly detect the populations $N_i(t)$ themselves. They have to be evaluated from counts of the numbers of photons spontaneously emitted by the gaseous medium we are studying.

Photon-Counting Statistics

The coefficient K_{ij} corresponding to an atomic transition to a lower energy level ($E_j > E_i$) may be written in the form of a sum $K_{ij} = A_{ij} + R_{ij}$, where A_{ij} is the Einstein coefficient for spontaneous emission on the $j \to i$, transition and R_{ij} is the probability of a nonradiative transition between the same states. Here $K_{ji} = R_{ji}$, $R_{ji} = D_{ij} R_{ij}$, and $A_{ji} = 0$, where the numbers D_{ij} are usually known in advance. For simplicity we shall assume that each detectable level (i) corresponds to one and only one detection channel (k_i, i) for counting photons on the $i \to k_i$ transition, where $E_{k_i} < E_i$. We shall rewrite the probability of the corresponding spontaneous transition in the form $A_{k_i i} \equiv \mathcal{A}_i$ for brevity. Spontaneous emission by atoms at each of these transitions forms a stochastic Poisson flux. From considerations similar to those in the section on the stochastic Poisson flux [cf. Eq. (8)] it follows that over a time $(t - \theta, t)$ the atoms in volume W emit random numbers of photons $S_i(t)$, given by

$$S_i(t) = \theta \mathcal{A}_i \omega [\langle N_i \rangle + \eta_i(t)] + \sqrt{\theta \mathcal{A}_i \omega \langle N_i \rangle} \psi_i(t), \tag{27}$$

where

$$\langle \psi_i(t) \rangle \equiv 0, \quad \sigma^2 \psi_i(t) \equiv 1, \quad \langle \psi_i(t) \psi_i(t+\tau) \rangle = 0 \quad \text{for} \quad |\tau| \gg \theta. \tag{27'}$$

In [1] the following were discussed in a similar way: the number $s_i(t)$ of photon counter pulses over the interval $(t - \theta, t)$, the resultant signal s_{i*} entering a computer as an estimate of $s_i(t)$, and the quantity $\mu_i(t) \equiv [s_{i*}(t) - s_{i*}(t)/Q_i$, which is an estimate for $\eta_i(t)$ (the meaning of Q_i is given below).

The observed fluctuations in the populations $\mu_i(t)$ can be written as a sum of two independent random processes, the internal fluctuations $\eta_i(t)$ and additive noise $\xi_i(t)$, i.e., $\mu_i(t) = \eta_i(t) + \xi_i(t)$. An estimate of the ratio of their dispersions, $\varkappa_i \equiv \sigma^2\xi_i/\sigma^2\eta_i$, is the final goal of an examination of the photon counting statistics. We shall use the notation of [1]. Let β_i be the fraction of the photons falling on the photomultiplier cathode of the i-th channel; p_i, the probability of detection of a photon incident on this photocathode; $Q_i \equiv \beta_i p_i \theta \mathcal{A}_i$, the coefficient of proportionality between the average value of the population $\langle N_i \rangle$ and the average signal from the photomultiplier $\langle s_{i*}(t) \rangle$; $\gamma_i = 2/\beta_i p_i \theta K_{ij} = 2/Q_i \alpha_i$, with $\alpha_i = \mathcal{A}_i/K_{ii}$; b_i, the standard deviation of the relative errors on going from $s_i(t)$ to $s_{i*}(t)$; and Θ, the maximum measurement time over which the experimental conditions may be still regarded as constant. We shall assume that the conditions for "sufficient slowness" are fulfilled; that is,

$$\Theta > 10 \max_{\lambda=1,J} \{1/\omega_\lambda\},$$

where ω_λ are the characteristic frequencies of Eq. (1) and we assume that the dark current does not exceed the external signal. Then (cf. [1]) under optimal conditions with

$$\langle N_i \rangle_{\text{opt}} \sim \frac{1}{Q_i b_i^2} \tag{28}$$

the ratio of the dispersions, $\varkappa_i$, is

$$\varkappa_i \equiv \frac{\sigma^2\xi_i}{\sigma^2\eta_i} \simeq \frac{2}{Q_i} = \frac{\gamma_i}{\alpha_i}. \tag{29}$$

Setting $K_{ii}\theta \leqslant 0.1$, $p \simeq 0.1$, $\beta_i \leq 1/3$, and $\alpha_i \simeq 1/2$, we obtain

$$Q_i \sim 3 \cdot 10^{-3}, \quad \gamma_i \gtrsim 700, \quad \varkappa_i \gtrsim 10^3. \tag{30}$$

Assuming $b_i \sim 10^{-2}$-10^{-5}, we find from Eqs. (28) and (30) that

$$\langle N_i \rangle_{\text{opt}} \sim 3 \cdot 10^6 - 3 \cdot 10^{12}. \tag{31}$$

We now evaluate the effect of the finiteness of the volume W on $\varkappa_i$ assuming the gas to be a sphere of radius R. If $\bar{c}$ is the average speed of the particles, then $\langle N_+ \rangle$ and $\langle N_- \rangle$, the average number of particles leaving and entering the sphere over time θ, are equal, with

$$\langle N_- \rangle = \langle N_+ \rangle \simeq \left(4\pi R^2 \frac{1}{6} \bar{c}\theta\right) \frac{\langle N_i \rangle}{W} = \frac{1}{2} \frac{\bar{c}\theta}{R} \langle N_i \rangle. \tag{32}$$

The processes of particle exchanges between the volume W and the surrounding gas may be regarded as stochastic Poisson fluxes and to take them into account we add another term to Eq. (26). Now

$$s_i(t) = \theta \mathcal{A}_i [\langle N_i \rangle + \eta_i(t)] + \sqrt{\theta \mathcal{A}_i \langle N_i \rangle} \psi_i(t) + \sqrt{\langle N_+ \rangle + \langle N_- \rangle} \psi_i^*(t),$$

where $\langle \psi_{i*}(t) \rangle \equiv 0$, $\sigma^2 \psi_i(t) \equiv 1$, and $\langle \psi_i^*(t) \psi_{i*}(t+\tau) \rangle = 0$ for $|\tau| > 0$. Clearly, as long as

$$\langle N_+ \rangle + \langle N_- \rangle \leqslant \theta \mathcal{A}_i \langle N_i \rangle, \tag{33}$$

the effect of finite W on $\varkappa_i$ may be neglected. From Eqs. (32) and (33) we obtain

$$R \gtrsim \bar{c}/\mathcal{A}_i. \tag{34}$$

Under ordinary conditions $\bar{c} \sim 10^5$ cm/sec and $\mathcal{A}_i = 10^7$-10^8, and it is required that $R \geq 10^{-2}$-10^{-3} cm. If the ratio of the populations $\langle N_i \rangle$ to the ground-state density N_0 is $\langle N_i \rangle / N_0 \simeq 10^4$, $R \simeq 10^{-2}$ cm, and $\langle N_i \rangle \simeq 10^{10}$ cm^{-3}, then the required maximum density for the experiment is $5 \cdot 10^{14}$ cm^{-3}.

A System with One Isolated Level

First we shall discuss the situation in which it is enough to describe the kinetics of the averaged population $N_1(t)$ of a single detectable state. Ignoring the fluctuations, we write

$$\frac{dN_1}{dt}(t) + K_{11} N_1(t) = H_1, \qquad K_{11} = K_{01} = A_{01} + R_{01}.$$

Here R_{01} is the sum of the "probabilities" of radiationless $1 \to 0$ transitions from state 1 into all others and A_{01} is the sum of the Einstein coefficients for spontaneous radiative transitions from state 1 (into all lower energy states). For simplicity we shall assume here that the probability of spontaneous decay, A_{01}, is known in advance with sufficient accuracy while the value of R_{01}, which must be determined here, is known only as to order of magnitude. In addition, we shall assume that the Einstein coefficient $\mathcal{A}_1 \equiv A_{k_1 1}$ for the transition $1 \to k_1$ at the photon counter frequency (in this example, the only detection channel) is also known to the required accuracy. Then, of course, $a_1 \equiv \mathcal{A}_1 / A_{01} < 1$. We note that by choosing the experimental conditions (primarily the density N_e and temperature T_e of the electrons) it is usually possible to vary the ratio $\alpha_1 \equiv \mathcal{A}_1 / K_{11} = a_1 (1 + R_{01}/A_{01})^{-1} < a_1$ over fairly wide limits.

The basic purpose of this preliminary discussion is to evaluate (as to order of magnitude) the observation time T_v required for the measurement $\hat{R}_{01}$ of the magnitude of R_{01}, with a given variability $v_R \equiv \sigma \hat{R}_{01}/R_{01} = v$. (Roughly speaking, v is the relative error of the estimate R_{01}.) For the fluctuations in the population of the detectable level smoothed over an interval θ, we have, according to Eq. (25),

$$\frac{d\eta_1}{d\theta} = -K_{11}\eta_1 + q_1\Phi_1(t);$$
$$\langle\Phi_1(t)\rangle \equiv 0, \quad \sigma\Phi_1(t) \equiv 1, \quad \langle\Phi_1(t)\Phi_1(t+\tau)\rangle = \begin{cases} 1 - |\tau|/\theta & \text{for } |\tau| < \theta, \\ 0 & \text{for } |\tau| > \theta; \end{cases} \tag{35}$$
$$q_1 = \sqrt{\frac{2}{\theta} K_{11} \langle N_1\rangle}.$$

It may be shown (for example, [7]) that it follows from this with relative accuracy θK_{11} that

$$\sigma^2\eta_1 = \theta q_1^2/2K_{11}.$$

Substituting the value of q_1 in this formula we obtain the estimate

$$v_{N_1} \equiv \frac{\sigma\eta_1}{\langle N_1\rangle} = \frac{1}{\sqrt{\langle N_1\rangle}}[1 + O(\theta K_{11})],$$

which improves upon Eq. (26) for this example. Hence, Eq. (29) which is based on the asumption (26) is adequate for the ratio of the dispersions of the additive noise $\xi_1(t)$ and the useful signal. Thus, the observed fluctuations in the population $\mu_1(t)$ can be written in the form of a sum of independent processes,

$$\mu_1(t) = \eta_1(t) + \xi_1(t); \quad \varkappa_1 \equiv \frac{\sigma^2\xi_1}{\sigma^2\eta_1} = \frac{\gamma_1}{\alpha_1}, \quad \gamma_1 \gtrsim 700. \tag{36}$$

From the accepted signal $\mu_1(t)$ it is possible to find the sampling dispersion $s^2\mu_1(t)$ and the autocorrelation function

$$r_\mu(t) = \overline{\mu_1(t)\mu_1(t+\tau)}/s^2\mu_1.$$

From Eqs. (22) and (23) it is easy to find for the actual autocorrelation function $\rho_\mu = \langle\mu_1(t)\mu_1 \times (t+\tau)\rangle/\sigma\mu$ that

$$\rho_\mu(\tau) = \frac{1}{1+\varkappa_1}\exp(-K_{11}\tau) \simeq \frac{\exp(-K_{11}\tau)}{\varkappa_1} \ll 1 \quad \text{for} \quad |\tau| \geqslant \theta. \tag{37}$$

Thus, because of the large additive noise, even for shifts θ the values of $\mu_1(t)$ seem almost stochastically unrelated. This means that for the full duration T of the measurements the effective sampling volume is $n_* = T/\theta$. From Eq. (37) it is clear that for an estimate of K_{11} we may use the quantity

$$\hat{K}_{11} \equiv \frac{1}{\tau_2 - \tau_1} - \ln\left[\frac{r_\mu(\tau_1)}{r_\mu(\tau_2)}\right], \quad \tau_2 > \tau_1 \geqslant \theta. \tag{38}$$

For $r_1 \simeq \theta$ and $\tau_2 \simeq 1\text{-}3/K_{11}$ this estimate has a variability given by [1]

$$v_{\hat{K}_{11}} \equiv \frac{\sigma\hat{R}_{11}}{K_{11}} \simeq \frac{3\varkappa}{\sqrt{n_*}}. \tag{39}$$

We now relate the accuracy of the estimates of the total probability K_{01} of a $1 \to 0$ transition to the unknown probability R_{01} of a radiationless transition assuming $\hat{R}_{01} = \hat{K}_{01} - A_{01}$. For the variability of the estimate of $\hat{R}_{01}$ we have, according to Eqs. (39) and (36),

$$v_{\hat{R}} = \frac{\sigma\hat{R}_{01}}{R_{01}} = \frac{K_{11}}{K_{11} - A_{01}} \frac{\sigma\hat{R}_{01}}{K_{11}} = \frac{1}{\alpha_1(1-\alpha_1)} \frac{3\gamma_1}{\sqrt{n_*}}. \tag{40}$$

It follows immediately from this equation that the optimum value of α_1 is $\alpha_{opt} = 0.5$. According to the same formula, we obtain for the total duration of a measurement with a given spread in the estimates

$$T_v = \left[\frac{3\gamma_1}{\alpha_1(1-\alpha_1)}\right]^2 \frac{\theta}{v^2}.$$

In the optimal case with $\gamma_1 \sim 700$, $\theta K_{11} \sim 0.1$, and $\alpha_1 = 1/2$, we have

$$T_v \simeq \frac{5\cdot 10^6}{K_{11}v^2}.$$

This means, for example, that for a spontaneous decay probability of $A_{01} \sim 10^7$ sec ($K_{11} = A_{01}/\alpha_1$) the time needed for a measurement with accuracy $v_{\hat{R}} \sim 0.05$ is $T_{01} \simeq 2$ min, which is fully acceptable.

A System with Several Isolated Levels

From the differential equations (25) for the fluctuations of the smoothed populations we now go to the correlation equations by multiplying both parts of these equations by $\eta_i(t-\tau)$ and averaging them over $\tau \geq \theta$. As a reult we obtain

$$\frac{d}{d\tau}\rho_{ji}^{\eta}(\tau) = -K_{ii}\rho_{ji}^{\eta}(\tau) + \sum_{s=1,\, s\neq i}^{J} \frac{\sigma\eta_s}{\sigma\eta_i} K_{is}\rho_{js}^{\eta}(\tau), \qquad i, j = 1, J, \tag{41}$$

where $\rho_{ik}^{x} \equiv \langle x_i(t)x_k(t+\tau)\rangle / \sigma x_i \sigma x_k$.

In the transition to the correlation equations for $\mu(t)$ we note that

$$\rho_{ik}^{\mu} = \rho_{ik}^{\eta} / \sqrt{(1+\varkappa_i)(1+\varkappa_k)} \simeq \rho_{ik}^{\eta} / \sqrt{\varkappa_i \varkappa_k}, \tag{42}$$

and, after substituting this in Eq. (41), obtain

$$\frac{d}{d\tau}\rho_{ji}(\tau) = -K_{ii}\rho_{ji}(t) + \sum_{s=1,\, s\neq i}^{J} \frac{\sigma\mu_s}{\sigma\mu_i} K_{is}\rho_{js}(\tau), \qquad i, j = 1, J, \tag{43}$$

where the superscript on μ has been left out for brevity. System (43) allows us to find the matrix of the kinetic coefficients $\|K_{ij}\|$ from the correlation matrix $\|\rho_{ij}\|$. In evaluating $\|K_{ij}\|$ it is best to avoid using the derivatives $d\rho_{ij}/d\tau$ since they typically have large scatter when they are evaluated using the sampling correlation function $r_{ij}(\tau)$. The derivatives are best avoided by integrating Eqs. (43). Then we obtain a new system of correlation equations without derivatives,

$$\rho_{ji}(\tau_2) - \rho_{ji}(i) = -K_{ii}\int_{\tau_1}^{\tau_i} \rho_{ji}(\tau)\,d\tau + \sum_{s=1,\, s\neq i}^{J} K_{is}\frac{\sigma\mu_s}{\sigma\mu_i}\int_{\tau_1}^{\tau_2} \rho_{js}(\tau)\,d\tau. \tag{44}$$

The optimum choice of the limits of integration τ_1, τ_2 for estimating the coefficients $\rho_{ij}(\tau)$ is determined by the form of the correlation functions $\rho_{ij}(\tau)$. As a first approximation in the general case we may use

$$\tau_1 = \theta, \qquad \tau_2(i, j) = 2\sqrt{(1+\varkappa_i)(1+\varkappa_j)} \int_{\theta}^{\infty} \rho_{ij}(\tau)\, d\tau. \tag{45}$$

Transforming from the actual correlation coefficients $\rho_{ij}(\tau)$ to the sample values $r_{ij}(\tau)$, we obtain, by analogy with Eq. (44),

$$r_{ji}(\tau_2) - r_{ji}(\tau_1) \simeq - K_{ii} \int_{\tau_1}^{\tau_2} r_{ji}(\tau)\, d\tau + \sum_{s=1,\, s\neq i}^{J} K_{is} \frac{\sigma\mu_s}{\sigma\mu_i} \int_{\tau_1}^{\tau_2} r_{js}(\tau)\, d\tau. \tag{46}$$

By specifying the parameters τ_1, τ_2 and numerically integrating the functions $r_{ii}(\tau)$ and $r_{is}(\tau)$, it is possible to reduce the evaluation of the coefficients K_{ij} to a linear algebra problem.

We now turn in more detail to second-order systems. Solving the system of equations (43) for J = 2 for the unknown kinetic coefficients K_{ij}, we find

$$K_{ij} = \frac{(-1)^{i+j}}{\omega(\tau)} \frac{\sigma\mu_i}{\sigma\mu_j} \left[\rho_{\bar{j}\bar{j}} \frac{d\rho_{\bar{j}i}}{d\tau} - \rho_{\bar{j}\bar{j}} \frac{d\rho_{ii}}{d\tau} \right], \qquad \tau \geqslant \theta, \tag{47}$$

where $\omega(\tau) \equiv \rho_{11}(\tau)\rho_{22}(\tau) - \rho_{12}(\tau)\rho_{21}(\tau)$; $\bar{i} = 2^{2-i}$; $i, j = 1, 2$.

Adding K_{11} and K_{22} Eq. (47) yields

$$K_{11} + K_{22} = \frac{1}{\omega(\tau)} \left[\rho_{12} \frac{d\rho_{12}}{d\tau} - \rho_{22} \frac{d\rho_{11}}{d\tau} \right] + \frac{1}{\omega(\tau)} \left[\rho_{21} \frac{d\rho_{22}}{d\tau} \right] - \rho_{11} \frac{d\rho_{22}}{d\tau} \Big] +$$
$$= - \frac{1}{\omega(\tau)} \frac{d}{d\tau} \left[\rho_{11}(\tau)\rho_{22}(\tau) - \rho_{12}(\tau)\rho_{21}(\tau) \right] = - \frac{d}{d\tau} \ln \omega(\tau).$$

Thus, by analogy with Eq. (38), as an estimate of the trace of the matrix we may use the quantity

$$\begin{aligned} \widehat{(K_{11} + K_{22})} &= \ln [w(\tau_1)/w(\tau_2)]/(\tau_2 - \tau_1), \\ w(\tau) &= r_{11}(\tau) r_{22}(\tau) - r_{12}(\tau) r_{21}(\tau). \end{aligned} \tag{48}$$

Now taking $\tau_1 = \theta$, $\tau_2 - \tau_1 \approx (1\text{–}3)/(K_{11} + K_{22})$, we obtain

$$\frac{\sigma\widehat{(K_{11} + K_{22})}}{K_{11} + K_{22}} \simeq \sqrt{\frac{\sigma^2 w(\tau_1)}{\omega^2(\tau_1)} + \frac{\sigma^2 w(\tau_2)}{\omega^2(\tau_2)}} \leqslant \frac{3\sigma w(\tau_1)}{\omega(\tau_1)} \leqslant$$

$$\leqslant 3\sqrt{[\rho_{11}(\tau_1)]^2 + [\rho_{22}(\tau_1)]^2 + [\rho_{12}(\tau_1)]^2 + [\rho_{21}(\tau_1)]^2}/\omega(\tau_1)\sqrt{n_*} \leqslant \frac{3\sqrt{1+\rho_0^2}}{\sqrt{n_*}} \frac{\sqrt{\varkappa_1^2 + \varkappa_2^2}}{1 - \rho_0^2},$$

where $\rho_0 = \dfrac{\langle \eta_1(t)\, \eta_2(t) \rangle}{\sigma\eta_1 \sigma\eta_2}$.

As a rule it is convenient to choose the experimental conditions so that the matrix $\|\hat{K}_{ij}\|$ is nearly diagonal (more precisely, so that $\hat{K}_{12}/\hat{K}_{22} < 0.5$ and $K_{21}/K_{11} < 0.5$). Then ρ_0^2 is much less than unity and

$$v_{\widehat{K_{11}+K_{22}}} \equiv \frac{\sigma\widehat{(K_{11} + K_{22})}}{K_{11} + K_{22}} \leqslant \frac{3}{\sqrt{n_*}} \sqrt{\varkappa_1^2 + \varkappa_2^2}. \tag{49}$$

To some extent this allows us to evaluate the relative errors in $\hat{K}_{ij}$ Thus, for example, if the relative errors in $\hat{K}_{11}$ and $\hat{K}_{22}$ are roughly the same and $\varkappa_1 \simeq \varkappa_2 \simeq \varkappa$, then, as is clear from Eq. (48), the accuracy (variability) of the estimates of K_{ij} may be reduced to the form

$$v_{\hat{K}_{ii}} \leqslant \sqrt{2}\,\frac{3\varkappa}{\sqrt{n_*}}. \tag{50}$$

An increase in the standard deviation of the estimate by $\sqrt{2}$ on going from a first-order to a second-order system is fairly natural. The number of parameters being evaluated increases by a factor of 4 in this case while the volume of information increases twofold; hence, the effective ("specific") volume of the sample per parameter is reduced by a factor of 2, and the corresponding reduction in the accuracy of the estimate, by $\sqrt{2}$.

A transfer to a J-th order system of fluctuation equations does not require fundamental changes in either the general considerations or in the sequence of operations. During an experiment J^2 correlation functions are evaluated from the photomultiplier signals, and then system (46) consisting of J^2 correlation equations is solved. For a fixed (for each signal) sample volume the variability in the estimates of the coefficients increases roughly in proportion to $\sqrt{J}$. (Including the above a priori information on the relation between the coefficients K_{ij} and K_{ji} increases the accuracy of the estimate somewhat.) Thus compared with a single detected state, the measurement time required to ensure a given accuracy must be increased by roughly J times. In the problems of particular current interest there is no sense in discussing experiments in which J is much greater than 10. Then for $v \simeq 0.05$ we may expect a measurement time of the order of several hours, quite within attainable limits for similar physical experiments.

Numerical Experiments

Earlier semiqualitative results [7] indicate the practical feasibility of the method for evaluating the probabilities of radiationless transitions recommended here. The recording time required for acceptable accuracy of the estimates has been found to within an order of magnitude. It is important that we are speaking here of minutes and hours rather than decades and centuries. Since the fluctuations being analyzed are stationary, the signals may be recorded in intervals with interruptions for adjustment of the apparatus. From the above it is clear that the intended recording apparatus is complicated and unique. Before developing it and, in any case, before the physical measurements are set up, it is desirable to do some auxiliary studies on the accumulated stochastic material, as these might help rational planning of those cumbersome efforts.

Evidently, it is necessary to turn to some comparatively simple preliminary experiments for this purpose. Their immediate result must be to find optimal plasma parameters, to determine the number of channels for counting photons, and to refine the measurement time. It is practically impossible to solve these "organizational" problems analytically. Thus, we may speak here of a crude analog model or of numerical experiments. We shall dwell on the latter. By "playing back" the stages of the measurement on a computer, in particular, solving the kinetic equations for the signal (modeling the intrinsic fluctuations and the additive noises on the right-hand side of these equations), and finally computing the correlation functions after making a series of such machine tests, we can predict for given conditions the actual accuracy in the estimates of the kinetic coefficients K_{ij} for a rather large number J of detectable states. In these numerical experiments, by varying the number of detectable states, the form of the matrix $\|K_{ij}\|$, the density and temperature of the free electrons in the plasma, and the smoothing and recording times, we can find the optimum conditions for evaluating (or improving the accuracy of) the coefficients K_{ij}. Clearly, it is much easier to carry out preliminary experiments on a computer (faster and cheaper) than it is to do actual physical experiments using the correlation method. Furthermore, since in real physical experiments the ratio of the dispersions of the external noise and the population fluctuations, η_i, is very large ($\varkappa \geq 10^3$) and all the correlation functions of the photon counts are very much less than unity, it is possible to reduce all the ratios K_{ij} by the same number of times during counting. This leads to a substantial speeding up in the calculations, permitting a scaling (increase by the same number of times) of the relative random errors in the estimates of the coefficients K_{ij}. Using this similarity, it is possible to do mathematical experiments with $\varkappa_i \ll 10^3$ and at the same time reduce the volume of measurements in the numerical experiment by millions of times.

The setup of the mathematical experiment can conveniently be divided into the following stages:

1. Creating and recording on magnetic tape (or magnetic drums) of G independent series of white Gaussian noise α.
2. Specifying the parameters of the experiments (including the averaging time θ and the maximum number J_0 of isolated levels; computing as accurately as possible the matrix $\|K_{ij}\|$, the coefficients q_i, the correlation coefficients $\langle \Phi_i(t)\,\Phi_j(t)\rangle \equiv -0.5 \times [K_{ij}\langle N_j\rangle + K_{ji}\langle N_i\rangle]/\sqrt{K_{ii}K_{jj}\langle N_i\rangle\langle N_j\rangle}$, and the roots $(-\lambda_i)$ of the characteristic equation of system (2); fixing the length T of the measurements in one series, $T > 50/\lambda_{min}$, where $\lambda_{min} = \min_i \lambda_i$.
3. Forming the random processes Φ_i and ξ_i for each series $g = 1, G$ with the aid of α.
4. Obtaining the response η_i of the system (2) to the random noise Φ_i for $g = 1, G$; adding the additive noise ξ_i to η_i; and forming the signal μ_i in imitation of the readout from the photon counters.
5. Computing the correlation functions $r^{\mu}_{ij}(\tau)$ and $\tau = \theta n$ for $g = 1, G$, where $n = 1, 2, \ldots, N_{max}$, and $N_{max} \simeq 0.1T/\theta$; analyzing the correlation functions and writing them in abbreviated form on magnetic tape or drum.
6. Isolating $J \leq J_0$ fixed levels; specifying the limits of integration for the correlation functions, $\tau_1(i, j)$ and $\tau_2(i, j)$; and, obtaining estimates of K_{ij} ($g = 1, G$) starting from the system (46).
7. Computing the mean values and standard deviations of $\hat{K}_{ij}$ for $g = 1, G$.
8. Searching for optimal $\tau_2(i, j)$ for estimating $\hat{K}_{ij}$ by the recurrence method with the aid of iterative calculations involving Eqs. (6) and (7).
9. Successively reducing J [repeating all calculations (6)-(8)] and finding J_{min} for which the condition

$$(\bar{\hat{K}}_{ij} - K_{ij})\ G/s(\hat{K}_{ij}) < 3, \tag{51}$$

   where $\bar{\hat{K}}_{ij}$ and $s\hat{K}_{ij}$ are the sample mean and standard deviation, is still satisfied.
10. Estimating (for J_{min} and optimal τ_1 and τ_2) the relative errors in $\hat{K}_{ij}$.
11. Repeating calculations of the types (2)-(10) for different variants of the matrix $\|K_{ij}\|$ and apparatus constant θ allowed in the experimental setup.

The calculation may be done by recording the intermediate results on a magnetic tape or drum. Naturally, it is not always necessary to carry out the scheme to its full extent. As noted earlier, the similarity principle is used in numerical experiments. In transferring from the physical to the mathematical experiment it is possible to reduce the ratio of the dispersions $\varkappa_i = \sigma^2\xi_i/\sigma^2\eta_i$ by K times (provided $\varkappa_i' \equiv \varkappa_i/K > 1$). Then the accuracy obtained in a mathematical experiment with duration T is achieved in a physical experiment with $T = (1 + \varkappa_i)/(1 + \varkappa_i')$. It is also natural to expect that when the recording duration is increased by a factor of Q in the physical experiment the standard deviations in $\hat{K}_{ij}$ are reduced by a factor of $(Q)^{1/2}$; however, these estimates may be shifted due to the limited number of levels J. Because of this it is not recommended that the accuracy of the estimates of $\hat{K}_{ij}$ in the mathematical experiment (where systematic errors are controlled) be scaled to the case with a long recording time T and the same $\varkappa_i$, since in the latter case the "modeled" errors may be much larger than the random errors.

We now consider the individual parts of this scheme in more detail. The first four states of the calculations imitate the observations. The main requirement in doing these calculations is that they not take up too much time. A single calculation and recording of several million dual-valued random numbers is no difficulty for a computer now. It is only important to note that heightened requirements then exist on the noncorrelation of sequences of pseudorandom numbers† and it is necessary to do correlation analyses on noises generated for control purposes. At the beginning it is convenient to obtain uniformly distributed uncorrelated quantities and then, with the aid of a small table, to convert these quantities into Gaussian quantities with zero average and unit dispersion. The normalization requirements on quantities in this problem are not very rigid, but it is necessary to make sure that the distribution be symmetric and that the dispersion of a random variable be equal to unity.

The transformation from J independent random sequences $\alpha_i(t)$ to $\Phi_i(t)$ $(i = 1, J)$ may be realized with the formulas

$$\Phi_i(t) = \alpha_i(t)\sqrt{1 - \sum_{k=1}^{i-1} \rho_{ik}^2} + \sum_{k-1}^{i-1} \alpha_k(t)\,\rho_{ik}, \tag{52}$$

where $\rho_{ik} = \langle \Phi_i(t)\, \Phi_k(t) \rangle$.

The numerical solution of system (2) may be obtained in several ways. For example, it is possible to use a difference scheme with first-order accuracy, approximating the derivatives by first differences,

$$\frac{\eta_i(t_{n+1}) - \eta_i(t_n)}{\theta} + K_{ii}\eta(t_n) - \sum_{j=1,\, j\neq 1}^{J} K_{ij}\eta_j(t_n) = q_i\Phi_i(t_n), \qquad t_{n+1} - t_n = \theta. \tag{53}$$

The choice of such a crude system is due to the statistical character of the problem and, specifically, to the fact that on the right-hand side there is a short-time correlated random function. The extreme case of a solution of Eq. (2) is to write the solution of the system in explicit form and evaluate the integrals numerically by a recurrence formula with step size θ. The choice of one or another scheme depends on the accuracy with which $\hat{K}_{ij}$ is determined. The regular (systematic) errors in $\hat{K}_{ij}$ caused by the crudity of the scheme for calculating $\eta(t)$ must be smaller than the statistical errors.

As initial conditions for solving system (2) we may take $\eta_i(0) \equiv 0$ since after the lapse of a time $t_{char} \simeq 3/\lambda_{min}$ the system "forgets" the initial conditions. In a preliminary analysis of the solution of system (2) the sample mean and standard deviations of η_i are calculated. The sample means must not differ significantly from zero, and they serve as a criterion for testing the correctness of the calculations. Knowing the approximate values (a relative accuracy of 10-30% is fully adequate here), we can, with the aid of the random sequences $\alpha_i(t)$, form a sequence $\xi_i(t)$ of external noise,

$$\xi_j(t) = \sigma\eta_j\varkappa_j'\alpha_j(t), \quad \varkappa_j' = \varkappa_j/K,$$

where K is the similarity multiplier which can significantly reduce the duration of the numerical experiment.

†It is impossible, for example, to use the RANDOM routine from the collection of algorithms [8] without appropriate modifications.

The fifth stage of the scheme begins with a calculation from the signal $\mu_i(t)$ of the correlation functions $r^{\mu}_{ij}(\tau)$ and then estimates the characteristic decay times of the correlation functions,

$$t_{ij} = 2\sqrt{(\varkappa_i' + 1)(\varkappa_j' + 1}\int_{\theta}^{\theta N_{max}} r^{\mu}_{ij}(\tau)\,d\tau,$$

where the integration is by simple summation of $r^{\mu}_{ij}(\tau)$. The values $r^{\mu}_{ij}(\theta)$ and the sequences

$$r^{\mu}_{ij}(\tau_{ij}),\ \int_{\theta}^{\tau_{ij}} r^{\mu}_{ij}(\tau)\cdot d\tau \tag{54}$$

are stored on magnetic tape for $\tau_{ij} = t_{ij} \pm m\Delta\tau_{ij},\ m = 0,1,\ \ldots,\ N_r,\ \Delta\tau_{ij} = 0.8\ t_{ij}/N_r,\ N_r \simeq 5\text{–}20$. The number G of series of trials that is sufficient for estimating the sample means and standard deviations of $\hat{K}_{ij}$ is G = 20-50. The total volume of the correlation function storage is $\simeq 4N_r J^2 G$, and for $J \leq 10$ it can always be done with less than 40,000 numbers. Thus, the material for later analysis is fairly compact.

In the initial calculations of the sixth stage it is possible to take $\tau_1(i, j)_0 = \theta$, and $\tau_2(i, j)_0 = t_{ij}$ as a first approximation.

The optimal values of $\tau_2(i, j)$ for evaluating K_{ij} may be found as follows. A measure of the errors $\varepsilon > 0$ (e.g., the sum of the sample dispersions weighted for the estimates of all the coefficients K_{ij}) is introduced and this approximation is used:

$$\begin{gathered}\varepsilon\,[\tau_2(i, j)_1] \equiv \varepsilon\,[\tau_2(i, j)_0] + \sum_{i,j}\frac{\partial\varepsilon}{\partial\tau\,(i, j)_0}[\tau_2(i, j)_1 - \tau_2(i, j)_0] = 0,\\ \frac{\partial\varepsilon}{\partial\tau}(i, j)_0 \simeq \varepsilon\,[\tau_2(i, j)_0 + \Delta\tau_{ij}] - \varepsilon\,[\tau_2(i, j)_0]/\Delta\tau_{ij};\end{gathered} \tag{55}$$

$\varepsilon(\tau_2(i, j)_0 + \Delta\tau_{ij})$ and $\partial\varepsilon/\partial\tau\,(i, j)$ are calculated from storage (54) and then system (55) is solved and the next approximation for $\tau_2(i, j)_1$ is found. The process is repeated anew and terminated when $\tau(i, j)_{n+1} - \tau(i, j)_n < \Delta\tau_{ij}$.

The next procedure, successively reducing J and determining J_{min}, reduces to repeating calculations of type (6)-(8). Despite the awkwardness of the description of procedures (6)-(8) given here, the time spent on them is small compared to the time for calculating the correlation functions (5) and forming the signal μ_i [in stages (3) and (4)].

We shall not specify computations (10) and (11) as their execution is substantially tied to the individual properties of the experiment (primarily to the form of the matrix K_{ij}). Clearly this scheme is only one of the possible versions of the mathematical experiment.

In conclusion we discuss the results of a preliminary numerical experiment for estimating the coefficients K_{ij}. A system with three isolated levels was examined. For definiteness the parameters of the problem were for a hydrogen plasma with detection of the 4th, and 5th, and 6th energy levels of the H atoms (for $N_e = 7 \cdot 10^{11}$ cm^{-3} and $T_e = 0.5$ eV). The time unit was chosen to be the computational step size $\theta = 10^{-9}$ sec. In this system of units we obtained

$$10^2\,\|K_{ij}\| = \begin{pmatrix} 3.95 & 1.68 & 0.171\\ 0.675 & 3.60 & 5.28\\ 0.0014 & 1.28 & 9.58 \end{pmatrix} \tag{56}$$

$$1/\lambda_1 \simeq 9,\ 1/\lambda_2 \simeq 22,\ 1/\lambda_3 \simeq 48.$$

The signal μ_i was generated using Eq. (2) and the matrix $\|K_{ij}\|$ (56) with ratios of the dispersions of the additive noise and population fluctuations given by $\varkappa_i' = 4$ (i = 1, 2, 3). With this statement of the problem, we neglected the question of systematic errors in K_{ij} due to an insufficient number of isolated levels.

After the signal μ_i established, 50 independent readings (G = 50) of the signal μ_i were taken, each lasting 5000θ.

The quantity $\hat{K}_{ij}$ was estimated from these readings. (Everywhere $\tau_1(i, j) = \theta$, $\tau_2(i, j) = 19\theta$, $\Delta\tau = 2\theta$, and the integrals of the correlation functions were computed from ten points.)

The total time for this kind of calculation on a BÉSM-6 computer was 16 minutes. The coefficients K_{11}, K_{22}, K_{33}, K_{12}, K_{23}, and K_{32} were reproduced on the average with an accuracy of up to 20% (with $\hat{K}_{ij} - K_{ij}/K_{ij}$ taken as the relative error) and the variability of the estimates of the coefficients were equal at $v_{ij} = 30\%$. The overall time for the corresponding physical experiment with $\varkappa = 2 \cdot 10^3$ is $T_{30\%} = 5000\theta \times 50[\varkappa/(1 + \varkappa')]^2 = 40$ sec. This could be reduced by several times on optimizing the experiment. To reduce the variability in the estimated coefficients to 5% would require an extension of the duration of the experiment by 36 times, that is, to $T_{5\%} \simeq 0.5$ hour, in accordance with the estimate of the previous section.

Literature Cited

1. L. I. Gudzenko, V. E. Chertoprud, and S. I. Yakovlenko, Preprint FIAN, No. 13 (1974).
2. L. I. Gudzenko, Radiofizika, 5, No. 3 (1962).
3. L. I. Gudzenko and V. E. Chertoprud, Radiofizika, 10, No. 3 (1967).
4. L. I. Gudzenko and T. M. Makhviladze, Radiofizika, 14, No. 11 (1971).
5. H. Cramer, Mathematical Methods of Statistics, Princeton University Press (1946).
6. E. S. Venttsel, Probability Theory [Russian translation], Nauka, Moscow (1969).
7. V. E. Chertoprud, Nauchn. Inform. Astrosoveta, No. 9 (1968).
8. M. I. Ageev et al., Algorithms [in Russian], Nauka, Moscow (1968).

CHAPTER VII

CORRELATION METHODS IN MEDICAL DIAGNOSTICS

General Considerations

Of the numerous and varied problems which are naturally analyzed using the methods of inverse problems, diagnostics, including medical diagnostics, is perhaps the closest to the now traditional cybernetic schemes. Medicine, as opposed to physiology, does not have as its immediate purpose the biological explanation of the functioning of organs. The achievements of pathophysiology, of course, aid in reliably setting up diagnostics for illnesses and in choosing treatments; nevertheless, properly speaking, medical studies often stop at the level of identifying life-threatening deviations from normal and searching for ways to eliminate them. The complexity of the interacting processes in a living organism is not comparable with the customary problems of physics or chemistry. In biological research it is not generally possible to use simplified descriptions in the form of models with a small number of parameters. Limiting ourselves solely to the aims of diagnostics, we can attempt to use this sort of simplification and describe the functioning of organs in both their normal and various pathological states by ordinary differential equations of low order n,

$$\Phi\left(u, \frac{du}{dt}, \ldots, \frac{d^n u}{dt^n}; \quad a_1, a_2, \ldots, a_Q\right) = 0. \tag{1}$$

Having taken this solution, it is appropriate after the above remarks to turn to an already prepared cybernetic scheme intended for studies of complicated objects, specifically, to the "black box" scheme. The purpose of this scheme is to organize the research and to choose economical ways of doing it. The object being studied is in effect placed in a "black box." The "box" is opaque (the object cannot be seen; the researcher is not diverted by seeing it and does not go off on sporadic searches) but has a number of previously determined "outputs" and "inputs" which link the observer with the object. Using first one, then another "input" it is possible to "excite" the corresponding degrees of freedom of the object and observe its responses to these stimuli at the "outputs." A comparison of the responses with the perturbations which produce them makes it possible in principle to find the characteristic parameters of the object, for example, the coefficients $a_1, a_2, \ldots, a_Q$ of the equations which describe it. However, the study of fairly complex objects, which certainly include living organisms, is conveniently done in another way, under conditions of, so to speak, "normal operation." In fact, the idealization in which the object is described by the simple equations (1) has a rather narrow range of applicability. The boundaries of this range are easily disrupted, thereby distorting the measurement scheme; when the perturbations are small, the responses are poorly reproducible and change chaotically during successive tests of the object. This is due to the appearance of many degrees of freedom which were not taken into account in the model. When, however, the perturbations are large, they take the object out of its normal condition and may present it with a direct danger. In addition, the parameters $a_1', a_2', \ldots, a_Q'$ evaluated in this manner are noticeably distorted. Thus, the traditional cybernetic "black box" scheme is too pragmatic and must be substantially modified. Genuinely complicated objects are best studied by examining their autonomous behavior under normal conditions, only supplementing this study with an analysis of the response to isolated perturbations when necessary.

But an observation made without involvement in the motion of the object would seem to give too little information. The steady motion of a model does not permit evaluation of the coefficients or determination of the type of equations (1) for this model. Taking this into account, Gudzenko [1] discussed a method for studying an uncontrollable object by statistical analysis of the steady-state fluctuations of the detected signal. This scheme was then called a "black box without an input." There a method was proposed for analyzing a comparatively simple dynamic model (1) with a small number k of parameters which included the effect (on the motion of the object) of numerous additional degrees of freedom by semiqualitative additions to the model.

Starting from a generalization of various physical problems, we proposed to describe the effect of these degrees of freedom which are not in the dynamic model by adding to Eq. (1) a broad-band fluctuating force F(t), whose autocorrelation interval is small compared to the smallest characteristic time of the dynamic model (1). The stochastic equations for the signal u(t) from such a system that has been perturbed by the noise F(t),

$$\Phi\left(u, \frac{du}{dt}, \ldots, \frac{d^n u}{dt^n};\ a_1, a_2, \ldots, a_Q\right) = F(t), \tag{2}$$

make it possible to go to normalizing equations which relate the correlation functions $\left\langle \frac{d^q u}{dt^q}(t) \frac{d^r u}{dt^r}(t+\tau) \right\rangle$ of the signal from the object to its dynamic parameters $a_1, a_2, \ldots, a_Q$. In the following we shall explain these generalizations by a discussion of two specific examples.

In medicine the recorded time-varying signal u(t) may be one of the widely different physiological quantities which determines the state of the organ being studied: instantaneous values of the concentrations of enzymes and oxygen in the blood, changes in the blood flow rate, the numbers of various blood corpuscles and microorganisms, the biological (electrical) potentials at various points, and so on. The choice of a dynamic model of an organ whose state is being diagnosed and the choice of a detectable signal are, of course, not made independently of one another. The suitability of this choice is determined both by the importance (in the functioning of this organ) of the quantity chosen as a signal u(t), and by the performance of the apparatus for recording the signal (primarily its sensitivity and response time). From the remarks made already there follows a need to obey the "principle of noninterference" of the entire signal detection process with the functioning of the organ and with the state of the entire organism. This excludes consideration, for example, of studies of the heart and blood vessels by inserting probes.

We now turn to a discussion of several possibilities for the diagnosis of various illnesses by means of correlation analysis of the simplest (most accessible to detection) characteristics of the mechanical motion of blood through the blood vessels. As is known, the cardiovascular system plays a vital role in the function of every human organ, supplying it with oxygen, nutrients, and immunological agents, transporting the products of the organ, removing decay products from it, and so on. It might thus be expected that the kinetics of the blood flow through the vessels associated with a given organ would reflect rather fully the state of that organ, even if only to the extent that one can, by beginning with the features of an observed transportation system, draw detailed conclusions about the life of the corresponding institutions. While we are not scrutinizing the scheme for mathematically analyzing the signal in detail yet, we must at once note the qualitative difference between the two types of major blood vessels, the arteries and the veins. The veins remove blood from the mechanically "passive" organs which mainly require only the transport of blood. The arteries go out from the heart, the main mover of the hemodynamic system.

These words about the mechanical passivity of the organs, the "users," should not be taken literally, as the actual situation is much more complicated. The intensity of blood circulation depends on the power of the cardiac pumps (determined by the ability of the myocardium to contract) as well as on the general resistance of the blood flow channel, on the tone of the peripheral vessels, and on the venous intake. Blood-pressure waves propagate along the vessels in accordance with the phase of the cardiac cycle. The speed of propagation of these waves is fairly high, about 5-20 m/sec. Since the maximum length of the vessels in a human is 1.0-1.5 m, it may be assumed in most problems that the pressure waves correspond quasi-statically to the phase of the cardiac contractions.

We may attempt to develop the technique for diagnosing ailments of an organ from the blood kinetics, of course, by mobilizing the most complete possible information. In that case, in the simplest approach to the problem it is necessary to record a set of time-varying quantities characterizing the blood flow in a large number of vessels entering and leaving this organ. Some simple estimates will show that it would be unrealistic to rely on the efficiency of the best available high-speed general-purpose digital computers for the numerical analysis of the immense mass of data which would arise from this. This difficulty is surmountable. Substantial economy results from the recording of, not the components of the multidimensional signal u(t) itself, but just their correlation matrices $\langle u^{(k)}(t)u_{(m)}(t+\tau)\rangle$ obtained using simple analog devices connected directly to the outputs of the sensors for the components $u^{(k)}(t)$. It seems to us, nevertheless, that it is best to begin the search with observations of the kinetics in a selected, single blood vessel and develop the prospects for diagnosis from a one-dimensional signal.

For the purpose of studying rapid (compared, at least, to the average period of the cardiac cycle, about 1 sec) fluctuations in some quantity characterizing the instantaneous state of a blood vessel, the most directly adapted technique at present is electrokymographic recording. A narrow x-ray beam is directed on the blood vessel during the recording process. The variable component of the x-ray absorption is roughly proportional to the instantaneous radius of the vessel. This type of recording lay at the basis of the cardiac diagnostic technique proposed in [1]. Since the material in the blood vessel differs fairly little in absorption from muscular tissue in the wavelength range of medical x-ray equipment, and since the vessel makes up only a small part of the x-ray path through the human body, the sensitivity of this method is rather low. Thus on ordinary electrokymographs the irradiation intensity over a single session is not negligibly small. The level of irradiation during the observation period of 3-5 min required for later statistical analysis certainly does not exceed the allowable single time dose. But simply to collect enough statistical data to develop dependable techniques for this diagnostic scheme would require that hundreds of people, both sick (with various pathologies of the organ involved, here the heart) and well, be subjected to this procedure. Radiation damage does not have clearly defined threshold behavior, and the variations in sensitivity and compensatory ability are rather large. All this was a barrier (perhaps mainly psychological) which kept Gudzenko [1] from testing the proposed method. In recent years new possibilities for nearly harmless observation of the transverse dimensions of large blood vessels involving ultrasonic irradiation have appeared. It is still perhaps early to settle completely on an ultrasonic measurement system since the available machines are not directly adapted to the requirements of the problems we are discussing and since they are, moreover, unique.

With all this in mind, in this article we shall take the signal u(t) to be the amplitude of the pulse oscillations of the outer parts of large blood vessels which are accessible to direct measurement. The sensor may be one of various mechanical or optical displacement (or pressure) measuring devices. The possibility of developing a sufficiently detailed diagnosis of very different diseases in this way is reinforced, to some extent, by the successful practice of Tibetan folk medicine. However, from the following it will be clear that the rigorous, but yet untested, analysis schemes described here differ greatly from the diagnostic procedures (not yet clearly formulated in the scientific literature) of the Tibetan doctors.

Diagnostics of a Passive Organ [2]

We now consider a scheme for analyzing the signal from a pulse displacement sensor to interpret the mechanical characteristics of the vitality of an organ from the pulsations in the veins taking blood from it. It is important that in this case the signal $u_0(t)$ formed somewhere at the exit from this organ is not accessible to direct detection. In the detected signal $u(x, t)$, which is picked up at a large distance x away from the organ $[u(0, t) \equiv u_0(t)]$, the information about the state of the organ has been distorted. Leaving out the presence of the cardiac rhythm here for simplicity (although of course of it has an effect on the entire cardiovascular system), we shall assume that the actual signal $u_0(t)$ is described by a differential operator of n-th order perturbed by a broad-band stationary noise, i.e.,

$$D^{(n)}[u_0(t)] = F(t),$$
$$\langle F(t)\rangle \equiv 0, \quad \langle F(t)\,F(t+\tau)\rangle = 0 \quad \text{for} \quad |\tau| \geqslant \tau_0, \tag{3}$$

where τ_0 is the correlation time of the noise and is much less than all the characteristic times of the dynamic system $D^{(n)}[u(t)] = 0$. The dispersion $\langle F^2\rangle$ is small so that under steady conditions the signal remains in a small neighborhood of the stationary point $u \equiv \bar{u}_0$ of this system given by $D^{(n)}[\bar{u}_0] = 0$.

The distorted signal u(x, t), observed at a number of points $x = x_1, x_2, \ldots, x_q$ along the vein (which we shall describe here as a delay line coming out of the organ), is recorded. The damping along this line is assumed to be nonnegligible, so reflection of the wave may be ignored. The signal, distorted by damping, obeys the telegrapher's equation

$$\frac{\partial^2 u}{\partial x^2}(x, t) = a\frac{\partial^2 u}{\partial t^2}(x, t) + 2b\frac{\partial u}{\partial t}(x, t) + cu(x, t), \tag{4}$$

where the positive constants a, b, and c are the initially unknown parameters of the delay line.

The smallness of the fluctuations in the signal $\xi_0(t) \equiv u_0(t) - \langle u_0(t)\rangle$ leads to linearization of the operator $D^{(n)}[u] \rightarrow L^{(n)}[\xi]$ near the stationary point $\bar{u}_0$. Thus, it is possible to write

$$L^{(n)}[\xi_0(t)] = F(t), \tag{5}$$

where

$$L^{(n)}[\xi(t)] = \sum_{k=0}^{n} D_k^{(n)} \frac{d^k\xi}{dt^k}(t), \qquad \langle u_0(t)\rangle = \bar{u}_0.$$

In the simplest case, according to the original model, a passive organ is characterized by n + 1 parameters, n constant ratio coefficients $l_k \equiv D_k^{(n)}/D_n^{(n)}$, $(k = 0, 1, \ldots, n-1)$, and the value $\bar{u}_0$ of the stationary point of the dynamic operator. When it is necessary to include the coupling of the fluctuations with the phase φ of the cardiac cycle, the constants l_k are replaced by the functions $l_k(\varphi)$ of the phase, and the constant $\bar{u}_0$, by the function $u_0 = \bar{u}_0(\varphi)$, where $0 \leq \varphi < T$ and T is the mean period of the cardiac oscillations. The stage of the analysis to be discussed now consists of evaluating these quantities from the distorted signal.

We use the notation

$$g(x, \omega) \equiv \frac{1}{2\pi} \int_{-\infty}^{\infty} \xi(x, t) \exp(-i\omega t)\, dt$$

for the amplitude spectrum of the fluctuations

$$\xi(x, t) \equiv u(x, t) - \langle u(x, t)\rangle$$

of the distorted signal recorded by the detector. Substituting these expressions in the linear telegrapher's equation (4), we find the spectrum obeys

$$\frac{\partial^2 g}{\partial x^2}(x, \omega) = [K(\omega)]^2 g(x, \omega).$$

Here $K(\omega) \equiv \sqrt{c - a\omega^2 + 2bi\omega}$. Thus, along the delay line the amplitude spectrum in general transforms according to the law

$$g(x, \omega) = g_+(\omega) \exp[K(\omega)x] + g_-(\omega) \exp[-K(\omega)x].$$

By assumption, the damping of the signal as it propagates along the line allows us to neglect reflections. In this case the functions $g_+(\omega)$ and $g_-(\omega)$ in this general form must be chosen so that the dispersion in the fluctuations $\xi(x, t)$ decreases monotonically with increasing x. Accordingly, we write

$$g(x, \omega) = g(\omega) \exp\{-x[\,|R(\omega)| + iI(\omega)\,\mathrm{sgn}\,R(\omega)]\}, \tag{6}$$

where we have used the notation

$$R(\omega) \equiv \mathrm{Re}\,[K(\omega)], \qquad I(\omega) \equiv \mathrm{Im}\,[K(\omega)],$$

$$g(\omega) = \begin{cases} g_+(\omega) & \text{for } R(\omega) < 0, \\ g_-(\omega) & \text{for } R(\omega) > 0, \end{cases} \qquad \mathrm{sgn}\,R = \begin{cases} 1 & \text{for } R > 0, \\ -1 & \text{for } R < 0. \end{cases}$$

If $g_0(\omega) \equiv \frac{1}{2\pi} \int_{-\infty}^{\infty} \xi_0(t) \exp(-i\omega t)\, dt$ is the amplitude spectrum of the fluctuations $\xi_0(t) \equiv u_0(t) - \langle u_0(t)\rangle$

in the signal which cannot be detected (i.e., the undistorted signal) at the outlet of the organ being studied, then from Eqs. (5) and (6) it follows that $g(\omega) = g_0(\omega)$.

It is easy to go from the amplitude spectrum to the correlation spectrum which is referred to as the spectral intensity. For the fluctuations in the distorted signal the spectral intensity is

$$G(x, \omega) \equiv \frac{1}{2\pi} \int_{-\infty}^{\infty} \langle \xi(x, t)\, \xi(x, t+\tau) \rangle \exp(-i\omega\tau)\, d\tau.$$

According to Eq. (6) it varies as (cf., for example, [2])

$$G(x, \omega) = \int_{-\infty}^{\infty} \langle g^*(x, \omega)\, g(x, \omega + \nu) \rangle \exp(i\nu t)\, d\nu =$$

$$= \int_{\infty}^{\infty} \langle g_0(\omega + \nu)\, g_0^*(\omega) \rangle \exp\{-x\,[|R(\omega+\nu)| + |R(\omega)| +$$

$$+ iJ(\omega + \nu) \operatorname{sgn} R(\omega + \nu) - iI(\omega) \operatorname{sgn} R(\omega)] + i\nu t\}\, d\nu.$$

Substituting here the spectral correlation function of the stationary fluctuations in the signal, $\langle g_0(\omega + \nu) g_0^*(\omega) \rangle = G_0(\omega)\delta(\nu)$, we obtain a simple expression for the variation along the delay line (without reflection) of the spectral intensity

$$G(x, \omega) = G_0(\omega) \exp[-2\,|R(\omega)|\,x], \tag{7}$$

where $G_0(\omega)$ is the spectral intensity of the undistorted signal $u_0(t)$. After determining $G(x, \omega)$ at two fixed points x_1 and x_2, Eq. (7) makes it possible in principle to calculate the spectral intensity of the fluctuations at the point $x = 0$, that is, at the vein outlet from the organ.† Then the parameters l_k of the dynamic model of the organ may be evaluated, for example, in the customary way, using the correlation function of its signal,

$$B(\tau) = \int_{-\infty}^{\infty} G_0(\omega) \exp(i\omega\tau)\, d\omega = \langle \xi(t+\tau)\, \xi(t) \rangle.$$

The ratios of the coefficients $D_k^{(n)}$ may afterwards be evaluated beginning with the equation

$$\sum_{k=0}^{n} D_k^{(n)} \frac{d^k}{d\tau^k} B(\tau) = 0 \quad \text{for} \quad \tau > \tau_0,$$

which is obtained in the obvious way from the linearized model (5).

Similar, although much simpler, considerations may be applied to estimating the stationary point $\bar{u}_0$ of the dynamic system of the model of the organ. From the telegrapher's equation (4) we find in succession that

$$\frac{d^2\bar{u}}{dx^2}(x) = c\bar{u}(x), \qquad \bar{u}(x) = \bar{u}_0 \exp(-\sqrt{c}\,x), \tag{8}$$

where $\bar{u}(x) = \langle u(x, t) \rangle$ is the mathematical expectation of the distorted signal.

† It should be noted, however, that the notion of large blood vessels as delay lines without reflections is very approximate at best. One may also speak very arbitrarily of the stationarity of the fluctuations in the signal. Then the spectral intensities $G(x_k, w)$ are evaluated by averaging the corresponding functions of the fluctuations over time. One should not count on high accuracy in such estimates. Thus, if conditions permit, it is best to detect the signal at an entire series of points of the vein and then to do the calculation using the method of least squares.

According to this model chosen the statistical averages $\langle u(x, t)\rangle, G(x, \omega)$ must be evaluated by time averaging $\langle w\rangle \sim \frac{1}{\theta}\int_t^{t+\theta} w(\eta)\,d\eta$ and proceeding from the ergodicity of the signal. The variability $v(\hat{z}) \equiv \frac{\sqrt{\langle(\delta\hat{z})^2\rangle}}{\langle\hat{z}\rangle}$ in the estimates $\hat{z}$ of the quantities z decreases with the length Θ of observation roughly as $\sqrt{\tau_{max}/\Theta}$, where τ_{max} is the maximum correlation time of the quantity $\hat{z}$ in the problem. If $\tau_{max} \sim T \sim 1$ sec (where T is the average period of the cardiac pulses) then to obtain $v(\hat{z}) \sim 0.05$ the signal must be recorded for more than 5 min. Clearly, even in this, the simplest statement of the problem, the question arises of accumulating and processing a fairly large amount of numerical data. It is thus convenient to specify that all the required statistical averages be evaluated during the observations by means of simple analog devices, rather than by direct recording of the signal $u(x_k, t)$ at several points x_k.

By assumption the set of estimates $\hat{u}_0, \{\hat{l}_k\}$ must give a fairly complete (for the diagnostic) picture which determines the state of the organ. Now it is primarily necessary to be able to distinguish an unhealthy from a healthy organ on the basis of this set and then to study further and distinguish the various ailments. This part of the problem must be solved by comparing the sets obtained for a large number of people with reliably established diagnoses. (Here it is also possible to use experiments on animals with various specially induced pathological changes in the appropriate organ.) Comparing large sets of numbers may seem rather cumbersome, and it is natural to apply computer "pattern recognition" techniques at this stage. We emphasize that it is usually unreasonable to use such techniques at earlier stages of the diagnosis. In the short record of a single pulse there is actually no information on the reaction (of the blood vessel system associated with the organ) to a broad-band perturbation, while the long untreated record of the signal which does contain the required (statistically hidden) information is too voluminous and thus is practically inaccessible to analysis by means of computer pattern recognition techniques.

Of course, it is possible that a reliable diagnosis may not be established from the set of numbers $\hat{u}_0, \{\hat{l}_k\}$. Then it is necessary to mobilize more detailed information about the organ, for example, by including in the diagnostic scheme some perturbation of its steady activity by certain actions which are more intense than the internal fluctuations F(t). It is possible, in particular, to evaluate the changes in $\hat{u}_0, \{\hat{l}_k\}$ due to a large physical stress produced in the patient after his taking some preparation. Much more complete information about the activity of the organ may be obtained by taking into account the nonstationarity of the processes taking place in the circulatory system due to the roughly periodic action of the heart. A statistical method which allows us to find the form of the functions $\bar{u}_0(\varphi)$ and $\{l_k(\varphi)\}$ (which arise when this periodic action is taken into account instead of the constants $u_0, \{l_k\}$) of the phase of the cardiac cycle is fully similar to the following scheme for diagnosing heart disease from fluctuations in the pulse.

Cardiac Diagnostics [3]

In the body the heart has the practically unique function of a hydraulic pump for the blood supply system. It is a double pump which drives blood though two "loops." The left side of the heart drives the "large circulatory loop," supplying all the organs of the body, including the heart muscle itself, through the aorta. The right side drives blood through the pulmonary artery to the lungs in the "small circulatory loop." Each side of the heart has two chambers, one which accepts blood from the "other" loop (the auricle) and the other, a more powerful "pump," which drives blood into its "own" loop (the ventricle). At the inlets and outlets of the ventricles there are heart valves whose role is to keep blood flowing in one direction through the vessels. The resistance of all the vessels of the large circulatory loop is roughly five times that of the small loop. Hence, the power and thickness of the musculature of the left ventricle are much greater than the right. The heart works rhythmically and its oscillations are nearly periodic. Each such period is a single cardiac "cycle." Within the cycle there are intervals of "systole" (contraction) and "diastole" (relaxation) of the chambers. The systoles of both auricles occur almost simultaneously, during the diastoles of the ventricles. The average period of the heartbeats in a state of rest is usually $T \sim 1$ sec, which frequency increases to about 200 per minute during strenuous exercise. The basic characteristic of the heart's operation is the amount of blood pumped by each ventricle during the systole, known as the "pulse volume." Often another quantity is used for this characterization, the amount of blood pumped through each of the circulatory loops over one minute, known as the "minute volume" of the heart. Evidently continuum models are needed for any kind of complete description of the heart's pump system. It is also important that the mechanical actions of the heart are not autonomous, in that they are initiated by the so-called conducting system, made up of the sinus node, the Hiss bundle, and the Purkinje fibers. The basic purpose of the latter is the generation of a sequence of comparatively rare, short, electric field pulses which lead to regular periodic interchanges of stimulation and relaxation (corresponding to the phases of the cardiac cycle) of the parts of the heart muscle. It is these pulses, propagating through the body as biopotentials, which are recorded at the various pickups of an electrocardiogram (ECG). As Van der Pol has already shown, the conducting system is adequately modeled by a chain of coupled, pulsed, autooscillators with several degrees of freedom.

This picture, as can easily be seen, relates directly to the basic function of the heart. And, nevertheless, in accordance with the remarks at the beginning of this chapter, we shall assume that for diagnostic purposes it is sufficient to model the heart as a self-oscillating system with a single degree of freedom that is perturbed by a broad-band noise $F(t)$:

$$\begin{gathered} \frac{d^2u_0}{dt^2}(t) = f\left[u_0(t), \frac{du_0}{dt}(t)\right] + F(t), \\ \langle F(t)\rangle \equiv 0, \quad \langle F(t)\ F(t+\tau)\rangle = 0 \quad \text{for} \quad |\tau| > \tau_0. \end{gathered} \tag{9}$$

As before, the correlation interval τ_0 of the noise is assumed to be small compared to the characteristic times of the dynamic model

$$\frac{d^2u_0}{dt^2}(t) = f\left[u_0(t), \frac{du_0}{dt}(t)\right]. \tag{10}$$

We now consider a scheme for estimating the parameters of the function $f(v, w)$ (the dynamic characteristics of the heart) from the signal $u(x, t)$ determined by the fluctuations in the outer parts of the major arteries. As in the above, we shall assume that it is sufficient to represent the arteries as delay lines attached to the output of a self-excited oscillator.

The signal $u(x, t)$, as distorted by the delay line, is recorded at several points $x = x_1, x_2, \ldots, x_N$, over an observation time Θ. The equation for the propagation of the signal along the artery has the same form (4) and, as previously, the point $x = 0$ corresponds to the point at which the signal leaves the organ and enters the delay line (the aorta or the pulmonary artery). Here it is also assumed that the fluctuating force $F(t)$ causes a weak deviation in the motion from the stable limiting trajectory in the phase plane, i.e., from the limiting cycle

$$\bar{u}_0 = \bar{u}_0(t), \quad \bar{u}_0(t+T) = \bar{u}_0(t) \tag{11}$$

of the dynamic model of the heart (10). Here this means that the deviations of the motion (9) from the normal to curve (11) (which describes the dynamics of the steady fluctuations at the output of the heart to the large or small circulatory loop) are small. We shall assume for now that the observation interval Θ is small so that the diffusive phase shift in the heartbeat during the analysis may be neglected. Then for the "actual" signal $u_0(t)$ we have

$$u_0(t) = \langle u_0(t)\rangle + \xi_0(t), \qquad \xi_0(t) = \int_{-\infty}^{\infty} g_0(\omega)\exp(i\omega t)d\omega, \tag{12}$$

where $\xi_0(t)$ are (sufficiently) small fluctuations in intensity and $\langle u(t)\rangle = \bar{u}_0(t)$ are the stationary dynamic fluctuations. According to Eq. (11) we may write further

$$\langle u_0(t)\rangle = \sum_{k=-M}^{M} q_{0k}\exp(ik\omega_0 t), \qquad \omega_0 = \frac{2\pi}{T}. \tag{13}$$

The distorted signal may thus be written in the form

$$\begin{gathered} u(x, t) = \langle u(x, t)\rangle + \xi(x, t), \\ \langle u(x, t)\rangle = \sum_k q_k(x)\exp(ik\omega_0 j), \qquad \xi(x, t) = \int_{-\infty}^{\infty} g(x, \omega)\exp(i\omega t)\,d\omega. \end{gathered} \tag{14}$$

Here $q_k(0) = q_{0k}$ and $g(0, \omega) = g_0(\omega)$.

We now expand the dynamic characteristic in a series of powers of the small deviations of the signal from the limiting cycle motion and drop the nonlinear terms. Then the fluctuations in the "actual" signal may be written [in accordance with Eq. (9)] in the form

$$\frac{d^2\xi_0}{dt^2}(t) = f_1(t)\frac{d\xi_0}{dt}(t) + f_0(t)\xi_0(t) + F(t), \tag{15}$$

where the quantities

$$f_0 \equiv \frac{\partial f(v, w)}{\partial v}\left(v = v_0, \quad w = \frac{dv_0}{dt}\right) = \sum_k f_{0k}\exp(ik\omega_0 t)$$

and

$$f_1 = \frac{\partial f(v, w)}{\partial w}\left(v = v_0, \quad w = \frac{dv_0}{dt}\right) = \sum_k f_{1k}\exp(ik\omega_0 t) \tag{16}$$

denote periodic time-dependent functions. These functions must in fact be regarded as dependent on the phase φ of the heartbeat as reckoned from the dynamically established motion along the limiting cycle (13). In a model formulated this way the dynamic characteristics of the heart are the limiting cycle (13) and two other functions of the phase of the heartbeat (16). The purpose of the following calculations is thus to evaluate 3(2M + 1) complex numbers $\{q_{0k}, f_{1k}, f_{0k}, (k = -M, M)\}$, where M is the maximum number of harmonics actually included in a given sensitivity of detection of the artery pulsations.

For simplicity we shall assume here also that we may neglect reflected waves because of damping of the signal as it propagates along the blood vessel. Then substituting Eq. (13) in the telegrapher's equation (4) we obtain the equations for the variation of the Fourier amplitudes of the mean statistical fluctuations along the vessel:

$$\frac{d^2 q_k}{dx^2}(x) = [K(k\omega_0)]^2 q_k(x), \qquad k = -M, M, \tag{17}$$

where $[K(\omega)]^2 = c - a\omega^2 + 2bi\omega$. Similarly, for the amplitude spectrum of the fluctuations in the signal we find

$$\frac{\partial^2 q}{\partial x^2}(x, \omega) = K(\omega)g(x, \omega). \tag{18}$$

This yields the general form of the variation in the amplitude characteristics with the coordinate x:

$$q_k^{(x)} = q_k' \exp[-K(k\omega_0)x] + q_k'' \exp[K(k\omega_0)x],$$
$$g(x, \omega) = g'(\omega) \exp[-K(\omega)x] + g''(\omega) \exp[K(\omega)x]. \quad (19)$$

The fluctuations in u(x, t) have no reflections by definition; hence,

$$q_k(x) = q_{0k} \exp\{-[|R(k\omega_0)| + iI(k\omega_0) \operatorname{sgn} R(k\omega_0)]x\},$$
$$g(x, \omega) = g_0(\omega) \exp\{-[|R(\omega)| + iI(\omega) \operatorname{sgn} R(\omega)]x\}, \quad (20)$$

where, as before, $K(\omega) \equiv R(\omega) + iI(\omega)$. From Eq. (20) it follows that with the pulsations measured at several points $x = x_1, x_2, \ldots, x_N$ ($N = 2$ being allowed as well) it is possible to evaluate the parameters q_{0k} of the limiting cycle at the outlet from the heart.

In cardiac diagnostics with this model it is necessary to include the periodic nonstationary behavior of the fluctuations both at the entrance to the aorta and at the entrance to the pulmonary artery. In either case the correlation function of the time representation of the signal has the form

$$B_0(t, \tau) \equiv \langle \xi_0(t)\, \xi_0(t+\tau) \rangle = \sum_k \psi_{0k}(\tau) \exp(ik\omega_0 t), \quad (21)$$

and the corresponding frequency representation has the correlation function

$$\langle g_0(\omega+\nu)\, g_0^*(\omega) \rangle = \sum_k G_{0k}(\omega)\, \delta(k\omega_0 - \nu),$$

where $G_{0k}(\omega) = \frac{1}{2\pi} \int_{-\infty}^{\infty} \psi_{0k}(\tau) \exp(-i\omega\tau)\, d\tau$ are the spectral components. Along the delay line the spectral intensity $G(x, \omega, t)$ of the distorted signal varies according to a simple law. Denoting its spectral components by $G_k(x, \omega)$, we obtain

$$G(x, \omega, t) = \sum_k G_k(x, \omega) \exp(ik\omega_0 t), \qquad G_k(x, \omega) = G_{0k}(\omega) \exp[-L(\omega, k\omega_0)x],$$
$$L(\omega, \nu) \equiv |R(\omega)| + |R(\omega+\nu| - iI(\omega) \operatorname{sgn} R(\omega) + iI(\omega+\nu) \operatorname{sgn} R(\omega+\nu). \quad (22)$$

Therefore, from the components $G_k(x, \omega)$ of the spectral intensity of the pulsations detected at N points $x = x_1, x_2, \ldots, x_n$ (in principle, $N = 2$ is sufficient) along the artery it is possible to evaluate the components $G_{0k}(\omega)$, then the spectral intensity $G_0(\omega, t)$, and finally the time correlation function $B(t, \tau)$ of the undistorted signal $u_0(t)$. The components $\psi_{0k}(\tau)$ of this correlation function yield the parameters $\{f_{0k}, f_{1k}\}$ of the dynamic characteristic of the heart linearized near the limiting cycle. Multiplying Eq. (15) by $\xi_0(t - \tau)$ for $\tau > \tau_0$ we obtain, after statistical averaging,

$$\frac{\partial^2 B}{\partial \tau^2}(t, \tau) = f_1(t) \frac{\partial B}{\partial \tau}(t, \tau) + f_0(t) B(t, \tau),$$

from which

$$\frac{d^2\psi}{d\tau^2}(\tau) = \sum_{l=-M}^{M} \left[f_{1l} \frac{d\psi_{k-l}}{d\tau}(\tau) + f_{0l}\psi_{kl}(\tau) \right]. \quad (23)$$

From the values of $\psi_k(\tau)$ at several shifts $\tau = \tau_1, \tau_2, ..., \tau_p$ we obtain the desired set of parameters from this system (23) of linear (with respect to $\{f_{1l}, f_{0l}, (l = -M, M)\}$) algebraic equations by using, for example, the method of least squares.

Having determined the dynamic characteristics of the heart from the pulsations of the large arteries, it is appropriate to simultaneously record the times of occurrence of a fixed phase of the heartbeat (coinciding, for example, with the passage of R-pulses in an electrocardiogram taken at a certain pickup point) along with the correlation functions of the signal $u(x, t)$ at the points $x = x_1, x_2, ..., x_N$. Let these times be $t_0, t_1, ..., t_S$; then $\hat{T} \equiv \frac{2\pi}{\hat{\omega}_0} = \frac{t_S - t_0}{S}$ is an estimate of the mean period and $T_s = t_s - t_{s-1}$, $(s = \underline{1, S})$ are the "instantaneous" periods of the heartbeat. From the discrete sequence T_s of instantaneous periods we go to a smooth function $T(t)$ for which $T(t_s) = T_s$. Replacement of the argument t (the time) by the phase $\varphi(t) \simeq (t_{s-1}) + \frac{\hat{\hat{T}}}{T(t)}(t - t_{s-1})$ makes it possible to do the above statistical evaluation of the heart parameters without including the phase diffusion of the heart cycle (heartbeats). To do this we have also to replace the derivative with respect to t (and with respect to the shifts τ), using the approximation $\frac{d}{d\varphi} \simeq \frac{T(t)}{\hat{T}} \frac{d}{dt}$. Statistical averaging of the periodically nonstationary processes in equations of this type must be done by harmonic-weighted averaging [4].

Literature Cited

1. L. I. Gudzenko and V. N. Orlov, in: Problems of Clinical Biophysics [in Russian], TsIU Vrachei (1966), p. 33.
2. L. I. Gudzenko and A. E. Sorkina, Kratk. Soobshch. Fiz., No. 8, p. 19 (1975).
3. L. I. Gudzenko and A. E. Sorkina, Krakt. Soobshch. Fiz., No. 1, p. 17 (1976).
4. L. I. Gudzenko, Radiotekh. Élektron., 3:1062 (1959).

CHAPTER VIII

A MODEL OF CYCLIC SOLAR ACTIVITY

Phenomenological Characteristics of the Simplest Model

The first application of the extremely time-consuming method of inverse problems in the theory of oscillations [1] was to analyze the nature of the cyclic activity of the sun, and this is not by chance. The problem of finding the physical mechanism which controls this phenomenon occupies a special place in astrophysics. The beginning of the study of sunspots is one of the very first steps in astrophysics. Since the development of the telescope a vast amount of observational material has been accumulated and many articles and books have been published on this topic. The timeliness of this group of problems is now assured. But it cannot be said that even the qualitative basis for the mechanism of solar activity has been reliably explained yet. Furthermore, the flow of reports filling the astronomical journals about measurements of various modifications of the activity indices or about arithmetic manipulations of these indices without a clear statement of the purpose of these manipulations or often without a minimal logical foundation is rather unusual in the modern exact sciences and is reminiscent of zoology in the pre-Darwin period. At the same time (and largely independently of experimental work) the journals include discussions of logically ordered theoretical models of the sun's activity which contradict one another even in the basic postulates. Thus there has naturally arisen a desire to investigate this group of problems and to first select anew and analyze the properties which are clearly inherent to cyclic solar activity while using as few a priori assumptions as possible. It is here that it is convenient to turn to the methods of inverse problems in the theory of oscillations or, more precisely, to a variant intended for the study of uncontrolled objects of unknown nature. This variant was called a "black box scheme without an input" in its own time although this name is now not fully correct.

In the traditional cybernetic "black box" scheme the radiating object acts as if placed inside an impenetrable box and is analyzed only in terms of its response to those actions with which the experimenter perturbs its established state. One then says that the perturbation is "fed in at the input of the box" and the response is "picked up at the output." Since one cannot expect a reaction from the sun or other astrophysical objects to any possible actions by the experimenter, this scheme must be changed. Here it is only possible to use the "black box without an input" scheme, which sets out to study the "response" of the object to internal perturbations. On the other hand, the very goal of the investigation, finding the physical nature of the phenomenon rather than an optimal mathematical description, already indicates that this problem belongs more to the theory of oscillations than to cybernetics. The two general assumptions about the structure of the object made during the solution of the problem are later subjected to verification.

These assumptions are given here:

1. The observed signal is adequately described by an oscillatory dynamical system with a small number of degrees of freedom being acted on by a broad-band noise perturbation. This noise is caused by microscopic fluctuations, that is, by the motions of a large number of elements of a single type which determine the observed effect.
2. The macroscopic properties of the effect, which are reflected in the dynamic description, are characterized first of all by the absence of a source of periodic oscillations (with a period of about 11 or 22 yr) lying outside the sun and capable of setting up the sun's rhythm. This means that the cyclic activity must be regarded as a manifestation of an autonomous oscillatory dynamic system located within the sun, with properties which must be analyzed carefully.

In a preliminary discussion [2, 3] it was assumed that the dynamic system which approximates the observed cyclic activity has a single, basically predominant degree of freedom. Neglecting the remaining degrees of freedom it is, therefore, a self-oscillating second-order system:

$$\frac{dx}{dt} = y, \qquad \frac{dy}{dt} = f(x, y), \tag{1}$$

where x is the signal which reflects the instantaneous state of the source of the activity and $f(x, y)$ is its dynamic characteristic. A more detailed description

$$\frac{dx}{dt} = y, \qquad \frac{dy}{dt} = f(x, y) + F(x, y, t) \tag{2}$$

takes into account the fluctuating deviations of the actual motion from its dynamic idealization (1). It was assumed that the fluctuations F ($\langle F \rangle \equiv 0$) are stationary and that the system (1) has an asymptotically stable periodic (with period T) solution $x = x_0(t)$. In the $\{x, y = dx/\partial t\}$ phase plane this solution corresponds to the closed curve $x = x_0(t)$, $y = dx_0/dt \equiv y_0(t)$, i.e., the closed limiting cycle

$$x_0(t + T) = x_0(t), \qquad y_0(t + T) = y_0(t). \tag{3}$$

Motion of a representative point of the system strictly along the cycle is equivalent to an established periodic motion of the dynamic system. Due to the fluctuating force a representative point of the real object (2) always deviates from the motion (3). An analysis of the motion (3) and then of these deviations makes it possible to find the characteristics [both dynamic $f(x, y)$ and statistical $F(x, y, t)$] of the mathematical model which correspond to the observations. Assuming that the action of the fluctuating force on the dynamic system is not too great, it is convenient to go from cartesian coordinates $\{x, y\}$ and the time t, respectively, to the local coordinates $\{n, \gamma\}$ and the phase of the motion θ, where the n are the deviations from motion (3) orthogonal to the limiting cycle, γ are the tangential deviations, and θ is the tangential coordinate of the representative point measured in a time scale. Limiting ourselves for simplicity to terms linear in the orthogonal deviation n we find that the following equations are valid (see [1, 4]) in place of Eq. (2) within a small neighborhood of the limiting cycle:

$$dn/d\theta + N(\theta)\, n(\theta) = F_n(\theta), \tag{4}$$

$$d\gamma/d\theta = \Theta(\theta)\, n(\theta) + F_\gamma(\theta). \tag{5}$$

Then the periodic (with period T) functions of the tangential coordinate (the phase of the oscillations) $N(\theta)$ and $\Theta(\theta)$ are known as the rigidity and the nonisochronism, respectively, of the cycle, and the fluctuations $F_n(\theta)$ and $F_\gamma(\theta)$ are the orthogonal and tangential components of the stochastic perturbing force F, which acts on the dynamic system (1). The rigidity characterizes the degree of asymptotic stability of the cycle and is inversely proportional to the time for orthogonal deviations from the cycle to disappear. The function $N(\theta)$ may go to zero for certain values of the phase of the oscillations or may be negative over entire intervals; however, in the case of average stability over the entire cycle the rigidity is positive. The nonisochronism shows by how much the orbital velocities of representative points $(x, y) \| t$ of the dynamic system differ for motions at different distances from the cycle. The limiting cycle, its rigidity, and its nonisochronism together determine the "linearized" dynamic characteristic $f(x, y)$ in the region of phase space adjacent to the limiting cycle where an analyzable, established motion of the object takes place.

The general properties of the perturbing fluctuating force (at first postulated, later proven) reduce to its having a small effect and a short-time correlation. Its having a small effect is assumed in the very statement of the problem approximating the dynamic system, and the short-time correlation of the fluctuations means that the correlation interval of F is small compared with both the time to establish a periodic motion and with the period of the important harmonics of the established dynamic motion. (For more detail, see [3, 4].)

We shall now note briefly the sequence of calculations which allow us to find the limiting cycle and then the functions $N(\theta)$ and $\Theta(\theta)$ from the signal x(t). First the "trajectory of the object's motion," $x = x(t)$, $y = dx/dt(t)$, is plotted on the (x, y) plane. It is assumed that the duration Q of observation of the signal is sufficiently large compared to the period of the motion ($Q/T = s_0 \gg 1$) so that the trajectory consists of a large number $s \simeq s_0$ of loops. When the fluctuation perturbation is small the limiting cycle coincides in the first approximation with the statistically averaged trajectory. This trajectory of the object is constructed from the stochastic trajectory plotted on the (x, y) plane by the method of successive approximations (see [1, 4]). Then the local coordinates n, γ are introduced in the neighborhood of the resulting limiting cycle and the equations of the object's trajectory are transformed to the new variables by the substitution $(x, y) \| t \to (n, \gamma) \| \theta$, where $\gamma = \theta - t$. The formulas

$$N(\theta) = -\frac{\left\langle \frac{dn}{d\theta}(\theta)\, n(\theta - \tau) \right\rangle}{\langle n(\theta)\, n(\theta - \tau) \rangle},$$
$$\Theta(\theta) = \frac{\left\langle \frac{d\gamma}{d\theta}(\theta)\, n(\theta - \tau) \right\rangle}{\langle n(\theta)\, n(\theta - \tau) \rangle}, \quad \tau > \tau_F, \tag{6}$$

where τ_F is the correlation interval of the random force, are used to compute the rigidity and nonisochronism.

Since all the random processes mentioned here may be regarded as ergodic [5], the correlation functions in Eq. (6) are determined, for sufficiently long recording intervals, by a single occurrence of the corresponding random processes.

The analysis of the structure of the self-oscillator for the cyclic solar activity began, of course, with a choice of signal x(t). This signal had to satisfy at least two major requirements: that it reflect the most important macroscopic properties of solar activity and that the signal observation interval Q cover a sufficiently large number of cycles (i.e., at least several hundred years). Here there was basically no choice. The only signal which satisfies both requirements to any extent is the Wolf index. The other indices of activity recorded nowadays, in particular, the indices characterizing the latitudinal distribution of sunspots, cannot serve as a basis for an analysis using a scheme from the inverse problems of the theory of oscillations because the longest recording interval is at best only about 90 years [that is, three times shorter than the list of annual Wolf numbers compiled by Waldmeier (1700-1974)] [6].†

The Wolf number W, introduced and evaluated since 1949 by Wolf, is given by the formula $W = k(10g + f)$, where g is the number of groups of spots on the visible disc of the sun and f is the number of spots of all types, both isolated and in groups; $k = 0.6-1.0$ is a normalizing factor to ensure uniformity of observations at different observatories and at different periods of time and also includes the characteristics of instruments, the visibility conditions, the observation techniques, the observer's skill, and so on. The daily Wolf numbers are not directly suitable for processing in this scheme as they are burdened by too large measurement errors (including, in particular, a large additive component). The simplest way of reducing the additive noise is to go over to the customary smoothed indices, the monthly-mean and annual-mean wolf numbers.‡ (For more detail on the problems associated with the additive noise, see [3].)

Preliminary processing of the annual-mean Wolf numbers [2] already yielded the main property of the cyclic solar activity, namely, that the variation in the Wolf numbers may be described by the signal from a relaxation oscillator with a single degree of freedom. In later work [3, 8, 9] this conclusion was confirmed, the more important assumptions were analyzed and justified, and several parameters of the cyclic activity were made more precise.

In [3] records of the monthly-mean smoothed Wolf numbers R(t) were used. The basis of the calculation was the object trajectory plotted on the (x, y) plane, a smooth curve $x_i = R(t_i)$, $y_i = dR/dt(t_i)$ drawn through 2504 points t_i separated by an interval $\Delta t = t_{i+1} - t_i = 1$. First the method of successive approximations was used to determine the statistical mean trajectory, with the estimate of the limiting cycle obtained in [2] from an analysis of annual-mean Wolf numbers being used as a first approximation. Because of the smallness of the sample for the orthogonal deviations from the limiting cycle (s = 19] the mean statistical trajectory could be determined to only low accuracy characterized by a dispersive $\sigma_n^2 = \langle n^2(\theta)\rangle/s$. This accuracy was obtained in the fourth approximation which was taken there as an estimate of the statistical mean trajectory and the limiting cycle. The two estimates of the limiting cycle, obtained in

† Naturally, these records will also be of great value in future analysis of the mechanism of solar activity.

‡ There are the "monthly mean observed Wolf numbers" obtained by simple averaging of the daily numbers and the "monthly-mean smoothed Wolf numbers" obtained from sliding averages over 13 neighboring points (12-month interval). By averaging the observed and smoothed monthly-mean numbers, we obtain the observed and smoothed annual-mean Wolf numbers, respectively [7].

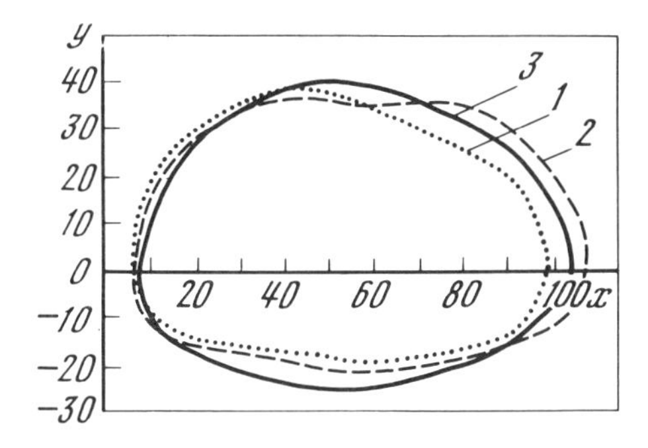

Fig. 1. The location of the limiting cycle in the phase plane.

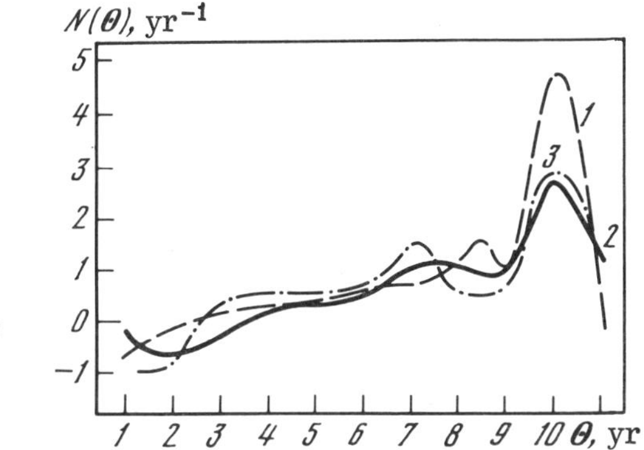

Fig. 2. Estimates of the rigidity (1) from the data of [2]; (2) from the data of [3]; (3) from the data of [8].

[2] from the annual-mean Wolf numbers and calculated in [3] from the smoothed Wolf numbers, are plotted in the phase plane in Fig. 1 (curves 1 and 2, respectively). It should be noted that in [3] the XIX-th activity cycle was used in the calculations but was neglected in [2]. Figure 1 shows that despite this and the slight difference in the signals employed there is fair agreement in the estimates obtained for the limiting cycle. The basic parameters of the limiting cycle in [3] are $x_{0\min} = 6.2$, $x_{0\max} = 180.0$, $y_{0\min} \simeq -22.5$, $y_{0\max} \simeq 36.0$, the time for the activity to rise from $x_{0\min}$ to $x_{0\max}$ is $T_1 = 4.2$ yr, the fall time is $T_2 = 7.0$ yr, and the period of the steady-state dynamic oscillations is $T = T_1 + T_2 = 11.2$ yr. The accuracy of the determination of the period is $\sigma_T \simeq 0.25$ yr. For the chosen coordinate scales ([x] = Wolf number, [y] = Wolf number $\times$ yr^{-1}) the standard deviation in the determination of the location of the limiting cycle varies within the limits $\sigma_n \simeq 1$-5 along the cycle. (Other details on the limiting cycle may be found in [3, 9].)

In [8] an attempt was made to take account of the slow fluctuations (which appear as "secular" variations in the amplitude of the eleven-year cycle). To do this a smooth plot of the variations in the relative intensity of the cycles $\alpha(t)$ was made and the index $x(t) = R(t)/\alpha(t)$ was chosen as a signal, where R(t) is the monthly smoothed Wolf number as before. Despite some differences in the use of the observational material and in the way of handling it the limiting cycle obtained (Fig. 1, curve 3) agrees qualitatively with the results of [2, 3]. The same may also be said of the variation in the rigidity $N(\theta)$ determined in [2, 3, 8]. In all three cases there is a positive spike in the rigidity in the phase interval $\theta = 9$-10.5 yr rising to values of 3-5 yr^{-1} (Fig. 2). The estimates of the nonisochronism $\Theta(\theta)$ in [2, 3] yielded values close to zero. Later [10] the assumption that there is a single significant degree of freedom in the solar activity source was confirmed and justified.

Prior to interpreting these results, we once again emphasize the stability of the qualitative conclusions about the properties of this system with respect to small changes in the observation interval and in the means for filtering the high-frequency fluctuations in the signal.

The existence of a sharp spike in the rigidity shortly before the phase of the activity minimum is one of the most important results obtained in [2, 3] and must be taken into account when constructing a physical model. In effect the intervals of heightened rigidity cut off the oscillations in the activity. As the system passes through each such interval the correlation between the orthogonal deviations from the limiting cycle almost completely disappears. Thus the eruption hypothesis of Waldmeier [11] is not only strictly justified but also takes on a much clearer significance. The signals of relaxation systems often have a discontinuous or pulsed nature. But the well-known property of the spot-forming latitudes which change almost discontinuously near the phase of minimum activity and the laws (Hale) of sign changes in the magnetic fields which appear to behave this way seem to be due to completely different aspects of the cyclic activity from the Wolf number.

Thus, the mechanism for solar activity has a relaxation character similar to a neon lamp oscillator. This means, in particular, that a description of the processes which produce the activity need not be related to periodic (for small changes in the total energy) transfer of one form of energy into another (for example, transfer of the energy of the poloidal field into toroidal field energy and back again). This relaxation oscillator is most simple interpreted by the fact that the mechanism for the cyclic changes in activity is the accumulation of some sort of "seeds," their rapid conversion into active regions of the photosphere, and then, with almost complete removal of the corresponding energy from the system, the development of activity.

Such is the main conclusion of [3] where it was noted that the properties of a phase of activity may be quite different from those proposed by us and may require additional information for their explanation.

The Effect of Planetary Motions

The eruption hypothesis of Waldmeier was one of the first which considered that the forces which cause cyclic activity come from within the sun rather than from outside it. A proof of the "internal origin" of the cyclic activity is vitally important to us for the following reasons. The analysis of the preceding section is based on the single principal assumption that we are examining the steady-state motion of a self-oscillating system. This assumption is quite natural if it is shown that the cyclic activity is generated within the sun. It is much simpler to demonstrate the latter point than it is to conclude that this regime of oscillations in the activity is conserved over a very extended time. Indeed the Wolf numbers span only 24 eleven-year cycles of solar activity. However, the Wolf numbers are not the only information on the cyclic recurrence of solar activity.

In [12] the energetics of cyclic activity and the motion of the planets was discussed and the possibility that the planets have a significant effect on the cyclic activity was rejected. According to current ideas the bulk of the energy of the cyclic activity leaves the sun (see [13-15] and [3]). The lower limit for the energy losses in cyclic activity during a cycle may be estimated by multiplying the number N_0 of groups of spots produced in the course of a cycle by the magnetic energy of a typical group of spots, W_M. According to [13, 16] $N_0 \simeq 3 \cdot 10^3$. By Alfvén's estimate $W_M = 10^{35}$ ergs [17]. He took $H = 3000$ G and assumed that the volume of a bipolar group is equal to the product of the area of a spot (taken to be $3 \cdot 10^{19}$ cm^2) and the distance between two primary spots (taken equal to 10^{10} cm). Using the model of a typical spot (see [18], Table 21), we obtained $W_M = 2 \cdot 10^{24}$ ergs and, therefore, $E_0 \sim 6 \cdot 10^{37}$ ergs. Thus, the average rate of energy release due to the cyclic activity is at least $\varepsilon_0 = E_0/T = 5 \cdot 10^{36}$ ergs/yr. The error in this estimate hardly exceeds an order of magnitude.

To estimate the upper limit on the energy of the tidal interaction with the planets we note that the maximum amplitude of these tides is $h < 0.1$ cm, while the speed of tidal motion is $v_{tidal} < 2\pi h/(T_\odot/2)$, where $T_\odot \simeq 2.2 \cdot 10^6$ sec is the average rotational period of the sun.

Thus, $v_{tidal} < 5 \cdot 10^{-7}$ cm/sec. The characteristic scale X of the spatial variations in the speed is much greater than 10^8 cm (the minimal height of the homogeneous atmosphere of the sun); thus, $v_{tidal}/X^2 \ll 5 \cdot 10^{-23}$ cm^{-1} · sec^{-1}. The dissipation of the energy of the planetary tides on the sun during a half-rotation is clearly less: $\nu \Delta v_{tidal} h M_{\odot} \simeq \nu v_{tidal} h M_{\odot}/X^2$. Setting $\nu = 10^{14}$ cm^2 · sec^{-1} and $v_{tidal}/X^2 = 5 \cdot 10^{-23}$ cm^{-1} · sec^{-1}, i.e., greatly increasing the estimate, we obtain 10^{25} ergs. Therefore, the perturbation due to the energy of the planetary tides during a cycle does not exceed $3 \cdot 10^{27}$ ergs, which is ten orders of magnitude less than the value of E_0 just given. Consequently, the tides on the sun due to planetary motion cannot be the source of the energy of the cyclic activity. Furthermore, the source of energy cannot be the kinetic energy of rotation of the sun (as assumed in the tidal hypothesis) or the energy of orbital motion of Jupiter. Both sources would require more than a million years. There remains only the energy of convective motions within the sun.

Wolf's hypothesis of the planetary origin of solar activity was one of the first explanations of this activity. At that time this hypothesis seemed fully plausible as the energetics of this activity were completely unknown (magnetic fields having not yet even been suspected on the sun) and the mean period of the activity cycle was known so inaccurately that it seemed the same as the period of Jupiter's orbit about the sun. Since then much has changed in our knowledge of solar activity. In particular, the "superposition hypothesis," which is based on fairly general ideas of external periodic effects on solar activity, suffered a complete failure. Discussions of the planetary origin of solar activity appear from time to time in the pages of journals and books (e.g., [7]). True, they now speak of the "nontidal" character of the interaction and no longer of an energy source modulating the activity, but of a "trigger," that is, of a mechanism which somehow synchronizes it. As will be explained a bit later, the transition to "trigger" schemes could reduce the energy required of the external source by two orders of magnitude, so in this case it would take a short time, just about 100 million years. It is, of course, possible to say that it is not known whether cyclic activity existed 100 million years ago and, therefore, that such a "phenomenological" statement of the problem does not of itself fully refute the hypothesis of a planetary effect on the cyclic activity. However, there is another, also simple, possibility for checking the importance of planetary effects on the sun [12]. To examine the reality of "nontidal" effects (of unknown nature) we turn to the elements of planetary orbits and use some data given in [20] in the following estimates. The total energy of interaction of planets with the sun is used up, thereby causing changes in the osculating elements of their orbits. This type of change, associated with real interactions, is not included in the theory of planetary motion (celestial mechanics, which, as is known, gives good agreement with observation) since these changes lie far beyond the limits of modern observational accuracy. As calculations show, the greatest uncertainty in evaluating the possible change in the energy of a planetary system is due to inaccuracy in determining the major semiaxes of the orbit. The relative random error in evaluating the interaction energy between the sun and the planets is equal (in absolute value) to the relative random error in the determination of the major radius. The gravitational interaction energy of Jupiter with the sun is $1.8 \cdot 10^{42}$ ergs and the standard deviation of the relative error in the measurement of the major semiaxis of this planet's orbit (1913 value) is $5.7 \cdot 10^{-9}$. Hence the uncertainty in the energy was 10^{34} ergs. Nowadays the major semiaxis of the orbit is determined with much greater accuracy. Taking this into account, we find that the upper limit to the monotonic variation in the Jupiter–sun interaction energy per cycle is $E = 10^{34}$ ergs/6 $= 1.5 \cdot 10^{33}$ ergs. Similarly, it is found that the upper limit E_1 for the "unaccounted-for" energy of saturn and uranus is half this amount while for the planets with orbits inside that of jupiter, it is smaller by two to three orders of magnitude. Including the interactions of planets with one another changes these estimates by no more than 5%. Thus, the change in energy of a planetary system over an eleven-year cycle is at least four orders of magnitude smaller than the energy expended during cyclic solar activity.

The "trigger" situation remains to be examined. Under these conditions the "trigger" would have to act over a period of $\Delta t \simeq (0.5\text{-}3)$ yr at best. The "trigger" cannot be faster since the relative position of the planets changes fairly slowly. With a larger time Δt, completely different states of activity would fall within this interval.

Strong variations in the monthly-mean values of the Wolf index with characteristic times of 0.2-0.4 yr (according to Tables 11 and 12 of [7]), as a rule, reach 0.5 times the average range of this index while the variability in these variations smoothed over a period of two years is roughly 0.1 [9]. In both cases the energy corresponding to the changes is no less than one percent of E_0 and we may assume that the standard deviation in the fluctuating energy is $\sigma E_f \geq 6 \cdot 10^{35}$ ergs. Over the eleven-year cycle somewhat stronger fluctuations occur. Either the fluctuations in the solar activity are generated by fluctuations in the external energy source and the energy expended by this source per cycle must be at least several times σE_f (not a single fluctuation per cycle), or the activity fluctuations are independent of an external source. But then, in order for the "trigger" to work against a background of such fluctuations it must have an energy greater than $\Delta E_c = 3\sigma E_f > 2 \cdot 10^{36}$ ergs, that is, an energy 300 times greater than the upper estimate of the observable changes in the energy of a planetary system. Therefore, a planetary system cannot possibly serve as a "trigger" for the eleven-year cycle.

There is then no basis for searching for an effect of planetary motions on the cyclic activity; there is, in fact, no such effect. We have given this proof (obvious enough even before) so much space because of the continuous year-by-year appearance of publications on various aspects of this irrelevant problem. The available observational data certainly also gives no foundation for the claim of a relationship between the motion of the planets and solar activity. The authors who appeal to such a connection most often have not even tried to justify their "conclusions," and their "statistical proofs" are usually the results of incorrect application of statistical methods.

The Phase of Cyclic Activity

A separate discussion of the orthogonal and tangential components of the signal from a self-oscillating system does not simply reduce to a convenient computational approach but has a greater conceptual significance. The character of the changes in the component orthogonal to the limiting cycle is due first of all to the coarse energy parameters of the oscillator, such as the energy input and the energy losses over a single period of the oscillations. As is known, this component is a generalization of the amplitude of the oscillations from a quasimonochromatic signal generator. From a comparison of the relaxation times of the orthogonal components (in the case of a system with one degree of freedom there is one such component, while this time is the reciprocal of the rigidity of the limiting cycle) with the period of the steady-state oscillations one finds the character of the self-oscillator, for example, whether it is relaxation (with intense energy pumping and rapid energy loss) or Thomson (in which there is almost no loss of energy from the oscillating system over a single period and accordingly no energy enters). The tangential component, which is associated with more subtle properties of the self-oscillator, generalizes the concept of the phase of the oscillations. The time variations of the phase and of the orthogonal deviations often occur over times differing by an order of magnitude in the case of a relaxation self-oscillator. For cyclic solar activity the relaxation time of the orthogonal deviations is close to the phase of the minimum of the Wolf index, about 0.3 yr, while the characteristic phase relaxation time is 20-30 yr. It is thus natural that the available recording interval for the Wolf numbers (about 200 years) allows us to establish with fair certainty that the cyclic activity is generated by a relaxation oscillator but is too short to clarify the properties of the phase. Over comparatively small time intervals a sort of diffusive spreading of the phase is observed. However, the behavior of the tangential component over large intervals was fairly unclear until recently. It was particularly necessary to study it since reports [21, 22] existed about some unusual behavior in the phase of cyclic activity. Thus, an analysis [23] was made of a longer series of dates of the activity maxima and minima. This series both included direct information on the dates of the extremal values of the Wolf index and relied heavily on extrapolation of indirect data. The dates of activity maxima from 300† through 1963 and of minima from 1610 through 1963 were used, taken mainly from [22, 24].

The phase ϑ of a periodic function coincides with the argument‡ to within an arbitrary (additive) component. The phase of a process which differs little from a periodic one can naturally be considered to be the same as the phase of the "nearest state" of this periodic process. The accuracy of this definition (and the appropriateness of introducing the concept of phase) is lost as the deviations of this process from periodic become greater. Finally, if the process consists of a sequence of qualitatively similar pulses the definition of phase retains a clear meaning only for certain characteristic points. In [23] practically only two states of activity, corresponding to the minimum and maximum Wolf numbers, were taken into account. In accordance with these remarks the phase difference between two such states was assumed equal to the average time interval between the occurrence of these characteristic points on the record of Wolf numbers. The phase of the first recorded minimum was taken to be zero ($\vartheta_1 = 0$), while the phase of the one N−1 cycles later (i.e., the N-th minimum) was $\vartheta_{2N-1} = (N-1)T$, where T = 11.1 yr is the average cycle length of the activity. The phase of the maximum immediately following the first minimum was $\vartheta_2 \simeq 0.4T$ (the mean interval from a minimum to the next closest maximum being about 4.2 years) and the phase of the N'-th maximum was $\vartheta_{2N''} = (N' - 0.6)T$. The principal value of the phase is given by the equations

$$\theta = \vartheta - k_\vartheta T, \quad k_\vartheta = \max\{k = 0, \pm 1, \pm 2, \ldots : kT \leqslant \vartheta\}, \tag{7}$$

and in [23] only two principal values were mainly used:

$$\theta_{2N'+1} = 0 \text{ and } \theta_{2N'} \simeq 0.4\ T.$$

The phase drifts were characterized by the quantities

$$\gamma(\vartheta) = \vartheta - t(\vartheta), \qquad \xi_{\tau,q}(\theta) = \gamma(\theta + qT + \tau) - \gamma(\theta + qT), \tag{8}$$

where $t(\vartheta)$ is the time at which the phase is ϑ.

Labeling the values of the shift τ used in the calculations, $0 < \tau_1 < \tau_2 < \ldots < \tau_m < \ldots$, and choosing nonoverlapping intervals $(\theta + qT, \theta + qT + \tau_m)$ we computed (for each m) the sample dispersion

$$s_m^2 = \frac{1}{q_m} \sum_{q=1}^{q_m} \xi_{\tau m,\, q}^2 \tag{9}$$

and the quantity

$$w_m \equiv \sqrt[3]{s_m^2}. \tag{10}$$

The values of s_m^2, w_m, and the sample volume q_m are given in Table 1.

† As in Russian original. – Consultants Bureau.
‡ For an arbitrary periodic function of t the behavior of the normalizing factor ($\vartheta - \vartheta_0 = 2\pi t/T$ or $\vartheta - \vartheta_0 = 360°\, t/T$, where T is the period) offers no advantage so it is best to write $\vartheta - \vartheta_0 = t$.

TABLE 1

m	τ_m	q_m	s_m^2	w_m	m	τ_m	q_m	s_m^2	w_m
1	0.4	30	2.0	1.26	11	5.0	17	13.7	2.39
2	0.6	30	2.1	1.28	12	6.0	14	8.0	2.00
3	1.0	16	2.9	1.42	13	7.0	16	8.8	2.07
4	1.6	10	4.6	1.66	14	8.0	16	8.0	2.00
5	2.0	10	4.8	1.69	15	9.0	15	8.4	2.03
6	2.6	20	8.4	2.03	16	10.0	14	11.7	2.27
7	3.0	19	11.3	2.24	17	12.0	11	15	2.27
8	3.6	18	9.3	2.10	18	13.0	11	14	2.41
9	4.0	19	10.4	2.18	19	19.0	8	7	1.91
10	4.6	24	11.3	2.24	20	22.0	8	11	2.22

Statistical analysis of Table 1 showed that the hypothesis of linear growth in the dispersion

$$\sigma_\tau^2 = A_1\tau, \qquad \sigma_\tau^2 = \langle s_\tau^2 \rangle \tag{11}$$

with τ and the hypothesis of constant $\sigma_\tau^2 \equiv A_2$ are rejected with a confidence level of less than 2% [23]. At the same time the available observational material does not permit rejection of the hypothesis that the dispersion grows in a more complicated way

$$\sigma_\tau^2 = A\ (1 - \exp\ (-\ a\ |\ \tau\ |)) \tag{12}$$

It was thus appropriate to do an autocorrelation analysis of the phase shift $\gamma(\theta)$ of the cyclic activity. The sequence of phase drifts γ_i, $i = 1, 2, \ldots, k$, calculated at the maximum activity points ($\theta \simeq 0.4T$) was examined. Also included was the fact that an inaccuracy in the determination of the period T leads to a regular variation in this sequence and, therefore, to a difference between $\sum_{i=1}^{k_{12}} \gamma_i$ and $\sum_{i=k_{12}+1}^{k} \gamma_i$. The method of least squares was used to improve the value of T until the difference in these sums ceased to be significant; specifically, from 156 points (γ_i chosen with the least probable errors) the period was estimated to be T = 11.084 yr, and

$$\sum_{i=1}^{78} \gamma_i = -3.7\ \text{yr}, \qquad \sum_{i=79}^{156} \gamma_i = -3.9\ \text{yr}.$$

The autocorrelation functions were evaluated from two groups of data, with sample volumes of k = 41 (observations from 1519-1954) and k = 121 (including still earlier dates for the maxima which seemed sufficiently reliable). It is shown that the sample autocorrelation function is adequately approximated by the exponential function

$$\rho\ (\tau) = \exp\ [-\ a\ |\ \tau\ |], \qquad a \simeq 1/(1.5 - 3)\ T. \tag{13}$$

Until these results were obtained it was assumed that the dispersion in the phase shift of an arbitrary autonomous oscillator increases according to an asymptotically linear (diffusion) law; that is, at large time shifts the fluctuations in the tangential coordinate always become stochastically independent. The self-stabilization of the phase in a relaxation oscillator was not yet known; hence, synchronization of the generator of cyclic activity with an external periodic source was proposed as an explanation of the unusual variation (12). This motivated us, on the one hand, to develop the inverse-problem method of the theory of oscillations for application to a more complicated oscillator [4] and, on the other, to find a source of highly stable oscillations and consider the mechanisms for its efficient interaction with the cyclic activity.

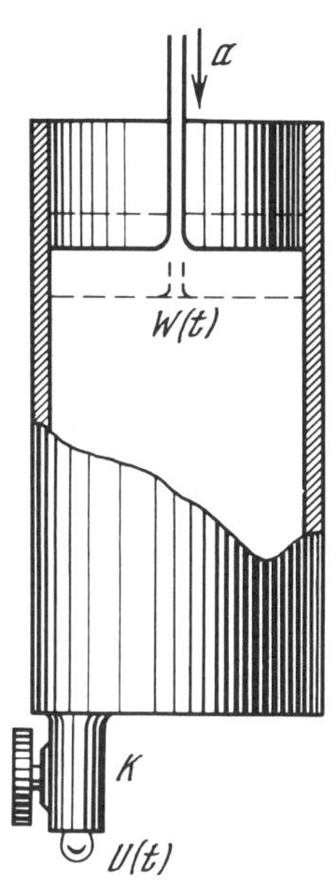

Fig. 3. A hydrodynamic self-excited relaxation oscillator.

No such external source of periodicity was found; furthermore, the autonomy of the cyclic activity generator was demonstrated comparatively recently. The search for a fundamentally new mechanism for the saturation in the growth of the phase dispersion, however, led to the discovery of phase self-stabilization in an autonomous ideal relaxation oscillator [25, 26].† It appears that in such an oscillator both the frequency and phase of the oscillations are dynamically stable without any synchronization by an external periodic force. The dynamic features of maintaining a stable phase in the case of external synchronization and (internal) "self-stabilization" are quite different. External synchronization is aimed at the removal of only small (compared to the period T of the autonomous oscillations) differences in the phases of the oscillations of the oscillator and of the synchronizing source. Clearly, when the phase difference is approximately T the dynamic situation is the same as when the phase difference is approximately 0. Phase self-stabilization is characterized by feedback within the autonomous relaxation oscillator itself which tries to liquidate any phase drifts away from the stable equilibrium. As will be obvious from the following, this difference in the method of dynamic phase maintenance may result in a substantial difference in the dependence of $\langle \xi_T^2 \rangle$ on the shift τ for a relaxation self-oscillator and for a synchronized oscillator of the common type (particularly when there is a high level of fluctuations).

When there is a small nonidealness in a relaxation oscillator the initial dynamic deviation in its phase is not completely suppressed. After the oscillations are set up there remains a phase shift which makes up a small part of the initial deviation. Correspondingly, the τ dependence of the phase shift dispersion σ_τ^2 does not exhibit complete saturation in the case of a stochastic model of a weakly nonideal self-oscillator. Phase self-stabilization appears here in that the initial portion of the relatively rapid increase in the dispersion of the phase shift which is proportional to τ goes into a slower (asymptotically linear with τ) rise. Clearly over a limited observation interval a sufficiently slow rise in the phase dispersion with τ is indistinguishable from complete saturation.

We now illustrate phase self-stabilization with a model of a hydrodynamic self-excited oscillator (Fig. 3). An incompressible fluid is poured into a sufficiently high cylindrical con-

† For more detail on the phase self-stabilization effect see Chapter IX of this volume.

tainer at a constant rate and leaves it only through a value K in the form of isolated drops of fixed volume. The rate of outflow is determined by the amount of fluid in the container and depends on the position of the valve. If its position is kept fixed the droplet outflow would have a strictly determined period T_0. Let the valve be rapidly (over times which are small compared to T_0) and randomly turned so the number of droplets breaking off from the valve during relatively small nonintersecting time intervals is practically independent. Here the dispersion in the number of droplets flowing out of the container over time τ is almost proportional to the interval τ. Over sufficiently large time intervals the flow takes place in a qualitatively different manner.

Let W_{max} denote the maximum volume of liquid which would accumulate in the container while the valve was kept minimally open (it is assumed that the valve is constructed in a way that it cannot be completely closed) and W_{min} denote the steady-state minimum volume of fluid in the container when the valve is fully open. For no observation time τ would the difference between the volumes of fluid flowing out through the valve and into the container during this time exceed $W_{max} - W_{min}$; thus the dispersion in the number of droplets in this scheme is always limited by the quantity

$$[(W_{\max} - W_{\min})/U]^2,$$

where U is the volume of a droplet.

Therefore, in the absence of leakage (around the valve) the growth in the dispersion of the phase of this hydrodynamic relaxation oscillator (which is practically proportional to the observation time for small τ) ceases completely as the time τ is increased.

This scheme for an ideal (leak free) relaxation oscillator may serve other purposes than as a graphic laboratory demonstration of the phase self-stabilization effect first found under astrophysical conditions during an analysis of cyclic activity. The closeness of this hydrodynamic arrangement to the model of cyclic activity developed here will be seen in the following discussion. In addition, the first results from an analysis of the Wolf numbers [2, 3] led us to another analogy with a neon lamp RC relaxation oscillator. It is in this type of arrangement that simplified analytic estimates were first made of the dependence of the dispersion in the phase of the oscillations on the observation time [25]. In Chapter IX we give a more complete discussion of the phase self-stabilization effect in a self-excited relaxation oscillator with a gaseous discharge lamp.

Here we should perhaps discuss briefly the idea of the model. A model is a schematic description of a phenomenon which takes certain aspects of the phenomenon which are isolated by experiments (or observations) of a given type into account. The words emphasize both the "modesty" of the model compared to the actual phenomenon and its limited purpose. In modeling phenomena it does not make sense to set out to find some kind of definitive solution which completely reveals the essence of the observed object. At best we may hope for a solution which, although tentative, adequately describes the (limited circle of) known features of the phenomenon at that time. Yet the model (speaking here of good models, of course) must always serve as a program for further research.

The search for the simplest analog model of the sun's cyclic activity was one of the main goals of our first papers [2, 3], in which a statistical analysis was made of the time variation in the Wolf index using the inverse-problem technique of the theory of oscillations. Among the main results of this work were the general conclusion that the self-excited oscillator for the solar activity is a relaxation oscillator and is isochronous, finding the specific form of the pulsed dependence of its rigidity on the phase of the activity, and, finally, proposing an analog for this oscillator in the form of a self-excited relaxation oscillator with a neon lamp. It is natural to compare this analogy with another relaxation analog proposed much earlier, the eruption hypothesis of Waldmeier [11, 28], in which the independence of the separate activity cycles was stressed and the unique development of each one was analyzed as a function of the maximum Wolf index in that cycle. Waldmeier's scheme essentially proposes no recipe for the transition from one cycle to the next. On the other hand, the neon lamp self-oscillator scheme, besides having a pulsed value of the rigidity which cuts the orthogonal deviations from the limiting cycle into weakly coupled segments, also predicts some sort of definite phase relationship between the successive cycles generated by the corresponding relaxation mechanism. At the same time the phase properties of the ideal neon lamp oscillator had not yet been studied. After studying these properties it was appropriate to draw some conclusions about the cyclic solar activity although it was, of course, unclear in advance how far it would be possible to follow the proposed analogy. Here general considerations could hardly be of help so it was necessary to turn to experimental data.

In this work it has already been mentioned that a statistical analysis of the phase properties of the cyclic activity made from a fairly long list of dates of strong arctic aurora clearly demonstrated saturation in the growth (with the observation time) of the phase dispersion. Starting from traditional concepts in the theory of oscillations we had to assume that this generator of the cyclic activity, which is similar to a neon lamp self-oscillator over a comparatively short time interval (a few cycles), is in fact not autonomous but is synchronized by some force external to it. Of course, we took this as a blow to the model which was beginning to become more concrete. On the one hand, the analysis of the time variation of a number of parameters of the activity (the Wolf numbers, the qualitative variation of the sunspot latitudes in the photosphere, the way the polarity of the magnetic field of groups of spots changes) outlined a unified relaxation model. On the other hand, it was now necessary to seek yet another mechanism for the cyclic activity besides this one which enters independently into the picture of its development. We have already noted that the energy of the tidal or any other interaction of the planets with the sun is insufficient for this by at least several orders of magnitude. The assumption of a hidden (unobservable from the earth through any other phenomenon), stable, sufficiently powerful, periodic process within the sun seemed both too arbitrary and unreal considering the sensitivity of mankind to a deep periodic modulation with such a noticeable period as 11 years (or multiples thereof). An "explanation" of the new properties by introducing some unprovable hypothesis is, of course, a big flaw in the model. Thus, when, as a result of an examination of the simplest scheme for an ideal relaxation self-oscillator, it seemed that it has limited phase dispersion even in the autonomous regime (without any external synchronization), we naturally had a feeling of satisfaction. The model was found to be successful and did not require arbitrary "additions." The search for a secret source of stable oscillations acting on the cyclic solar activity could be stopped. The analogy of the cyclic-activity source with the simple neon lamp RC-oscillator was deeper than its authors expected.

We now dwell on questions about both the methods and the psychology of scientific research since their significance was very appreciable during the course of this work. The relationship between the model and the effect itself is given enough attention in physics courses on the theory of oscillations, and a brief discussion of this topic has long been included in physics textbooks. In addition it is necessary from time to time to deal with some confusion in understanding the initial concepts of this subject, particularly as a misunderstanding of the role of models as approximations and of the meaning of their evolution.

The Autonomy of the Cyclic-Activity Oscillator

The phase self-stabilization effect was discovered as a result of three factors: (1) experimental observations (from indirect data on the dates of the maxima of the Wolf index) of saturation in the growth of the dispersion in the phase shift of solar activity, (2) the analogy of the activity generator (oscillator) with a relaxation oscillator, and (3) direct calculations of the phase drift in an ideal autonomous relaxation oscillator. All this is a convincing orientation which, together with the obvious absence of an external stable source acting effectively on the cyclic activity, makes a conclusion that the activity generator is autonomous seem probable, but not (strictly speaking) proven. In [29] a numerical experiment was done to choose one of two competing hypotheses: (1) the cyclic activity oscillator is synchronized by an external periodic source, and (2) this oscillator is fully autonomous but the phase self-stabilization effect is realized in it.

To clarify the reasons for saturation in the phase dispersion of the cyclic activity it is important that the ratio of the steady-state standard deviation of the fluctuations ($\sigma\gamma$) in the phase σ_∞ to the period of the activity not be small. It is

$$\sigma_\infty/T \sim 0.2-0.3, \qquad \sigma_\infty = \lim_{\tau\to\infty} \sqrt{\langle\delta\gamma^2\rangle_\tau}.$$

For a self-oscillator of the general type that is controlled by an external interaction such a large steady-state standard deviation is unrealistic since there would be large phase fluctuations $\delta\gamma \geq T/4$ which could not be removed by synchronization. This may be shown analytically with the simplest example, when a noisy Thomson self-oscillator is synchronized by a sinusoidal interaction. In the general case an analytical treatment is difficult. A mathematical experiment was done to obtain a conclusive analysis of the situation in the case of the self-oscillator for the cyclic solar activity. The equations of the relaxation activity oscillator obtained in [9] were modeled numerically by processing direct observational data. The right-hand side of these equations contains a sinusoidal force $f \sin(2\pi t/T)$ which is synchronized with the fundamental frequency of the oscillations as well as with a short-time correlated noise (with an intensity determined by analyzing the same observational data). The result of the computer calculation was a mean (over time) estimate $\overline{\xi_\tau^2}$ of the variation in the dispersion $\langle\xi_\tau^2\rangle$ in the phase of the oscillations of such an oscillator for various amplitudes f of the synchronizing force. Various periods of the synchronizing force close to the mean period T of the cyclic activity were also tried.

Let $n_f(t)$ and $n_0(t)$ be the deviations from the limiting cycle (of the autonomous dynamic system) in the phase plane of the trajectory of the oscillator being studied with and without the synchronizing force. The relation between this external force and the internal noise of the oscillator may be characterized by the ratio of the mean squares of these deviations,

$$\alpha(f) = \sigma^2 n_f / \sigma^2 n_0.$$

Figures 4 and 5 show part of the results of the numerical experiments. The smooth curves of

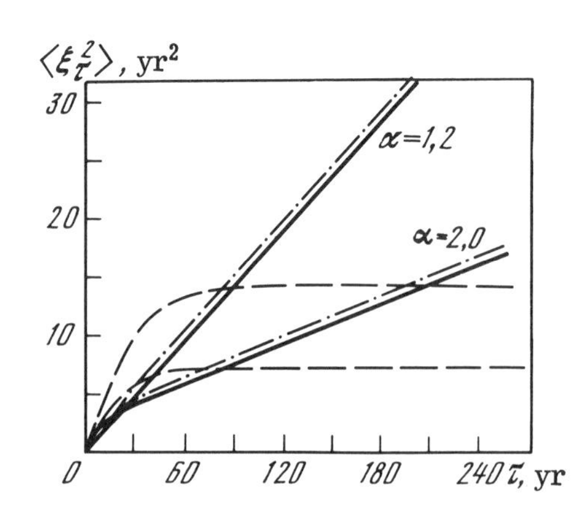

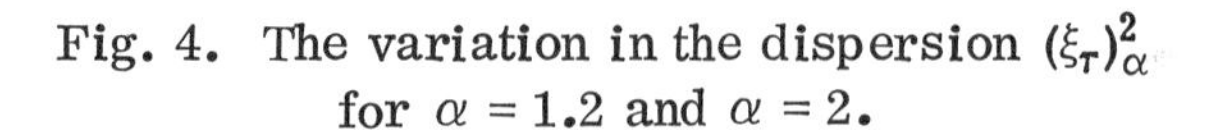
Fig. 4. The variation in the dispersion $(\xi_T)^2_\alpha$ for $\alpha = 1.2$ and $\alpha = 2$.

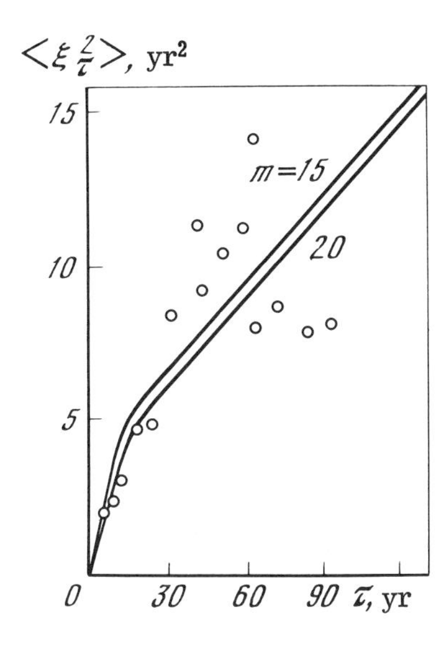

Fig. 5. The 95% confidence limits for the maximum sample value of ξ_τ^2.

Fig. 4 are the values of $\overline{(\xi_\tau)_\alpha^2}$ for α = 1.2 and α = 2.0. The step size (and correlation interval of the internal noise) was 1 yr. The external noise was neglected. The accuracy of the estimates of $\overline{(\xi_\tau)_\alpha^2}$ was

$$\sigma^2 \overline{(\xi_\tau)_\alpha^2} = \begin{cases} 0.07 \text{ yr}^2 & \text{for } \tau = 6 \text{ yr.} \\ (0.2-0.3) \text{ yr}^2 & \text{for } \tau = 40 \text{ yr,} \end{cases}$$

Including the external noise yields only small shifts in these curves. (The displaced curves are shown in dot-dashed curves.) The dashed curves show the boundaries of the 95% confidence region for the observed variation in $\overline{(\xi_\tau)_\odot^2}$ for the cyclic solar activity. This region was determined starting from a model of the experimental values of $\overline{\xi_\tau^2}$ in Table 1 in the form of Eq. (12) with Eq. (13) taken into account. From Fig. 4 it is clear that for α = 2 the function $\overline{(\xi_\tau)_\alpha^2}$ differs significantly from $\overline{(\xi_\tau)_\odot^2}$, particularly in the parts with small shifts at τ = 30-60 yr. As the amplitude f of the synchronizing force (and thus of the parameter α) increases the mismatch between $\overline{(\xi_\tau)_\alpha^2}$ and $\overline{(\xi_\tau)_\odot^2}$ increases in the large-shift region. This is illustrated by the $\overline{(\xi_\tau)_\alpha^2}$ curves for α = 1.2 and 2.0. Choosing another correlation time Δ of the internal noise (equal to a computational step size Δ = 0.2 instead of Δ = 1.0 yr) yields practically the same $\overline{(\xi_\tau)_\alpha^2}$ curves. Small changes in T also do not improve the agreement between $\overline{(\xi_\tau)_\alpha^2}$ and $\overline{(\xi_\tau)_\odot^2}$.

We may convince ourselves in yet another way of the contradiction between observation and the assumption that the cyclic-activity generator is synchronized without turning to a model for the variation in $\langle\xi_\tau^2\rangle_\odot$ Starting with the estimate $\overline{(\xi_\tau)_{\alpha=2.0}^2}$, we plotted the 95% confidence limits of the maximum sample value of $(\xi_\tau)_{\max}^2$ (the phase dispersion) for sample volumes q = 15 and q = 20 in Fig. 5. The values of $\overline{(\xi_\tau)_\odot^2}$ from Table 1 are shown here in the form of separate points. For 36 yr < τ < 100 yr the sample volume for calculating them exceeds 15-20. Clearly these points lie outside the probable limits. Other types of synchronization may be rejected in a similar fashion.

Therefore, the large value of the steady-state standard deviation in the phase shift of the activity makes it possible to definitively reject the assumption that the cyclic activity of the sun is synchronized with some kind of external stable oscillations.

We note in conclusion that thanks to the self-stabilization of the phase of the cyclic activity it is theoretically possible to have a definite correlation between the motion of the planets in the solar system and the level of activity over large time intervals. However, contrary to the just rejected hypothesis of planetary effects on the activity, we must reverse the cause and effect in this case.

The phase stability of the activity over long time intervals could, over many millions of years, lead to synchronization of an almost conservative planetary system due to, for example, the action of a solar wind that is correlated with the activity. It is uncertain whether the lifetime of the solar system is enough for such a synchronization to be established. To identify such a correlation would require a recording period of at least thousands of years. Over smaller intervals the phase of the cyclic activity fluctuates strongly.

A Kinematic Model

Our conclusions, derived from an analysis of the time variation of the Wolf index, that the rigidity depends on the phase of the oscillations and that there is no synchronization mechanism for the cyclic activity served as a key for constructing the first rigorous relaxation model. This model was an analogy with an autonomous relaxation oscillator with one degree of freedom, an RC-oscillator with a neon lamp. In the simplest variant used by us a direct current supplies a condenser and lamp in parallel. Oscillations occur because when the lamp is off the voltage on the condenser increases until the lamp breaks down, after which the lamp is again shut off. Already with this variant, described by a system of two first-order, ordinary differential equations, it was possible to perceive a new effect which did not enter in the original considerations for the model: a limitation on the dispersion of the phase fluctuations. Of course, from the very beginning it was clear that this model, which did not include the spatial development of the activity, was one-sided. It was now necessary to go to a second stage of evolution of the model which reflected the kinematics of the cyclic-activity mechanism but did not yet pretend to explain the physical causes (forces) which set this mechanism in motion.

An analysis of the spatial development of the activity is naturally carried out in the coordinate system Λ, Φ, z, where Λ and Φ are the heliographic longitude and latitude and z is in the direction of the radius. Each of these directions has specific features. We begin with the fact that both energy considerations and the results of earlier observations indicate that the processes which produce the activity propagate outward from some depth within the photosphere, and into higher (accessible to observation) regions of the solar atmosphere. The "cutting" of the activity cycles by the bursts of rigidity preceding the minima in the Wolf number means that in a physical model which relates the oscillations in activity to vertical displacements of carriers of these oscillations it is necessary to speak of their rise and loss, followed by succeeding generation of new carriers in the depths of the sun rather than of smooth up-and-down oscillations.

We note further that during the time the distribution of sunspots over the photosphere has been recorded there have been no cycles in which the maximum in activity in the northern hemisphere has coincided in time with the minimum in the southern hemisphere (or vice versa). The passage of a process outward from the depths, the joint occurrence of activity in both hemispheres, the onset of each cycle at comparatively high heliographic latitudes, the gradual drift of the mean latitudes of spot formation toward the equator (the Spörer law), and the specific way the dispersion of these latitudes varies, depending on the development during the cycle (the Maunder "butterflies"), have led to the idea of modeling the kinematic features of cyclic activity by belts made up of cells of a single type (Fig. 6) which break up randomly into their elements as they float periodically to the surface. Each of these belts covers the equator and both "royal zones" (in which sunspots are observed). The absence of spots in the latitudes

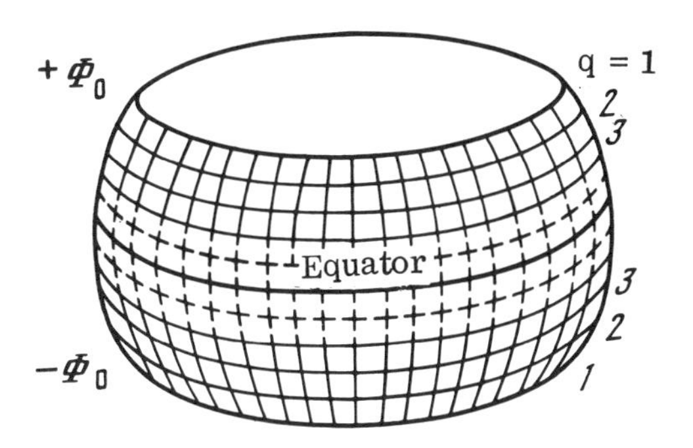

Fig. 6. A belt made up of cells.

adjacent to the equator is presumably due to the loss of the corresponding part of a belt under the photosphere. Any cell which has broken off from the belt rises to the surface toward the photosphere much more rapidly than the entire belt. If the cell has not been destroyed, then, having reached the photosphere, it is observed in the form of an incipient and evolving group of spots and later in the form of a residual active region. Underneath the comparatively slowly rising and gradually disintegrating cellular belt a new belt develops which, when the first belt is finally destroyed, rises and begins to break up into individual cells thus forming the next activity cycle, and so on. A very sketchy kinematic model of solar activity was examined in [30] and used to demonstrate the feasibility of having the basic global properties of cyclic activity agree with the decay of a belt (according to a simple stochastic law) into cells. In this model it was assumed that each belt, with latitudes $-\Phi_0 < \Phi < \Phi_0$, consists of closely adjacent identical "rectangular" cells. Before the onset of disintegration the cells in the belt occupy 2k rows with g_0 cells in each row parallel to the equator. Since the belt is initially symmetric with respect to the equatorial plane and the averaged statistical parameters remain symmetric during the decay as well, it is sufficient to analyze the deoay into cells of one of the halves of the belt.

The number q of the row is reckoned from the corresponding edge ($\Phi = \Phi_0$ or $\Phi = -\Phi_0$) of each half of the undecayed belt. It is assumed that during the entire decay period only the outer (at a given time) cells can break away from the belt. A cell which has a nonzero probability $\lambda(t)dt$ of leaving the belt over the time interval $(t, t + dt)$ was referred to as "perturbed at time t." We now restate these remarks about the decay of the belt from its edges more precisely as follows. An arbitrary cell of the q-th row is regarded as perturbed if the neighboring (having the same longitude) cell in the $(q - 1)$-st row has already left the belt by time t. Thus at the initial moment of decay all the cells of the first row, and only they, are perturbed. During the activity cycle the perturbation propagates latitudinally, while the randomness of the decay leads to smearing out of the latitude distribution of the perturbed cells. For simplicity it was postulated that $\lambda(t)$ depends only on the degree of decay of the belt at time t and is the same for all perturbed cells. The number $2kg_0$ of cells in the undecayed belt was assumed to be sufficiently large. The function $\lambda(\theta)$ was chosen so that the parameters of this essentially mathematical model would be close to the observed characteristics of the cyclic activity. A published comparison [30] of the computed and observed characteristics is shown in Figs. 7-9, in which the smooth curves correspond to statistical estimates of the characteristics obtained from observations of the cyclic activity and the dashed curves are computed by using this mathematical model of the decay of a belt. Figure 7 shows the limiting cycles obtained by analyzing the records of the Wolf index and from the model. Figure 8 shows the activity phase θ dependence of the mean reduced latitude $\overline{\eta}_{\Theta}$ of sunspot formation (obtained from the data of [31]) and the corresponding result for the model. Figure 9 shows a comparison of the observed and theoretical variation in the rigidity with θ. The scatter in these estimates of the "observed" and theoretical characteristics does not exceed the observational limits of accuracy. Hence we may say that without introducing new numerical results on the observations or new qualitative considerations there is no sense in complicating the decay model. In this model the func-

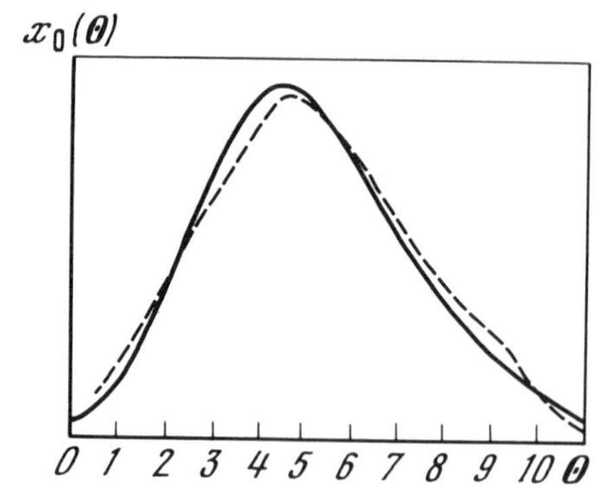

Fig. 7. The limiting cycles obtained from an analysis of observational data and from the model.

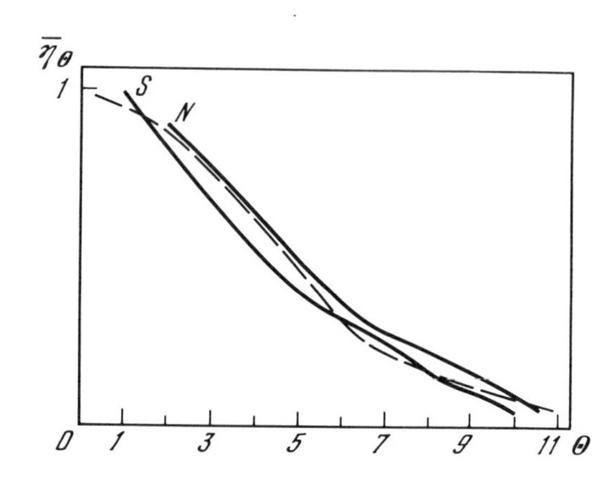

Fig. 8. The observed and theoretical variation of the mean reduced width $\bar{\eta}_\Theta$ with the phase of the 11-year cycle Θ.

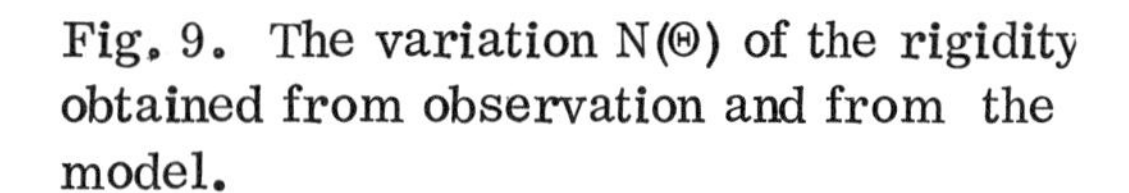

Fig. 9. The variation $N(\Theta)$ of the rigidity obtained from observation and from the model.

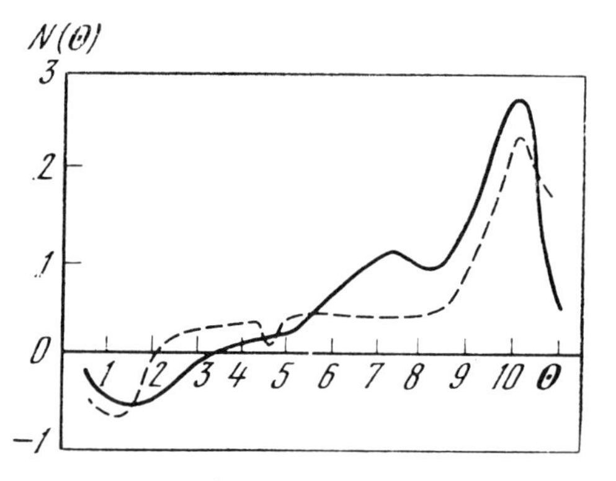

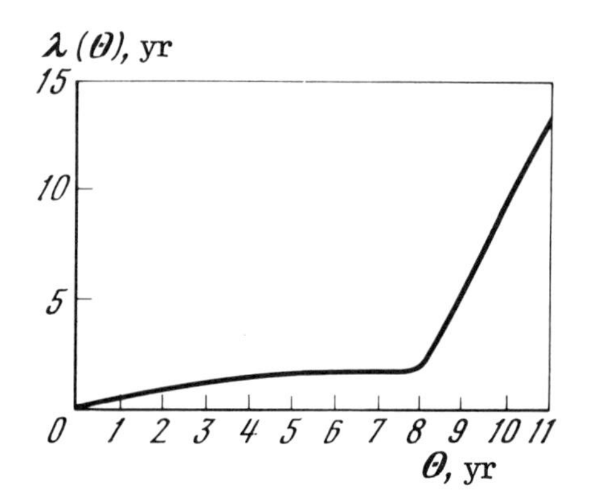

Fig. 10. The function $\lambda(\Theta)$ chosen for the model.

tion $\lambda(\theta)$, for which we have obtained such good agreement with experiment, is apparently close to the actual effective probability density for decay of a "belt" into its elements. The variation of this function with the phase θ is shown in Fig. 10, where $\lambda(\theta)dt$ is the probability that a perturbed element (cell) will break off over the time interval dt at the activity phase θ. Assuming the random process of the cells' breaking off from the belt to be a Poisson process, we obtain a simple expression for the probability distribution $F_\theta(t)$ that a perturbed cell will break off at cycle phase θ over time t:

$$F_\theta(t) = \exp[-\lambda(\theta)t]. \quad (14)$$

Thus, the mean lifetime of a perturbed cell in the belt is $\tau = 1/\lambda(\theta)$. As can be seen from Fig. 10, in the first year of a new cycle ($\theta \lessapprox 1$ yr) this time is fairly large ($\tau > 2$ yr), that is, the belt is practically stable. For $\theta < 8$ years the stability remains ($\tau \gtrsim 8$ months), but toward the end of the cycle the belt breaks up rapidly ($\tau < 1$ month). This means that the interpretation of this mathematical model must reflect the growth in the instability of the belt as it rises toward the photosphere.

The Laminar-Convective Region Hypothesis

We now consider the conditions under which such a belt of cells is formed. We begin with the fact that the agreement between the orientation of the magnetic fields of the groups of sunspots and the rotation of the sun indicates that the cyclic activity is produced by regular motions. The existence of "active longitudes" in the distribution of spots over the photosphere allows us to state more accurately that these motions must take place sufficiently deeply within the sun, where there is still no nonuniformity in the angular velocity of the sun (specifically, where the velocity does not yet depend on the heliographic latitude). It is natural to associate the source of such motions with thermal convection. But at what depth under the photosphere does the plasma move regularly? This question is important not only for explaining the physical mechanism of cyclic solar activity.

The dimensionless similarity criteria (such as the Rayleigh number) for the entire subphotosphere convective zone as a whole are so large as to guarantee, with room to spare, a turbulent, random convective motion. Immediately below the convective zone lies another massive zone with small Rayleigh numbers which is gas-dynamically stable with respect to thermal perturbations. Motions with relatively small (compared to the dimensions of the activity regions in the photosphere) spatial and time scales must be rapidly damped in this region, the more rapidly the deeper below the convective region they originate (for whatever reason). There is no place for the source of regular motions in the stable zone. Only direct experimental evidence could justify the development of clever models in which the regular motions originated more deeply, within the central convective zone of the sun, after which they somehow broke through the stable and turbulent zones while retaining their regularity. Such evidence, however, is lacking.

The way out of this contradictory situation is, we feel, related to the inhomogeneity of the subphotosphere convective zone, where the plasma density and the degree of ionization change significantly with height (reckoned from the depths of the sun toward the photosphere). The traditional reliance (in solar and stellar research) on unified parameters for the entire zone, which is based on Rayleigh's rigorous solution of the problem of convection in a homogeneous plane layer, is far from justified under these conditions. After breaking the convective zone down into comparatively uniform (with height) portions it is appropriate to first see how the corresponding parameters vary with z. Without studying the z dependences of the plasma temperature and density and of the degrees of ionization of hydrogen and helium in detail, it is possible, in view of the monotonicity and continuity of these parameters, to conclude that there exists a range of heights $0 < z < \Delta z$ near the lower boundary ($z = 0$) of the convective zone in which the parameters correspond to stable laminar convection. However, it is not clear at first whether the interval Δz is sufficiently large for a regular laminar structure which would cause a horizontal (spherical) layer of Benard cells to be formed and maintained.

As noted above, the conclusion on the basis of the Spörer law that the self-oscillation mechanism is a relaxation process led to the idea of a belt of decaying cells. Thus the basic (for our model of cyclic activity) hypothesis gradually developed: In the lower portion of the subphotosphere convective zone there is a stable region of laminar motions which form the cellular structure. We emphasize that the assumption that the subphotosphere convective zone (mainly turbulent) has a laminar boundary layer continues to be basic even if some significant changes have to be introduced in the details of the model of cyclic activity discussed below. The existence of such a boundary region is not dictated solely by the rather unique pattern of the processes involved, which agrees with observations and has already led to the prediction of several previously unknown features of the cyclic activity. It is also important that when there is a region of laminar cellular motions inside the convective zone any concrete mechanism for the cyclic activity discussed previously becomes impossible.

The necessary presence of a magnetic field (also with a cellular structure) in the laminar convective region cannot draw objections from astrophysicists as it is a special case of the interaction of the magnetic field with the long-term regular motions of the plasma. The magnetization of the convective cells leads to a series of radical consequences. The cyclic change in the orientation of the magnetic fields of the groups of spots (the Nicholson–Hale laws) is convincing evidence of a relation between the motions of the cellular belts which follow one another in time. It is thus already natural to suppose that the laminar convection region is stratified, that is, it consists of several horizontal layers of Benard cells. The cells of a particular layer are almost similar to one another, while in the adjacent layers they are the inverse; that is, in any pair of immediately adjacent (in height z) layers one layer is always "ordinary" and the other is an "inverse" layer. In the "ordinary" layer convective upwelling of the plasma takes place in the center of the cells and the cooler plasma flows down around the periphery. In the "inverse" layer, however, the heated plasma rises around the periphery of the cells and flows downward in the central portion after being cooled. Thus there arises a three-dimensional, periodic, convective plasma "crystal" in which the cells of one layer are located exactly beneath those of another (Fig. 11).

In this model the activity cycle may now be related to the decay of the uppermost layer into individual cells and the floating of these cells to the top, where, when any of them go out into the photosphere, they are seen as a group of spots at that location. From the first it is clear that the intense random flows of plasma in the turbulent part of the convective zone lying above the "crystal" do not allow the cells to float vertically to the top. This must cause significant dispersion in the coordinate positions at which the active regions appear in the photosphere. Such stochastic scattering may almost completely smear out the "traces" in photosphere of the cell locations in the upper layer (belt) of the "crystal" which is periodic in the heliographic longitude and almost periodic in the latitude. But the development of a periodic structure would be an important argument in favor of both the hypothesis that a laminar-convective region exists and the appropriateness of modeling the activity cycle by the decay of convective cells. Thus for the sake of verifying these fundamental questions we decided to risk a futile, painstaking analysis of the available photoheliographic data [32] in order to try to identify longitudinally and latitudinally periodic components in the distributions of the places of first appearance of groups of sunspots on the photosphere. Surprisingly to us, even the first "rough" stage (done only for the latitudinal distribution and only over part of the available observational material) of this analysis (described in more detail in the next section) yielded positive results. Calculations [33, 34] indicate the definite presence of a quasiperiodic com-

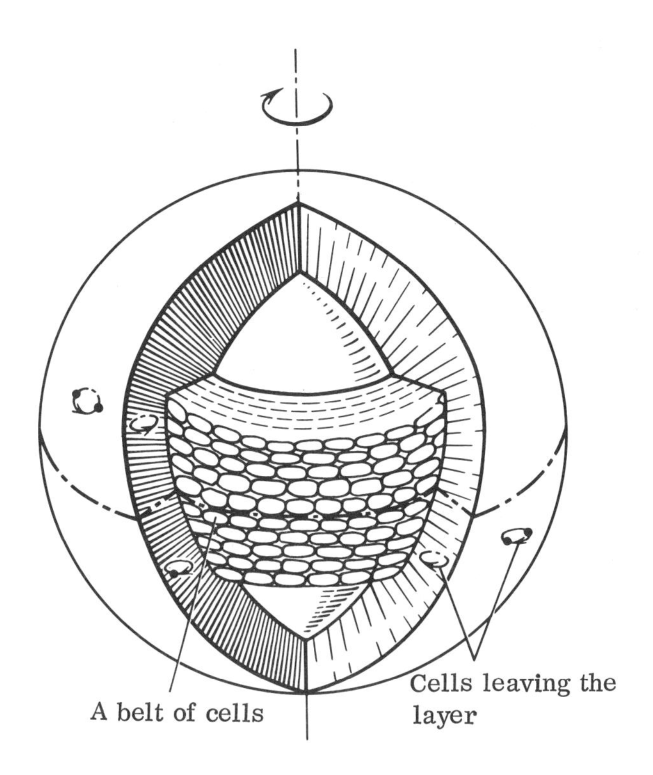

Fig. 11. The periodic cell structure.

ponent in the distribution of latitudes of the places where regions of activity first appear on the photosphere. This distribution has an oscillatory component whose "period" apparently changes slowly with latitude.

The order of magnitude of the expected angular period $D_\Phi \simeq D_\Lambda$ is easily found from the data on the number N_0 of groups of cycles created during a cycle and the known area of the zone of activity $S_{ac} \simeq S_\odot/4$ (where ($S_\odot$ is the area of the sun's surface). The period in degrees is

$$D_\Phi \simeq D_\Lambda \simeq \frac{360°}{2\pi R_\odot} \sqrt{S_c}, \tag{15}$$

where S_c is the area of a cell and $R_\odot$ is the solar radius. The number of cells in the belt, N_c, must be somewhat larger than N_0 (as not every cell produces a group of spots and not every group is observed); thus,

$$N_c \simeq (1-4)N_0,$$
$$S_c \approx \frac{S_{ac}}{N_c} \simeq \left(\frac{1}{16} - \frac{1}{4}\right) S_\odot/N_0. \tag{16}$$

From Eqs. (15) and (16) we obtain

$$D_\Phi \simeq D_\Lambda \simeq \frac{360°}{2\pi R_\odot} \sqrt{\frac{S_\odot}{N_0}} \left(\frac{1}{4} - \frac{1}{2}\right) \simeq (50° - 100°) N_0^{-1/2}. \tag{17}$$

Setting $N_0 = 3000$ (cf. [13, 16]) in Eq. (17) we find

$$D_\Phi \approx D_\Lambda \approx 0.9° \text{ to } 1.8°. \tag{18}$$

The Development of Spatial Periodicity

The observed differential rotation of the sun consists of a regular dependence of the angular rotation velocity of the photosphere on the latitude. Since at a sufficient depth the sun must rotate as a unitary solid body, it follows that the angular velocity depends on the depth z as well. From the existence of "active longitudes" it first follows that in this description of the cyclic activity the entire crystal, including its upper layer (the belt), must be located in the region of solid rotation. In this problem it is also important that the depth and latitude dependence of the differential rotation (which is hardly reflected in the average values of the periodic set of discrete latitudes of the "trace" of the belt in the photosphere) results in a continual blurring of the behavior in the middle latitudes. From this it is clear that the method for developing the latitudinal periodicity is much simpler than that of the longitudinal. Thus the analysis [33, 34] of the Greenwich catalogs [32], which contain the heliographic coordinates of the groups of spots, was begun with an analysis of the distribution of latitudes. In this preliminary work we used data on the latitudes of groups of spots for four activity cycles (Nos. 15-18). From the latitudes Φ of the centers of the first groups of spots to appear sixteen histograms $\psi(\Phi_i)$ of the number of groups to appear in intervals of 0.1° were compiled separately for the northern and southern hemispheres, for each of the four chosen cycles, and for two intervals of the phase of the cyclic activity ("maximum activity" and "average activity"). To reduce the effect of nonuniformities in the material due, in particular, to a reduction in the average number of spots near the borders of the "royal" zones, the sliding average (with a period of 4.6°) was deducted from each histogram, where

$$\overline{\psi}(\Phi_i) = \frac{1}{47} \sum_{k=-23}^{23} \psi(\Phi_{i+k});$$

that is, we used the function

$$f(\Phi_i) = \psi(\Phi_i) - \overline{\psi}(\Phi_i). \tag{19}$$

At first [33] these functions $f_l(\Phi_i)$ (l = 1-16, where l is the number of one of sixteen cases) were examined over the latitude range Φ = 17.5°-30°. Thus each of the histograms had 125 points. For simplicity instead of $f_l(\Phi_i)$ the following normalized quantities were used:

$$f_l^{\circ}(\Phi_i) \equiv f_l(\Phi_i)/s_l,$$

$$s_l^2 = \overline{f_l^2} = \frac{1}{n} \sum_{i=1}^{n} f_l^2(\Phi_i).$$

Values of the period D from 0.4° to 2.9° with a step size of 0.025° were tested. For each of these values of D and l = 1-16 the periodogram

$$a_l^2(D) = 4\left\{\left[\overline{f_l^0(\Phi)\sin\left(2\pi\frac{\Phi}{D}\right)}\right]^2 + \left[\overline{f_l^0(\Phi)\cos\left(2\pi\frac{\Phi}{D}\right)}\right]^2\right\} \tag{20}$$

was calculated, where, as before,

$$\overline{Z} = \frac{1}{n} \sum_{k=1}^{n} Z_k, \qquad n = 125.$$

If the function $\psi(\Phi)$ does not contain an oscillating quasimonochromatic component then the mathematical expectation of a_l^2(D) should not depend on D. It may also be shown that if $f_l^0(\Phi_i)$ and $f_l^0(\Phi_{k\neq i})$ do not correlate pairwise and the distributions f^0 are Gaussian, then these exact relations hold:

$$\langle a_l^2 \rangle = \frac{4}{n}, \qquad \sigma a_l^2 = \langle a_l^2 \rangle = \frac{4}{n}. \tag{21}$$

We shall prove the first of these. Introducing the notation

$$x_i = f_l^0(\Phi_i), \qquad s_i = \sin 2\pi \frac{\Phi_i}{D}, \qquad c_i = \cos 2\pi \frac{\Phi_i}{D} \qquad (c_i^2 + s_i^2 \equiv 1)$$

for brevity, we obtain from Eq. (20)

$$\frac{\langle a_l^2(D)\rangle}{4} = \left\langle \left(\frac{1}{n}\sum_{i=1}^{n} x_i s_i\right)^2 + \left(\frac{1}{n}\sum_{i=1}^{n} x_i c_i\right)^2 \right\rangle = \frac{1}{n^2}\sum_{i=1}^{n}\langle x_i^2\rangle + \frac{1}{n^2}\sum_{i=1}^{n}\sum_{\substack{j=1\\ j\neq i}}^{n}\langle x_i x_j\rangle (s_i s_j + c_i c_j) = \frac{1}{n}\sigma^2 x_i = \frac{1}{n}.$$

The second of Eqs. (21) is proven similarly (only with more cumbersome manipulations). We note that the normalization of the distribution f^0 is required only for the derivation of the second equation and that the presence of a weak exponentially falling correlation between x_i and x_j hardly changes the results (21) if

$$\Phi_{i+1} - \Phi_i \ll D, \qquad \tau_* \ll D, \tag{22}$$

where τ_* is the correlation interval of x. The first inequality of Eq. (22) is ensured by the method of calculating a_l^2(D) (the test period D is always much greater than the observational step size), while the second inequality is verified by computing the autocorrelation function for f^0. From a_l^2(D) it is convenient to go to the average over the sixteen cases,

$$\tilde{a}^2(D) = \frac{1}{16}\sum_{l=1}^{16} a_l^2(D).$$

It follows from Eq. (21) that

$$\langle \tilde{a}^2(D)\rangle = \frac{4}{n}, \qquad \sigma\tilde{a}^2(D) = \frac{\langle a_l^2\rangle}{\sqrt{16}} = \frac{1}{n}.$$

Figure 12 shows the function

$$A(D) \equiv [\widetilde{a^2}(D) - \langle\widetilde{a^2}\rangle]/\sigma(\widetilde{a^2}) = n\,[\widetilde{a^2}(D) - 4/n] \tag{23}$$

calculated from the observational data. When there is no periodic component in the histogram of $\psi(\Phi)$ this function must have a zero mathematical expectation and unit standard deviation. The graph of this function in Fig. 12 indicates the existence of a spatial period lying within

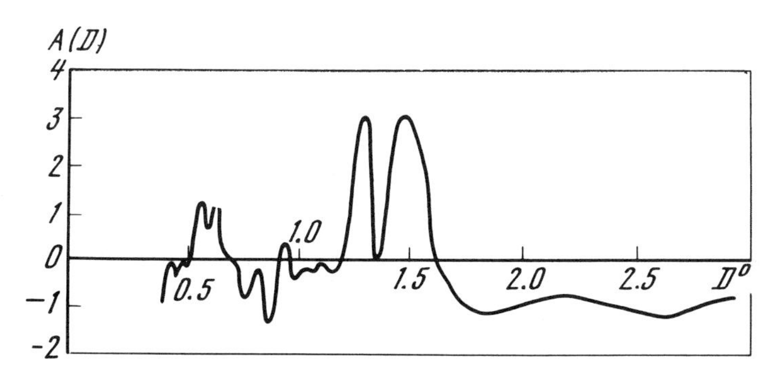

Fig. 12. The normalized function A(D).

the range

$$D = 1.2\text{–}1.6°, \tag{24}$$

which agrees with the model [see Eq. (18)].

We now introduce some new [34] (but still preliminary) estimates of the period D using the same data but, in view of the importance of this question, a slightly different method. There [34] the histograms of $f(\Phi_i)$ were reduced to a single form using the normalizing factor $[\overline{\psi}(\Phi_i)]^{-1/2}$, and the behavior of the function

$$f^1(\Phi_i) = f(\Phi_i)[\overline{\psi}(\Phi_i)]^{-1/2}$$

was analyzed in three latitude ranges:

$$5° \leqslant \Phi_i < 11°,$$
$$11° \leqslant \Phi_i < 17°,$$
$$17° \leqslant \Phi_i < 23°.$$

These operations do not introduce a periodicity but make the dispersion $\sigma^2 f^1$ almost constant and the functions f^1 nearly stationary (in the broad sense) random functions. For both the northern and southern hemispheres and for each of the two phase intervals the average periodograms were calculated over four cycles similarly to Eq. (20) for each of the three latitude intervals. As opposed to [33] twelve periodograms were now examined with $n = 240$ points in each. Without stating that the idea of a periodic component in the spatial distribution over the last latitude range is somewhat arbitrary, it is clear beforehand that the dimensions of the cells in the crystals as well as their average drift during the decay of the belt must depend on the latitude, although little, Thus, the "period" D must vary weakly, rather than strongly, with the latitude and phase of the activity. This fact complicates the computational scheme. The way the calculations were done in [34] reduces to the following: (1) A sufficiently wide interval of tests of the quasiperiod D(1°, d°) was chosen. The step size for the tests was taken to be 0.1° everywhere. (2) The locations of the maxima in the periodograms D_l^* were sought for fixed d. (3) The interval (g_1°, g_2°) within which the unknown quasiperiod D was expected to lie was fixed; that is, $1° \leqslant g_1^\circ \leqslant D \leqslant g_2^\circ < d°$. (4) For fixed d, g_1, and g_2 the number of maxima N lying in the interval (g_1, g_2) was calculated. Here the overall number of maxima was $M = 12$ or 6 (either 12 or 6 periodograms having been examined). (5) The probability P_M^N of the appearance of $K \geq N(d, g_1, g_2)$ maxima in the interval (g_1, g_2) if there is no quasiperiodicity was evaluated. A small value of P_M^N would indicate quasiperiodicity with a period lying within the interval (g_1, g_2).

In the absence of quasiperiodicity the fact that the maximum of one periodogram falls within the interval (g_1, g_2) does not depend on other periodograms having maxima which fall in this interval, and the probability distribution of the number of maxima falling in this interval is binomial. The probability of N or more maxima of M periodograms falling within (g_1, g_2) is

$$P_M^N = \sum_{l=N}^{M} C_M^l p^l (1-p)^{M-l}, \tag{25}$$

where p is the probability of the maximum of a periodogram which fills the interval (1°, d°) falling into the range (g_1, g_2) and

$$C_m^l = \frac{m!}{l!\,(m-l)!}\,.$$

For evaluating p we note that in the absence of a quasiperiodicity the functions $f^1(\Phi_i)$ are close to white noise and the probability that the maximum of the periodogram will fall into the chosen frequency interval is proportional to the length of the interval. On going from frequencies to periods D the factor D^{-2} appears so that the probability that the maximum of the periodogram filling the interval (1°, d°) will fall in the region (g_1, g_2) is

$$p = \frac{\sum\limits_{D_k \in (g_1^0,\, g_2^0)} \frac{1}{D_k^2}}{\sum\limits_{D_k \in (1^\circ,\, d^\circ)} \frac{1}{D_k^2}}, \qquad (26)$$

where the summation is over all D_k falling in either the interval (g_1°, g_2°) or the interval (1°, d°).

The results of some calculations using this method to find the quasiperiod of the oscillations, D, are shown in Table 2 (g_2 = 1.6°). They fairly reliably exclude randomness in the agreement of the positions of the maxima of the periodograms in the latitude range 1.2°-1.6°, which indicates that there are oscillations in the distribution $\psi(\Phi)$ with the corresponding quasiperiod. The value of D apparently changes slowly with the phase and latitude. This conjecture is in accordance with the fact that the periodograms, which refer only to the phase of the maximum activity, make it possible to reject chance more surely as the cause of the agreement in the maxima since the range of periods has been narrowed to 1.4°-1.6°.

No search has yet been made for the longitudinal periodicity. We note that the method for this search must differ considerably from that described above. The difficulty is the following. (1) The longitude of the center of a group of spots (even at the moment it first appears) is determined very much less accurately than the latitude and (2) the differential rotation causes smearing of the longitudinal period; i.e., the displacement along the longitude of a cell which has floated to the top depends on the depth at which it floats up (i.e., on the phase of the 11-year cycle) and on the latitude at which the groups of spots is observed. In a preliminary analysis these dependences could be assumed to be linear, but even then the problem is made difficult (compared to finding the latitudinal period) by the need to evaluate a number of supplementary parameters. In addition, it is hardly possible to expect that the belt is strictly axially symmetric over the observation interval. The existence of active longitudes indicates, rather, that the depth of the belt slowly fluctuates at different longitudes and thus the angular longitudinal dimensions of the cells may also vary. Therefore, at least one further parameter is needed to describe the variation in D_Λ with latitude.

The Magnetic "Crystal"

In these sketchy discussions of the variation of the Rayleigh number with depth, whose purpose was to explain the presence of a laminar boundary region in the convective zone, the magnetic field has been ignored. However, it has a very noticeable effect on the stability of

TABLE 2

d	g_1	p	M	N	P_M^N	Remarks
1.9°	1.2°	0.49	12	11	$3 \cdot 10^{-3}$	Both phase intervals of the cyclic activity used
2.3°	1.2°	0.42	12	11	$6 \cdot 10^{-4}$	
2.9°	1.2°	0.37	12	10	$2 \cdot 10^{-3}$	
3.2°	1.2°	0.35	12	9	$5 \cdot 10^{-3}$	
2.9°	1.4°	0.19	6	6	$5 \cdot 10^{-5}$	Only the interval of maximum activity used
3.2	1.4°	0.18	6	5	$9 \cdot 10^{-4}$	

this region in two ways. We begin with the fact that the field strengthens the local structure of the laminar crystal and greatly expands its thickness Δz. In the upper layers the comparatively strong field suppresses attempts by the turbulent plasma to destroy the central regions of these layers; thus the cells break off only from the edge. At the lower boundary of the crystal the weak magnetic field penetrating from the convective layers into the stable zone makes it easier for the laminar convective motions to enter the upper part of this zone. The cellular structure of the magnetic field, which gradually increases with z and is in accordance with the periodic distribution of the regular plasma velocity, results in a certain hierarchy in the stabilities of the elements of the convective crystal. Most stable are the magnetized cells themselves. Due to the field they continue to exist even after the conditions for laminar convection completely vanish as they rise and their orientation changes. The coupling of the cells in the single horizontal layer is not so stable and the layers of the crystal have the weakest coupling with one another. It may be said that it is the periodic structure of the magnetic field (and not immediately the structure of the mean values of the plasma velocity) which is the basis of the crystal. In the simplest description the lines of force of the magnetic field are situated in loops around the center lines of the nearly toroidal cells. A small part of the lines of force twist alternately around the central lines of one cell, then of its neighbors. These lines are the scattering field. This field lies mainly in one layer, but a small fraction of its loops go out from each cell into the closest cells in neighboring layers.

The magnetic field in the toroidal region of the cell may be broken down into axial and azimuthal components, which are its projections onto the central and generating circles of the torus. The field direction may be characterized by two numbers (σ_1, σ_2), for example, by setting $\sigma_1 = +1$ if, when looking upon the layer from above, the axial component is directed clockwise ($\sigma_1 = -1$ otherwise) and $\sigma_2 = +1$ if the loops of the lines of force rotate about the central circle of the torus according to the right-hand rule. If we assume that the Coriolis force is negligible here, then the cells themselves and their fields are the same in the northern and southern hemispheres of a given layer (n); i.e.,

$$\mathbf{H}_{(n)}(\Phi + D_\Phi,\ \Lambda + D_\Lambda) = \mathbf{H}_{(n)}(\Phi,\ \Lambda), \tag{27}$$

where, as before, Φ and Λ are the heliographic latitude and longitude, and D_Φ and D_Λ are the latitudinal and longitudinal periods. It is easy to see (Figs. 11 and 13) that after development of the cells this leads to the same polarization in each group of sunspots in the northern hemisphere of the sun and to the opposite (in view of the coincidence of the axial components) polarity in the groups in the southern hemisphere. Under these conditions we may write in both (S and N) hemispheres that

$$\sigma^{(N)}_{(n)1} = \sigma^{(S)}_{(n)1}, \qquad \sigma^{(N)}_{(n)2} = \sigma^{(S)}_{(n)2}. \tag{28}$$

If in fact the direction of the increased field in the cells changes with the direction of the Coriolis force, then the symmetry of the axial and azimuthal components of the field is different; i.e.,

$$H^{\text{ax}}_{(n)}(\Phi + D_\Phi,\ \Lambda + D_\Lambda) = H^{\text{ax}}_{(n)}(\Phi,\ \Lambda); \quad \sigma^{(N)}_{(n)1} = \sigma^{(S)}_{(n)1} \tag{29}$$

$$H^{\text{az}}_{(n)}(\Phi,\ \Lambda + qD_\Lambda) = -H^{\text{az}}_{(n)}(-\Phi,\ \Lambda);\ q = 0,\ \pm 1,\ \pm 2,\ \ldots;\ \sigma^{(N)}_{(n)2} = -\ \sigma^{(S)}_{(n)2}. \tag{30}$$

In this case within a given activity cycle the sign of the loop of the lines of force as well as the polarity of the groups of sunspots must change from one hemisphere to another.

The magnetic field does not only cause a reinforcement of the local structure of the convective crystal. Its second (almost contradictory) role is equally important. As the field is

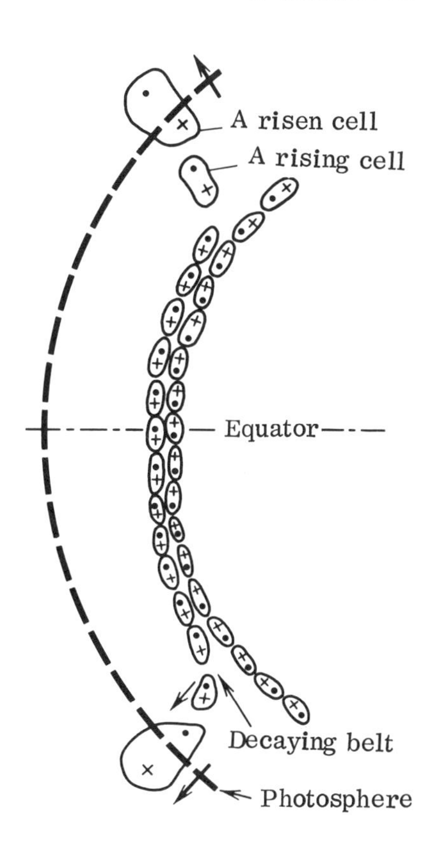

Fig. 13. The predominant directions of the magnetic fields in the cells.

frozen in the plasma it produces an Archimedean buoyancy and therefore destroys the static stability of the entire crystal. The distributed magnetic buoyancy increases with height z, thereby speeding up the upward motion of the higher layers and, evidently, causing the uppermost layer to break off from the crystal. This layer (the belt) is gradually destroyed by the surrounding turbulent plasma. The cells which are at the edge of the belt up to this moment are broken off from it and rapidly carried to the photosphere after the volume force develops in a vertical configuration. In place of the disintegrating n-th (top) layer (the belt) the (n + 1)-st layer floats up and later, during the roughly 11-year cycle, it too disintegrates. At the same time as the decay ("evaporation") occurs at the top, a new layer is created under ("frozen to the bottom") the crystal.

Therefore, in the presence of a magnetic field the crystal becomes dynamically, rather than statically, stable. From a purely gas-dynamic standpoint the sequentially located layers within it are not fully equivalent. The convective layer in which the plasma rises in the central portions of the cells transfers heat somewhat more effectively than the layers adjacent to it where the plasma rises in the periphery of the cells. When the plasma is magnetized this difference may be unimportant. Then as the n-th layer is replaced at height z by the next, (n + 1)-st layer (i.e., after about 11 yr), the speed at these same heliographic points changes sign, that is,

$$\mathbf{V}_{(n+1)}(\Phi, \Lambda, z) = -\mathbf{V}_{(n)}(\Phi, \Lambda, z). \tag{31}$$

With this symmetry the field transformation

$$\mathbf{H}_{(n+1)}(\Phi, \Lambda, z) = -\mathbf{H}_{(n)}(\Phi, \Lambda, z), \tag{32}$$

which reduces to a sign change at the same points, is a possible solution. Considering the controlling role of the magnetic field in the crystal it is possible to regard Eq. (32) as fairly plausible. From this it follows that both the axial field component and the orientation of the loop change sign on going from one cycle to the next; i.e.,

$$
\begin{aligned}
H^{\mathrm{ax}}_{(n+1)}(\Phi, \Lambda, z) &= -H^{\mathrm{ax}}_{(n)}(\Phi, \Lambda, z),\\
H^{\mathrm{az}}_{(n+1)}(\Phi, \Lambda, z) &= -H^{\mathrm{az}}_{(n)}(\Phi, \Lambda, z),\\
\sigma^{(\alpha)}_{(n+1)k} &= -\sigma^{(\alpha)}_{(n)k}, \quad \alpha = S, N, \quad k = 1, 2.
\end{aligned}
\tag{33}
$$

These transformation laws for the axial component correspond exactly to the well-known observational Nicholson–Hale laws for the direction of the magnetic fields in sunspot groups. The results of any observations of the sign of the force loops of the active regions, or even the existence of any such observations, are unknown to us. It seems that observing the sign of the field loop of the active regions would be of considerable interest.

These schematic discussions do not pretend to be rigorous and are qualitative in nature. In a number of problems they clearly cannot be relied upon, in particular, when analyzing the enhancement of the magnetic field in the layer as it rises. For a given velocity field $\mathbf{V}$ the change in the magnetic field is given by [35]

$$
\partial \mathbf{H}/\partial t = \mathrm{curl}\,[\mathbf{V} \times \mathbf{H}] + \nu_m \Delta \mathbf{H},
\tag{34}
$$

where $\nu_m = c^2/4\pi\lambda$ is the magnetic viscosity. For identical heliographic coordinates (Φ, Λ, z) let the plasma velocity and magnetic field distributions in two successive layers (i.e., separated by about 11 yr) differ from the above scheme as follows:

$$
\begin{aligned}
\mathbf{V}_{(n)} &= \mathbf{V} + \mathbf{V}', & \mathbf{V}_{(n+1)} &= -\mathbf{V} + \mathbf{V}',\\
\mathbf{H}_{(n)} &= \mathbf{H} + \mathbf{H}', & \mathbf{H}_{(n+1)} &= -\mathbf{H} + \mathbf{H}'.
\end{aligned}
\tag{35}
$$

Small deviations are fully natural. The nonazimuthal component $\mathbf{V}'$ of the velocity may be due to both weak axial flows within the cells along the central circles and radial flows which mix material from neighboring cells. The component $\mathbf{H}'$ of the magnetic field may be due, in particular, to a small dissymmetry between the even and odd layers. In the general case Eqs. (34) and (35) may be rewritten in the form

$$
\begin{aligned}
\partial \mathbf{H}/dt &= \mathrm{curl}\,[\mathbf{V}' \times \mathbf{H}] + \mathrm{curl}\,[\mathbf{V} \times \mathbf{H}'] + \nu_m \Delta \mathbf{H},\\
\partial \mathbf{H}'/\partial t &= \mathrm{curl}\,[\mathbf{V} \times \mathbf{H}] + \mathrm{curl}\,[\mathbf{V}' \times \mathbf{H}'] + \nu_m \Delta \mathbf{H}'.
\end{aligned}
\tag{36}
$$

These equations must be solved jointly with the equations for the time variation of the plasma velocity. In order to evaluate the situation simply we shall assume that the coordinates Φ and Λ are fixed and that the height $z = z(t)$ changes in accordance with the way the layer rises in the crystal. First we come to the conclusion that for strict fulfillment of the conditions ($\mathbf{V}' \equiv 0$, $\mathbf{H}' \equiv 0$) the field strength in the layer would change with time according to

$$
\partial \mathbf{H}/\partial t = \nu_m \Delta \mathbf{H}.
$$

But then the enhancement of the magnetic field by the laminar plasma motion in the convective cells is impossible. From this it follows that the deviations $\mathbf{V}'$ and $\mathbf{H}'$ from the above simple scheme must be taken into account in the field-enhancement problem.

While the magnetic field is still comparatively small and has almost no effect on the plasma motion the increase in its strength may be estimated using only one of Eqs. (36) and setting $[\mathbf{V}' \times \mathbf{H}'] = 0$. Furthermore, since the characteristic damping time of the field due to finite conductivity $t_{\mathrm{damp}} \approx r^2/\nu_m \sim 2 \cdot 10^{11}$ sec (where it is assumed that the characteristic dimension of the spatial variation of the field is $r \approx 0.2\ R_{\odot} \sin D_{\Phi} = 2 \cdot 10^8$ cm and that $\nu_m = 2 \cdot 10^5\ \mathrm{cm^2/sec}$) is much greater than the characteristic time for enhancement of the field, $t_{gain} \sim 0.1T = 3 \cdot 10^8$ sec, the magnetic viscosity may be neglected in Eqs. (36). Assuming for the calculation that in the initial stage of enhancement the functional dependences of the fields H and H'(t, r) on time and position **r** are separable, that the velocities **V** and **V'** are constant in time, and that the growth in the magnetic field strengths is exponential, i.e.,

$$\mathbf{H}(t) = \mathbf{H}(t_0) \exp [k(t - t_0)], \quad k > 0,$$
$$\mathbf{H}'(t) = \mathbf{H}'(t_0) \exp [k(t - t_0)],$$

then from Eq. (36) we obtain

$$k\mathbf{H} = \mathrm{curl}[\mathbf{V} \times \mathbf{H}'] + \mathrm{curl}[\mathbf{V}' \times \mathbf{H}],$$
$$k\mathbf{H}' = \mathrm{curl}[\mathbf{V} \times \mathbf{H}], \quad \mathbf{V}(t) = \mathbf{V}(t_0), \quad \mathbf{V}'(t) = \mathbf{V}'(t_0). \tag{37}$$

If the field **H** is force-free then the value of the field given by Eq. (37) is reasonable until the interaction energy density $(\mathbf{H}' \cdot \mathbf{H})/4\pi$ of the magnetic fields **H'** and **H** approaches the kinetic energy density $E_L = \rho V^2/2$ of the plasma and the field begins to have a significant effect on its motion.

In our situation the question is whether Eqs. (37) agree with the conditions†

$$\tilde{H} \gg \tilde{H}', \tag{38}$$
$$\mathrm{curl}\ \mathbf{H} = \alpha \mathbf{H}, \quad \alpha = \mathrm{const} > 0, \tag{39}$$

where for simplicity we have chosen the simplest form of a force-free field and neglected the curvature of the torus. From Eqs. (36) and (39) we obtain equations relating **V** and the velocity **V'**, **H**, **H'** (equal to the characteristic scale $w = k/\alpha$ of the spatial variation in the field H divided by its evolution time 1/k):

$$w\mathbf{H} = \mathbf{V} \times \mathbf{H}' + \mathbf{V}' \times \mathbf{H},$$
$$w\mathbf{H}' = \mathrm{curl}\ [\mathbf{V} \times \mathbf{H}]/\alpha. \tag{40}$$

We shall assume the order of magnitude of the velocity **V** to be more or less known,

$$V \simeq (1 - 5) \cdot 10^4 \quad \mathrm{cm/sec}. \tag{41}$$

The order of magnitude of w and the ratio of the velocities $q \equiv w/\tilde{V}$ is evaluated from the characteristic dimension of the convective cell and the characteristic time for the magnetic field to grow, i.e.,

$$\alpha \simeq 1/r \sim 5 \cdot 10^{-9} \quad \mathrm{cm^{-1}},$$
$$k \simeq 10/T = 3 \cdot 10^{-8} \quad \mathrm{sec^{-1}},$$
$$w = k/\alpha \simeq 6 \quad \mathrm{cm/sec}, \tag{41a}$$
$$q = w/\tilde{V} \simeq (1 - 6) \cdot 10^{-4}.$$

For the other calculations it is convenient to go to a local cylindrical coordinate system $\{s, \varphi, z\}$

† We shall characterize the order of these quantities by their root mean squares calculated over the volume of the part of the cell in question and denote them by a tilde above.

Fig. 14. The local coordinate system.

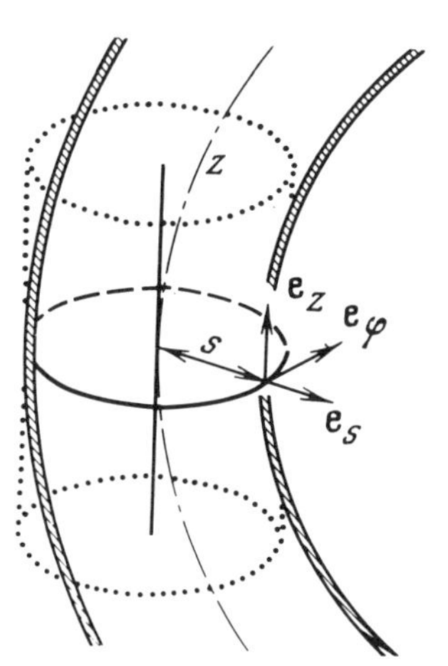

(corresponding to the unit vectors $\mathbf{e}_s$, $\mathbf{e}_\varphi$, and $\mathbf{e}_z$) with the z axis oriented along the tangent to the central circle of the torus (Fig. 14). In these coordinates

$$\begin{aligned} \mathbf{H} &= \mathbf{e}_\varphi H_\varphi + \mathbf{e}_z H_z, \\ \mathbf{V} &= \mathbf{e}_s V_s + \mathbf{e}_\varphi V_\varphi. \end{aligned} \tag{42}$$

The small additions $\mathbf{e}_s H_s$ and $\mathbf{e}_s V_z$ are neglected as an examination shows that they are unimportant for qualitative considerations. From Eqs. (40) and (42) it is easy to obtain

$$\begin{aligned} \alpha w H_s' &= \frac{1}{s}\frac{\partial}{\partial \varphi}(H_\varphi V_s) + \frac{\partial}{\partial z}(H_z V_s), \\ \alpha w H_\varphi' &= -\frac{\partial}{\partial s}(H_\varphi V_s) + \frac{\partial}{\partial z}(H_z V_\varphi), \\ \alpha w H_z' &= -\frac{1}{s}\frac{\partial}{\partial s}(s H_z V_s) - \frac{1}{s}\frac{\partial}{\partial \varphi}(H_z V_\varphi), \\ w H_\varphi &= -H_z' V_s - H_z V_s', \\ w H_z &= H_\varphi' V_s + H_\varphi V_s' - H_s' V_\varphi, \\ H_z' V_\varphi &+ H_z V_\varphi' - H_\varphi V_z' = 0. \end{aligned} \tag{43}$$

It is natural to assume that

$$\tilde{V}_\varphi \simeq \tilde{V}, \qquad \tilde{H}_\varphi \simeq \tilde{H}_z \simeq \tilde{H}, \qquad \tilde{H}_s' \simeq \xi \tilde{H}, \tag{44}$$

where ξ is much less than unity. Then by comparing the orders of magnitude of the quantities and their derivatives in Eq. (43) we obtain

$$\tilde{V}_s \simeq \xi w, \qquad \tilde{V}_s' \simeq w, \qquad \xi \simeq q. \tag{45}$$

Thus,

$$\frac{\widetilde{H^2}}{8\pi} \approx Q\frac{(\widetilde{H'H})}{4\pi} \simeq Q E_L, \tag{46}$$

$$Q = 1/2\xi \simeq 1/2q \simeq (1-5)\cdot 10^3, \tag{47}$$

and the energy density of the force-free field $\widetilde{H}^2/8\pi$ may exceed the energy density E_L of the laminar motions by several orders of magnitude. When the field **H** differs from force-free the factor Q is reduced.

It seems reasonable that estimates of the orders of magnitudes in Eq. (43) still do not constitute a proof that the force-free field is enhanced until the magnetic field energy density greatly exceeds the energy density of the laminar motion. But this assumption is supported by indirect estimates of the energy of the magnetic fields in the cellular belt (see the next section).

Observations of the magnetic fields in sunspots might answer several questions about the structure of the magnetic fields in the belt. In particular, observations of the sign of the loop could help us understand several important questions. Among these we must first include the questions about the validity of the ideas on the layered-periodic, magnetic plasma crystal and on the loop model of the cellular belt. From the unity of the belt it follows that during each activity cycle the sign of the loop must be the same in all bipolar groups in a given hemisphere of the sun and must change in the next cycle. Two possible theoretical variants of the distribution of signs (of the loops over the hemispheres during a single cycle) combined with observational data might clarify the important features of the scheme whereby the magnetic field is enhanced. If the signs are the same, $\sigma^{(N)}_{(n)^2} = \sigma^{(S)}_{(n)^2}$, the change in the direction of the Coriolis force does not change the field direction and the layer of cells (belt) is the same in both hemispheres. Here the observed appearance of active regions only in two nonintersecting bands of latitude ("the royal zones") may be related to the fact that the part of the decaying uppermost layer (belt) near the equator is lost without reaching the photosphere or breaking up into individual cells. In the future we hope to discuss this final stage in the life of the belt in detail, analyzing its possible connection with the solar wind. In a certain sense the second possibility is closer to traditional models of the cyclic activity. In it the sign of the loop in the northern and southern hemispheres is opposite in a given cycle. We shall not analyze this case, which would greatly complicate the structure of the magnetic crystal, at this time (until observations are made).

As already indicated, the magnetic field ensures the dynamic stability of the quasiperiodic convective structure by destroying its static state. The cells are located above one another, with boundaries above boundaries and centers above centers. At the longitudes of the crystal, periodicity is strictly preserved over times including many 11-year cycles because of the naturalness of the overall cylindrical symmetry of the sun's regular structure. Over small times (one to three cycles) the belt usually is not cylindrically symmetric. These fluctuations may explain the existence of active longitudes. It is also natural to expect that the latitudinal dimensions of the cells should vary with latitude. A somewhat more detailed (than described in the previous section) analysis of photoheliographic data might clearly answer this question. The results of such an analysis should also help a great deal in the analysis of the mechanism for field enhancement. The thickness of the layer (the third linear characteristic of the quasiperiodicity of the crystal) is a monotonically increasing function of the height z. As it rises continuously the belt enters ever less dense regions of the convective zone; hence, the cells must always be expanding. Because of the rigidity of the longitudinal-latitudinal frame of the crystal this expansion takes place as a growth in the thickness of the layer. From this it is already clear that the rise of the layers is accelerated. Such motion is ensured by the Archimedean buoyancy of the field distributed over the crystal, which increases with z and after enhancement of the field by the plasma ceases due to an increase in the ratio of the magnetic and gas kinetic pressures. This ratio decreases both due to the drop in pressure with z in the unmagnetized plasma outside the crystal and as a consequence of the partial transformation of the force-free component of the magnetic field into a force component. It is also evidently necessary to include the additional contribution to the buoyant force from the boundary cells of the upper layer (belt) of the crystal. In order to estimate the buoyant force of the belt and the rate of rise, we now turn to observational data and specify the parameters of the belt.

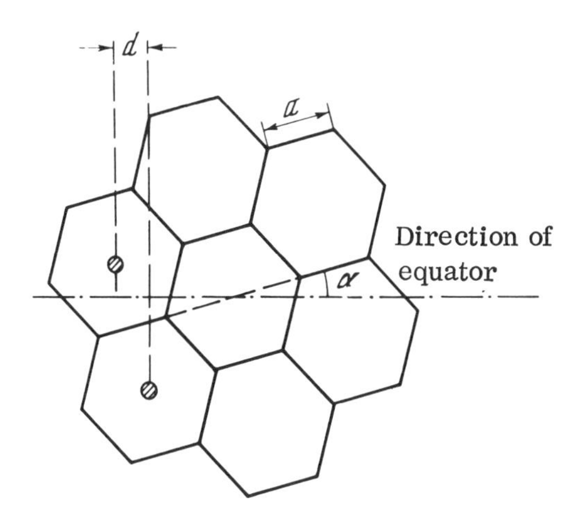

Fig. 15. The orientation of the cells with respect to the direction of the solar equator.

The Emergence of the Belt

We shall evaluate several characteristics of importance for the picture of a cellular belt floating to the top assuming for concreteness that the horizontal cross sections of the convective cells are regular hexagons with sides of length a (Fig. 15). The orientation of the hexagons with respect to the direction of the solar equator may be characterized by the minimum angle α between the sides of the hexagons and the direction of the solar equator (Fig. 15). It is easy to see that for $\alpha = 0$ and $\alpha = 30°$ the latitudinal period of the structure $d \simeq R_{\odot} \sin D_{\phi}$ is equal to

$$d = \sqrt{3}a \sin(30° + \alpha), \tag{48}$$

while the area of a cell is

$$S = 2.6\, a^2 = 0.9 d^2/\sin^2(30° + \alpha).$$

The angular width of the activity zone in each hemisphere of the sun is about 30° and the area of the entire activity zone is roughly a quarter of the sun's area $S_{\odot}$. Thus, the number of cells in the belt is about

$$N_c \simeq S_{act}/S_c \simeq 0.3 \sin^2(30° + \alpha)\frac{S_{\odot}}{d^2}.$$

Evidently the number of cells, N_c, is 1.5-2 times greater than the number N_0 of observed groups of spots created over an 11-year cycle ($N_0 \simeq 3000$) since not all cells (especially near the equator) form groups of spots and not all groups of spots are observed. Thus the condition

$$\sin^2(30° + \alpha) \simeq (1.5 - 2.0)\cdot 10^4\, d^2/S_{\odot} \tag{49}$$

must be satisfied. From observations of the latitudinal periodicity (period about 1.4°) it is found that $d \simeq 1.7 \cdot 10^9$ cm. Thus condition (49) is best satisfied for $\alpha = 30°$. The fact that condition (49), which relates observed quantities (the number of groups of spots created during a cycle, N_0) to the latitudinal period D_{ϕ} (or d), is satisfied is an argument in favor of the concept of a cellular belt. We note that going from regular hexagons to cells which are stretched along the equator would result in an increase in the right-hand side of Eq. (49) so that strongly elongated cells contradict this condition. Setting $\alpha = 30°$, we find from Eq. (48) that

$$a = 10^9 \text{ cm}. \tag{50}$$

Assuming that the radius R_{max} of the largest circle of the magnetic torus is roughly equal to the radius of the circle inscribed in a hexagon, we obtain

$$2R_{max} \simeq \sqrt{3}\,a = 1.7 \cdot 10^9 \text{ cm}. \tag{51}$$

The radius r of the generating circle of the magnetic torus may be assumed equal to $(1/4 - 1/8)a$, that is,

$$r \simeq (2.5 - 1.2) \cdot 10^8 \text{ cm}. \tag{52}$$

Comparing $2R_{max}$ with the average extent of a large group of spots ($1.2 \cdot 10^{10}$ cm, or about 10°) and r with the radius of a typical spot [18], we conclude that a volume element of a magnetized cell is magnified by roughly 100-350 times as it rises from the belt to the region where sunspots are observed and as the magnetic field strength falls by 10-50 times. If the matter in the magnetized torus is conserved throughout the process of emersion, then the ratio of the densities at these heights is also 100-350. If matter flows out of the torus region at the level of sunspot observation this ratio must be increased by a few times. The density of matter in a spot is not known exactly. Assuming that it differs little from the density in Mattig's model, $\rho_s \simeq 7 \cdot 10^{-7}$ g/cm³, we find that $\rho_0 \simeq 10^{-4}$ g/cm³ at the level of the decaying belt layer, and taking all the uncertainties into account we shall write $\rho_0 \simeq (10^{-4}-10^{-3})$ g/cm³. It is not yet possible to determine the depth z_0 of the decaying belt from this density since at the depths z corresponding to such values of ρ_0 the uncertainty in $\rho_0(z)$ is still large and the results obtained with various models of the solar atmosphere differ from one another by almost an order of magnitude. In any case we may assume that estimating the depth at which the belt disintegrates by the range $z_0 \simeq (3-8) \cdot 10^9$ cm does not contradict modern ideas about the structure of the convective zone. At these depths $T = (1-4) \cdot 10^5$°K and $p_g \simeq (10^9-10^{10})$ dyn/cm².

For conformity among the gas-dynamic quantities in the following calculations we shall take

$$\rho_0 = 10^{-3} \text{ g/cm}^3, \quad T = 10^{5}\text{°K}, \quad p_g = 10^{10} \text{ dyn/cm}^2. \tag{53}$$

Observations of the magnetic fields in the centers of sunspots yield the maximum value $H_{max} \sim 4 \cdot 10^3$ G [36]. Thus, at the level of the disintegrating layer the field reaches

$$H_0 = (4 \cdot 10^4 - 2 \cdot 10^5) \text{ G}. \tag{54}$$

Since the mean square of the magnetic fields at the centers of the spots is a fifth of the maximum value ($l_1 = 0.2$) we find the energy density at this level to be

$$E_M = l_1 (H_0^2/8\pi) \simeq (10^7 - 3 \cdot 10^8) \text{ ergs/cm}^3. \tag{55}$$

This estimate shows that the magnetic energy density is at least one and a half orders of magnitude less than the thermal energy density (roughly equal to p_g) at the same height. Taking the average velocity of the laminar motion to be $V_0 = 0.5$ km/sec and the density of the layer where the field is mostly produced to be $3\rho_0$ we obtain the energy density of the laminar motion

$$E_L = 3\rho_0 V_0^2/2 = 4 \cdot 10^6 \text{ ergs/cm}^3. \tag{56}$$

The resulting ratio of the energy density of the laminar motion to that of the magnetic field,

$$E_L/E_m \simeq 0.4 - 0.01, \tag{57}$$

favors the above arguments that the magnetic field generated in the cells is close to force-free.

In [37] we considered a magnetic field which is the sum of a cylindrically symmetric force-free field $\mathbf{H}^{(f)}$ and a field $\mathbf{H}^{(p)}$ parallel to the axis of symmetry which is also axially symmetric, i.e.,

$$\mathbf{H}^{(s)} = \mathbf{H}^{(f)} + \mathbf{H}^{(p)}. \tag{58}$$

As a measure of the deviation of $\mathbf{H}^{(s)}$ from the force-free field we introduced the parameter

$$\beta = \frac{[H^{(s)}(0)]^2 - [H^{(f)}(0)]^2}{[H^{(s)}(0)]^2},$$

where the values of the fields used in this equation are taken along the symmetry axis.

As will be shown in the next section [Eqs. (74)-(75)] the field $\mathbf{H}^{(s)}$ acts on the plasma $1/\beta$ times more weakly than if it were parallel but with the same strength. Assuming that the energy density of the laminar motion, E_L, and the energy density of the magnetic field interaction with this motion are roughly equal we obtain [cf. Eq. (57)].

$$E_L \simeq \beta\, [H^{(s)}]^2/8\pi,$$
$$\beta = 0.4 - 0.01, \tag{59}$$

and the deviation of the field $\mathbf{H}^{(s)}$ from force-free is of order

$$\bar{H}^{(p)} \simeq \frac{\beta}{2} \bar{H}^{(s)} \simeq (0.2 - 0.005)\, \bar{H}^{(s)}.$$

Thus the field has considerable magnetic buoyancy.

When the conditions for freezing-in are satisfied (i.e., the time for a layer of cells in a 10-year crystal to evolve is about 100 years while the damping time of the field due to finite conductivity is at least several thousand years) the magnetized belt may be regarded as a cylindrical barrier which is impermeable to the outside plasma (see also [38]). Due to magnetic buoyancy this cylinder increases in radius r_{cyl} (as the belt floats up) while pushing ahead and driving behind itself the field-free plasma. The rate of rise of the belt is automatically established so that the plasma is able to flow around it (Fig. 16). The effect of the stretching of the

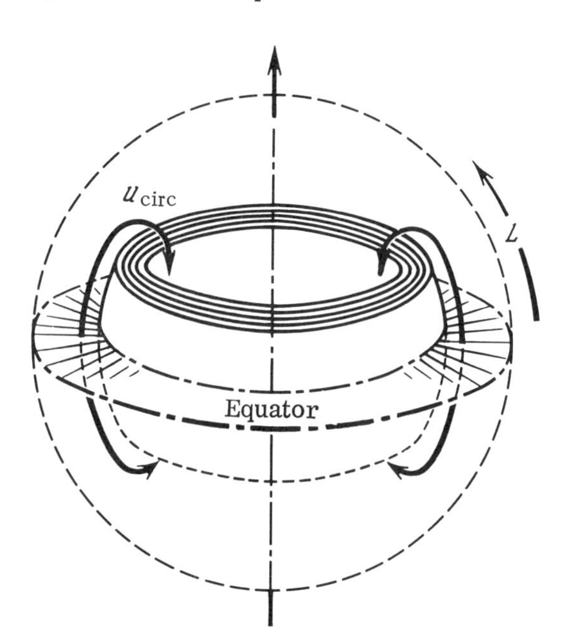

Fig. 16. The flow of plasma around the belt.

belt on its rise may be neglected (as the magnetic coupling forces between the cells are much less than the magnetic buoyancy). The velocity $\mathbf{V}_{circ}$ with which the plasma flows around the belt is determined by the Euler equation

$$\rho \frac{d\mathbf{V}_{circ}}{dt} = -\operatorname{grad} p + g_{\odot}\rho + \eta_{tur}\Delta\mathbf{V}_{circ},$$

where $g_{\odot}$ is the acceleration of gravity. For the horizontal component of $\mathbf{V}_{circ}$ in steady-state motion we have the approximation

$$P_{+} \simeq L\eta_{tur}\tilde{V}_{circ}/h_0^2. \tag{60}$$

Here L is the characteristic length of the flow around the belt, $L \simeq R_{\odot}$; P_{+} is the pressure drop along this length; and h_0 is the height of the uniform atmosphere given by

$$h_0^{-1} = -\frac{d \ln p}{d \ln z} = \frac{g_{\odot}\rho_0}{p_g}. \tag{61}$$

Here it was assumed that $\widetilde{\Delta V_{circ}} \simeq \tilde{V}_{circ}/h_0^2$. Setting $r_{cyl} \equiv R_{\odot}$ and $L = R_{\odot}$, from Eq. (60) we find the mass flow rate of the unmagnetized plasma in the form

$$\frac{dM}{dt} = 2\pi r_{cyl} h_0\rho_0\tilde{V}_{circ} = 2\pi \frac{\rho_0 h_0^3 P_{+}}{\eta_{tur}}.$$

The time for the belt to rise by one layer, that is, the duration t_{cyc} of a cycle, is

$$t_{cyc} = h_c S_{ac}\rho_0 \Big/ \frac{dM}{dt},$$

where, as before, $S_{act} \simeq \frac{1}{4} S_{\odot}$ is the area of the active zone and h_c is the thickness of the decaying layer.

Substituting the last but one equation into the previous equation, we obtain

$$t_{cyc} = \frac{\eta_{tur}}{P_{+}} \frac{h_c}{h_0^3} \frac{S_{\odot}}{8\pi} = \frac{\eta_{tur}}{P_{+}} \frac{h_c R_{\odot}^2}{2h_0^3}. \tag{62}$$

The pressure drop P_{+} is due to the presence of the buoyant force F_b in each of the cells and is given by

$$P_{+} \simeq F_b/S_c, \tag{63}$$

where S_c is the horizontal cross-sectional area of a cell.

The difference between the external and internal gas-kinetic pressures for a field of the form (58) (also see the following section) is

$$\Delta p = \beta l_1 H_0^2/8\pi,$$

where β is a parameter which determines how much the field differs from force-free and the factor $l_1 \simeq 0.2$ takes into account the fact that the mean square of the magnetic field is five times less than the maximum of H_0^2. The buoyant force F_c of the cell is given by

$$F_b = (\Delta p/p_g) M_T g_{\odot} = \beta l_1 g_{\odot} \frac{H_0^2}{8\pi} M_T/p_g, \tag{64}$$

where M_T is the effective mass of the magnetized region of the toroidal convective cell,

$$M_T = l_2 M_C = l_2 \rho_0 h_C S_C. \tag{65}$$

From Eqs. (61), (63), (64), and (65) we obtain

$$P_+ = l_1 l_2 \beta \frac{H_0^2}{8\pi} h_C \frac{g_\odot}{p_g} \rho_0 = l_1 l_2 \beta \frac{H_0^2}{8\pi} \frac{h_C}{h_0}.$$

Substituting this expression for P_+ in Eq. (62), we find

$$t_{cyc} = \frac{\eta_{tur} R_\odot^2}{2 l_1 l_2 \beta \frac{H_0^2}{8\pi} h_0^2} = \frac{\nu_{tur} R_\odot^2}{2 l_1 l_2 h_0^3 g_\odot} \frac{p_g}{E_L}. \tag{66}$$

Before using this formula we note that the emersion time is mainly determined by two quantities, the viscosity and the energy of the laminar flow. To evaluate t_{cyc}, the duration of the cycle, we have only to substitute several parameters of the convective zone of the sun in Eq. (66). Recalling that $g_\odot = 2.7 \cdot 10^4$ cm/sec (the difference in the acceleration of gravity at the depth where the belt disintegrates and at the surface of the sun does not exceed 10%), we obtain $h_0 \simeq 4 \times 10^8$ cm from Eq. (61). We note that the closeness of h_0 and r [Eq. (52)] is not accidental and that a large difference between them (especially with $r \gg h_0$) would cast doubt on these discussions. From Eqs. (56) and (53) we also have (to within an order of magnitude) $p_g/E_L = 3 \cdot 10^3$. The volume of the magnetized torus is $V_T \simeq 19.7(R_{max} - r)/r^2$ and the volume of a cell is $V_c \simeq 2rS_c = 5.2a^2r$. Thus

$$l_2 = V_T/V_C \simeq 3.8 \ (R_{max} - r/a)^2,$$

and from Eqs. (50)-(52) we have

$$l_2 = \begin{cases} 0.15 & \text{for } r = a/4, \\ 0.09 & \text{for } r = a/8. \end{cases} \tag{67}$$

Assuming $\nu_{tur} \simeq 0.1\bar{V}_0 h_0 \simeq 4 \cdot 10^{12}$ cm^2/sec and substituting the above quantities in Eq. (66), we find $t_{cyc} = 25$ yr. The accuracy of this estimate is mainly determined by the accuracy of p_g/E_L and ν_{tur} and is one or two orders of magnitude. Thus, the observed period of the cyclic activity, T ~ 11 yr, must be regarded as being in good agreement with the model discussed here.

We now estimate the energy losses from the magnetic belt as it rises. (This energy is expended in setting the plasma above the belt into motion and ultimately ends up as heat.) From Eqs. (50), (51), (55), and (67) we find that the magnetic energy of a cell is

$$\mathscr{E}_C = E_M l_2 V_C = \begin{cases} 2 \cdot 10^{33} \text{ ergs for } r = a/4, \\ 1.7 \cdot 10^{34} \text{ ergs for } r = a/8. \end{cases}$$

Assuming that the number of cells in a belt is $N_c = 10^4$, we find the average rate of change of the energy to be

$$\frac{d\mathscr{E}}{dt} = \frac{N_C \mathscr{E}_C}{T} = (7 \cdot 10^{28} - 5 \cdot 10^{29}) \text{ ergs/sec}. \tag{68}$$

which is 10^{-5}-10^{-4} of the total solar radiation. If we assume that the energy which is converted to heat hardly propagates latitudinally beyond the active zone and that the losses depend on the phase of the cycle, we find that the maximum changes in the solar constant associated with the motion of a magnetized belt may reach 0.01–0.1%. It is possible that such variations in the solar constant have already been observed [39].

The Formation of Sunspots

As is clear from the preceding discussion, the magnetic field of a convective cell in the region considered in this model has a complicated configuration and clearly is not the purely toroidal field of the cell. It is also important that this field cannot be purely force-free since such a field could not of itself lead to buoyancy. It is possible to reach a similar conclusion from direct observations of the magnetic field structure of the active region in the photosphere [36]. In [37] the field of the convective cell (or more precisely, of a small part of the magnetizing current inside the cell) was analyzed in the form of a sum of the two simplest cylindrically symmetric fields,

$$\mathbf{H}^{(s)} = \mathbf{H}^{(f)} + \mathbf{H}^{(p)},$$

where $\mathbf{H}^{(p)}$ is directed along the axis of the cylinder and $\mathbf{H}^{(f)}$ is a force-free field. A cylindrically symmetric field $\mathbf{H}^{(s)}$ that was uniform along the symmetry (z) axis and dropped off rapidly at a sufficient distance s from the symmetry axis was considered (i.e., $H^{(s)}(s) \simeq 0$ for $s \geq S_0$). It can always be written in the form

$$\mathbf{H}^{(s)}(s) = [a(s) + b(s)]\,\mathbf{e}_z + c(s)\,\mathbf{e}_\varphi, \tag{69}$$

where

$$\begin{gathered} b^2(s) = 2\int_s^{S_0} \left\{\frac{c(\xi)}{\xi}\frac{d}{d\xi}[\xi c(\xi)]\right\} d\xi, \\ c(0) = 0, \quad u(\infty) = b(\infty) = c(\infty) = 0. \end{gathered} \tag{70}$$

Here s, φ, and z are the cylindrical coordinates and $\mathbf{e}_s$, $\mathbf{e}_\varphi$, and $\mathbf{e}_z$ are their unit vectors. It is easy to see that field $b(s)\,\mathbf{e}_z + c(z)\,\mathbf{e}_\varphi$ is force-free; that is,

$$\begin{gathered} \mathbf{H}^{(s)}(s) = \mathbf{H}^{(f)}(s) + \mathbf{H}^{(p)}(s); \quad \mathbf{H}^{(f)}(s) = b(s)\,\mathbf{e}_z + c(s)\,\mathbf{e}_\varphi; \\ \mathbf{H}^{(p)}(s) = a(s)\mathbf{e}_z. \end{gathered} \tag{71}$$

The force determined by the field $\mathbf{H}^{(s)}(s)$,

$$\mathbf{f} = -\frac{1}{4\pi}\mathbf{H} \times \operatorname{curl}\mathbf{H},$$

is a potential force since it obeys the equation

$$\mathbf{f}(s) = -\frac{1}{4\pi}\left\{\frac{1}{2}\frac{d}{ds}[a(s) + b(s)]^2 + \frac{c(s)}{s}\frac{d}{ds}[sc(s)]\right\}\mathbf{e}_s = -\operatorname{grad}\psi(s), \tag{72}$$

where

$$\psi(s) = \frac{1}{8\pi}\left([a(s) + b(s)]^2 - 2\int_s^{S_0}\left\{\frac{c(\xi)}{\xi}\frac{d}{d\xi}[\xi c(\xi)]\right\}d\xi\right) = \frac{1}{8\pi}\left\{[a(s) + b(s)]^2 - [b(s)]^2\right\}. \tag{73}$$

(The indeterminate additive constant in the potential $\psi(s)$ has been chosen here so that $\psi(\infty) = 0$.) From Eqs. (72) and (73) it follows that the difference between the internal p and external

p^* pressures is

$$p - p^*(s) = \psi(s) = 1/8\pi\{[H_z^{(s)}(s)]^2 - [\bar{H}_z^{(f)}(s)]^2\}.$$

If the symmetry axis z is oriented perpendicular to the force of gravity and the temperature is the same inside and outside the cylinder, then for buoyancy of the magnetized region it is necessary that

$$\int_0^{s_0} [a(\xi) + b(\xi)]^2 d\xi > \int_0^{s_0} [b(\xi)]^2 d\xi,$$

or

$$\int_0^{s_0} [H_z^{(s)}(\xi)]^2 d\xi > \int_0^{s_0} H_z^{(f)}(\xi)]^2 d\xi.$$

Limiting ourselves only to configurations of $\mathbf{H}^{(s)}$ for which $a(0) \cdot b(0) > 0$, we find the maximum gas-kinetic pressure drop inside the cylinder to be

$$\Delta p = \beta p_m(0); \qquad \beta = 1 - \left[\frac{b(0)}{a(0) + b(0)}\right]^2, \qquad P_m(0) = \frac{(\mathbf{H}^{(s)}(0))^2}{8\pi}. \tag{74}$$

As can easily be seen the positive parameter β characterizes the degree of deviation of the field $\mathbf{H}^{(s)}$ from force-free,

$$\beta = \{[H^{(s)}(0)]^2 - [H^{(f)}(0)]^2\} / [H^{(s)}(0)]^2. \tag{75}$$

We now consider the evolution of the field during adiabatic expansion of this magnetized cylinder (cavity), assuming that the high conductivity of the plasma ensures practically complete freezing-in of the field. Let the cavity expand uniformly along the z axis and in the radial directions as

$$\frac{d\xi_z}{\xi_z} = A_z(t)\,dt, \qquad \frac{d\xi_s}{\xi_s} = A_s(t)\,dt, \tag{76}$$

while conserving the symmetry type of the field $\mathbf{H}^{(s)}(s)$. Here ξ_α is the corresponding element of length. We use the subscripts 0 and 1 to denote the initial and final states of the cavity as it expands adiabatically over a small time interval during which the ratio $K \equiv A_z(t)/A_s(t)$ may be assumed constant. Let the radial dimension s_0 of the cavity increase by q times ($s_1 = qs_0$) due to expansion. Then, according to Eq. (76), its scale along the z axis increases by q^k times and the plasma density $\rho^*(s)$ within the cylinder falls by q^{2+k} times, i.e.,

$$\rho_1^*(s) = q^{-(2+k)} \rho_0^*(s). \tag{77}$$

Then the z component of the total field $\mathbf{H}^{(s)}(s)$ in the cavity decreases by q^2 times and the φ component c(s) by q^{k+1} times. As can be seen from Eq. (70) the z component b(s) of the force-free field $\mathbf{H}^{(f)}$ varies in time in proportion to its φ component c(s). This means that $b_1(s) = b_0(s)q^{-(1+k)}$. The magnetic pressure on the symmetry axis, determined (since $c_1(0) = c_0(0) = 0$) by the z component of the field $\mathbf{H}^{(s)}$, decreases by q^4 times. From the definition (75) of β we find

$$1 - \beta_1 = q^{2(1-k)}(1 - \beta_0). \tag{78}$$

This formula describes qualitatively the variation in the ratio of the force-free and "parallel" components of the field $\mathbf{H}^{(s)}(0)$ on the axis of the cavity during expansion. Its qualitative significance is clear: If the cavity expands more along the z axis than along its radius (K > 1) then the deviation of the field $\mathbf{H}^{(s)}$ from force-free becomes greater. This situation is opposite to free expansion of the field when Joule losses are negligible. From Eq. (78) there follows, at last, an expression for the maximum pressure drop,

$$\Delta p_1 = q^{-4} [1 - q^{2(1-k)} (1 - \beta_0)] \Delta p_0. \tag{79}$$

It is known [40] that expansion (compression) of the plasma perpendicular to the direction ofthe "parallel" field frozen into it is described by the usual solutions of gas-dynamics without a magnetic field with the equation of state $p_m = \Delta p \propto \rho^{\gamma_m}$, in which $\gamma_m = 2$. If, however, a plasma that is magnetized by a field such as $\mathbf{H}^{(p)}$ expands uniformly and isotropically, then $\gamma_m = 4/3$. A quantity $\gamma_m^{(0)}$ analogous to the adiabatic index may be introduced in the present, more complicated case [with the field $\mathbf{H} = \mathbf{H}^{(s)}(s)$] by defining

$$\gamma_m^{(0)} = \frac{d \ln \Delta p}{d \ln \rho^* (0)} . \tag{80}$$

From Eqs. (77) and (79) we obtain

$$\gamma_m^{(0)} = \frac{2}{k+2} [2 + (1-\beta)(1-k)]. \tag{81}$$

Therefore, the nonuniform expansion of the magnetic field leads to a power-law dependence $\Delta p \propto \rho^{\gamma_m^{(0)}}$ with an exponent $\gamma_m^{(0)}$, which may be less than unity (as opposed to the ordinary adiabatic index γ). Equations (80) and (81) make it possible to close the system of gas-dynamic equations for the state of the cavity and to explain how its temperature changes during deformation of the magnetic fields.

The reason for cooling of the active regions and formation of sunspots is still undetermined despite the seeming validity of the explanation for cooling of magnetic tubes in the suppressed convection models of Babcock and Parker [35, 41-43]. In the case of our model, where an isolated active region is associated with the emersion of a magnetized cell into the photosphere, it is paradoxical that the sunspots (which are volumes of comparatively low-temperature gas) belong to convective cells which have brought up higher energy from below. Considering only the spatially smoothed parameters, we shall show that this is explained by expansion of the emerging cells or deformed magnetized tori.

Withoug analyzing the effect of a magnetic field, a number of authors, beginning even in the 1920's, offered a similar explanation (reviewed in [11]), but it was abandoned later due to contradictions. First, the cooled cell should have sunk at once and, second, it was not known why it did not get heated by the downward heat flux [44]. The role of the magnetic field in this problem is fundamental, as it causes a magnetized isolated cavity to develop in the plasma. Creating an additional pressure, the field expands the cavity and causes it to rise despite substantial cooling. The field also suppresses convection, weakening the flow of heat into the cavity from the surrounding plasma. The convection suppression mechanism has already repeatedly been drawn on to explain the cooling of the active regions [35, 41-43]. Alone it is not sufficiently effective in the case of an isolated convective cell. We shall therefore consider adiabatic expansion of the corresponding cavity.

The partially ionized gas in the "quiet regions" of the convective zone is heated from below as it moves in the downward (opposite to z) homogeneous gravitational field. We shall write the z variation of its temperature T and gas-kinetic pressure p in the form

$$\frac{d \ln T(z)}{d \ln p(z)} = [1 + \varepsilon(z)] \frac{\gamma(z) - 1}{\gamma(z)} \qquad \left(1 < \gamma \leqslant \frac{5}{3}\right), \tag{82}$$

where $\varepsilon(z) \geq 0$ characterizes the deviation from adiabaticity and $\gamma(z)$ is the effective adiabatic index, defined by the formal equality [35]

$$\frac{\gamma-1}{\gamma} \equiv \left(\frac{d \ln T}{d \ln p}\right)_s . \tag{83}$$

We shall neglect the differences in the degree of ionization and thus in the adiabatic index inside and outside the convective cell.

The time τ_{act} for evolution of the active region as a rule is $\tau_{act} \geq 2$ days, its size $r_{act} \leq 10^4$ km, and its sound speed $c \sim 8$ km/sec; that is, $\tau_{act} \gg r_{act}/c$ in the photosphere. We shall assume that the total pressure inside the cell is able to equilibrate with the external pressure $p(z)$ in the depths adjacent to the photosphere as well. The characteristic time τ_R for equilibration of the temperature inside a cell of radius r with the surrounding medium is roughly $\tau_R \simeq r^2/K_R$, where $K_R \sim (10^{10}\text{-}10^{11})$ cm^2/sec is the coefficient of radiant thermal diffusivity. Thus, for $r \sim 5 \cdot 10^8$ cm, $\tau_R \sim 1$ month. The time for a cell to emerge is of order τ_{act}, so the rise of extended vertical portions (Q) of the emerging torus near the photosphere (Fig. 17) is almost adiabatic. Inside Q the density $\rho^*(z)$, temperature $T^*(z)$, and gas-kinetic pressure $p^*(z)$ obey the equations

$$\frac{d \ln T^*(z)}{d \ln p^*(z)} = \frac{\gamma(z)-1}{\gamma(z)}, \qquad \frac{d \ln [\rho^*(z)/\mu(z)]}{d \ln p^*(z)} = \frac{1}{\gamma(z)}, \tag{84}$$

which have been written for an ideal gas with an effective molecular weight $\mu^*(z) \simeq \mu(z)$ which varies along z. Assuming, as before, that the magnetic field is the sum of a "parallel" $\mathbf{H}^{(p)} \parallel 0z$ and force-free $\mathbf{H}^{(f)}$ field where the contribution of the force-free field (as shown above) is large ($\beta \ll 1$), we write, by analogy with Eqs. (80) and (81) and including the variation of $\mu = \mu(z)$,

$$\frac{d \ln [p(z) - p^*(z)]}{d \ln [\rho^*(z)/\mu(z)]} = \gamma_m(z), \tag{85}$$

where

$$\gamma_m = \gamma_m^{(0)} \Big/ \left(1 - \frac{d \ln \mu}{d \ln \rho^*}\right) \simeq \frac{6-2K}{2+K}\left(1 - \frac{d \ln \mu}{d \ln \rho^*}\right)^{-1}.$$

We note that under these conditions in the subphotosphere depths [45]

$$|d \ln \mu / d \ln \rho^*| \approx 0.1$$

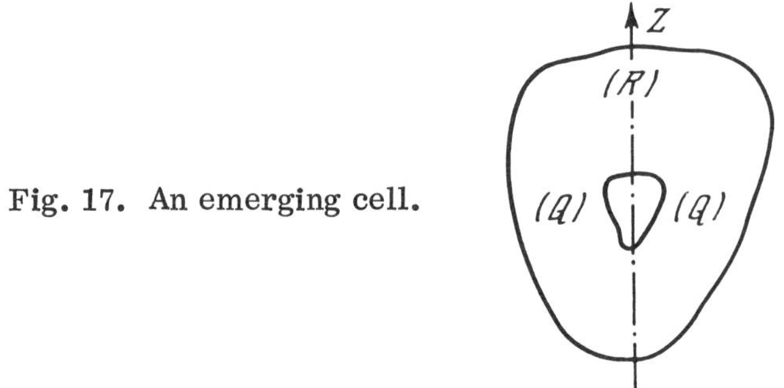

Fig. 17. An emerging cell.

and

$$\gamma_m \simeq 0.9 \frac{6-2K}{2+K}. \tag{86}$$

The buoyancy of an isolated magnetized cavity increases as it rises in the field-free "quiet" plasma since, as z increases, the effective magnetic pressure $p_m \simeq (H_z^{(f)} H^{(p)})/4\pi$ constitutes an ever larger fraction of the total pressure inside the cavity. Thus the upper part (R) of the edgewise emerging magnetized torus, since it has the bulk of the buoyant force, pulls the entire torus behind itself. Thus, both vertical parts (Q) of the torus expand primarily along the z direction and are characterized by the parameter $K_{(Q)} > 1$. Introducing the dimensionless quantities

$$U(z) = \frac{p(z)}{p^*(z)} - 1, \quad \Theta(z) \equiv \frac{T^*(z)}{T(z)}$$

and eliminating $p^*(z)$, $\rho^*(z)$, $T^*(z)$, and $T(z)$ from Eqs. (82-(85), we obtain

$$\frac{d \ln U}{d \ln p} + \frac{\gamma_m - \gamma}{\gamma} \frac{d \ln (1+U)}{d \ln p} = \frac{\gamma_m - \gamma}{\gamma}, \tag{87}$$

$$\frac{d \ln \Theta}{d \ln p} = -\frac{\gamma - 1}{\gamma}\left[\frac{d \ln (1+U)}{d \ln p} + \varepsilon\right]. \tag{88}$$

In Eqs. (87) and (88) it is convenient to change the argument as well, going from the linear coordinate z (subphotosphere depth) to the dimensionless reading $\xi = -\ln (p/p_0)$ of the "altitude in the free atmosphere."

Roughly estimating the cooling of the parts Q, we break up the interval of the ascent by the cell into two parts. We assume for simplicity that beginning from the time the cell breaks off from the comparatively slowly rising belt until the time it reaches some subphotosphere depth $\xi = \xi_0$ the magnetized cell is able to actively exchange heat with the surrounding field-free gas, i.e.,

$$T^*(\xi) = T(\xi) \quad \text{for} \quad \xi \leqslant \xi_0.$$

At the depth cutoff

$$\xi_0 < \xi < \xi_1, \quad \xi_1 \simeq \xi_{ph}$$

(where ξ_{ph} is the dimensionless coordinate of the level from which the sunspot emission is detected) the expansion and rise of the cell take place adiabatically. The variation in the magnetic adiabatic index $\gamma_m(\xi)$ with depth is determined by the rather complex structure of the magnetized deformed toroidal cell and the conditions under which it emerges. Although we cannot yet soundly analyze the nature of this dependence, we shall assume (starting from these considerations about the distribution of the buoyant force over the torus) that the ratio $K_{(Q)}$ of the axial and radial expansion coefficients of the parts Q is much greater than unity over the interval $\xi_0 < \xi < \xi_1$. Here it is sufficient to assume that $k_{(Q)} \geq 1.2$ and therefore [cf. Eq. (86)] $\gamma_m(\xi) \leq 1$. In the two limiting cases of weak and strong initial magnetic pressure $U_0 \equiv U(\xi_0)$, the solution of Eqs. (87) and (88) may assume the following comparatively simple form

$$U(\xi) = U_0 \exp\left[\int_{\xi_0}^{\xi} l(\eta)\, d\eta\right], \tag{89}$$

$$\Theta(\xi) = \exp\left\{-\int_{\xi_0}^{\xi} [l(\eta) - \varepsilon(\eta)] \frac{\gamma(\eta) - 1}{\gamma(\eta)}\, d\eta\right\}, \tag{90}$$

where

$$l(\eta)=\begin{cases}1-\gamma_m/\gamma & \text{for} \quad U_0 \ll 1,\\ \gamma/\gamma_m-1 & \text{for} \quad U_0 \gg 1.\end{cases}$$

In the case $U_0 \ll 1$, when the effective magnetic pressure at the beginning of the adiabatic rise of the cell is relatively small ($\Delta p \ll p^*$), we find, starting from detailed tables [45] of one of the typical "solar models" and assuming $p_0 = 4 \cdot 10^{10}$ dyn/cm^2, $p_1 = 1.2 \cdot 10^5$ dyn/cm^2, and $\gamma_m(\xi) = 1$, that

$$U_1 \simeq 10 U_0, \quad \Theta_1 \simeq 0.7. \tag{91}$$

Therefore, for the very moderate initial parameters of the convective cell used in these estimates, the temperature of the parts Q is lower than that of the surrounding gas by 30%, in accordance with the observed spatially smoothed temperature difference between a sunspot and the quiet photosphere. The need to choose large values of U_0 and K would lead to the conclusion that the subphotosphere parts of the active region cool much more rapidly than they actually do.

Conclusion

We would like once more to discuss briefly the general characteristics of this model of the processes responsible for cyclic activity, to show to what degree the conclusions drawn from it agree with observations, and to consider a plan for further work on the verification of the model, its refinement, and elaboration.

First we shall list the basic features of the model we have obtained in the course of analyzing a broad circle of observational data and determine the model's place among other models.

1. As in all modern physical schemes for cyclic activity which are at all justified, a source inside the sun is considered. This approach is natural for a number of reasons. We begin with the fact that within the solar system there is no stable energy source outside the sun which could long sustain the energy output generated by the cyclic activity (at least, it is unknown to modern physics and astrophysics). The motions of the planets (over the last century) are in such good agreement with theory that any changes in the energy of the planetary system permitted within the measurement accuracy are much less than required to maintain the cyclic activity. Thus, there is no place for any sort of hypothetical extrasolar mechanism for the activity in the solar system. The strong fluctuations in the solar activity also make the assumption of a triggering mechanism in the planetary system unrealistic. Even if we leave aside the question of energetics, from the standpoint of magnetohydrodynamics there would remain the riddle of how the magnetic fields of the active formations were produced by internal forces.

Here it is relevant to note that the occasional references to "statistical proofs" of a coupling between the motions of the planets and the level of cyclic activity are a consequence of incorrect statement of the statistical problems. (The authors of such proofs "forget" to monitor the fulfillment of the requirements imposed on the initial data by the statistical methods being used.)

2. In this model, as in practically all models of cyclic activity generation, the source of the magnetic fields lies in the subphotosphere convective zone or immediately below it. Alternative hypotheses, in which the magnetic fields are initially generated in unstable regions of the sun's core and perturbations propagate through the stable zone of the sun and the sub-

photosphere convective zone, finally entering the photosphere, now seem artificial. (The critique of the specific models of Alfvén and Walen is widely known.)

3. The basis of any model is the enhancement of magnetic fields by a magnetic dynamo. However, this mechanism is fundamentally different from the solar and earth dynamo considered by Parker [46]. The magnetic field is enhanced in an ordered system of convective cells. The field amplification within each cell is monotonic, similar to the enhancement (or maintenance) of the earth's field, rather than quasiperiodic as in Parker's solar dynamo. However, a cell cannot be considered in isolation from the other cells of the belt and, therefore, the nature of amplification differs in some sense from Parker's earth dynamo. The change in the sign of the fields is due to a transition from one layer of cells to another. The regular change in the sign of the field on going from one layer to the succeeding layer is ensured by a change in the direction of the motions inside the convective cells of neighboring layers due to friction.

4. As opposed to the model of Babcock, Leighton, and Parker, we have examined a depth model (with a sufficiently thick crystal of cells taking up about 10 layers of cells), and the conditions for creation and enhancement of the fields in the cells differ greatly from the conditions under which the cells leave the belt. Because of the large depth at which the cells are formed (a feature our model has in common with that of Piddington) the generation region assumes a spatial stability ("boil-off" of cells takes place at the top of the crystal and they "freeze-on" at the bottom). As Piddington has already pointed out [14], there is no such stability in the models of Babcock, Leighton, and Parker. There the entire system of magnetic fields must sooner or later float to the top and leave the sun.

5. The "overall field" observed on the sun's surface is not the basis for the mechanism of cyclic activity (it is not necessary for the model). This field is only the fluctuating residue of the magnetic fields determined by the activity regions and is thus only a consequence rather than a cause. Reliance on the overall field to explain the change of sign of the fields of the 11-year cycles is one of the weakest points in the models by Babcock, Leighton, and Parker. Thus, in Babcock's model there is an obvious (also noted by Babcock himself in [13]) time instability in the mechanism for generating the fields. (According to that model 99% of the generated fields leave the sun while 1% change sign and are amplified in the next cycle. If, let us say, 99.5% of the fields were to leave the sun due to fluctuations (instead of 99%) then field generation would be cut off after three or four cycles.) In general, a scheme involving downward penetration of the overall field to the site where fields are created for the next cycle seems very unstable.

6. In this model the activity elements are separated in the depths close to the photosphere. The magnetic field configuration consists of isolated toroids near the photosphere, rather than bundles of magnetic force tubes as with Parker and Babcock. To some extent this property brings it close to Alfvén's model (with the fundamental difference that with Alfvén the activity elements are separated at much greater depths and the elements are not emerging magnetized plasma cavities but vortical perturbations of the overall radial magnetic field of the sun) and leads to significant differences in the distribution of active regions, the structure of the magnetic fields, and the physics of processes in this model from the models of Babcock, Leighton, and Parker.

7. As opposed to the models of Babcock, Leighton, and Parker, the magnetic field strength at the depth of the convective zone (at the level of the decaying belt in our model) is much greater than the observed field strengths in sunspots.

8. The replacement of magnetic force tubes by rapidly rising, isolated, magnetized toroidal regions allows us to propose a very effective cooling mechanism in this case: adiabatic expansion of the magnetized cavity. In models which rely on slowly changing bundles (or yarns) of force tubes, cooling is generally taken to be due to suppression of convection.

But here we are fully allied with H. Zirin and consider it appropriate to quote him ([47], p. 388): to quote him [47]: "Unfortunately, we still cannot explain in detail why sunspots are cool, even with Biermann's model. We know that in the solar atomosphere magnetic fields are, in general, connected with higher temperatures, except in the sunspots. If the field suppresses the flow of energy from the interior of the sun to the surface at the sunspot, why does it not suppress the flow of energy to the corona above the spot? This is a question we are still a long way from answering."

9. This model involves stability of the phase of activity over long times, independently of the level of fluctuations in the activity. The stability of the activity amplitude (more precisely, the total Wolf number over a cycle) is ensured by the roughly equal number of cells in each layer of the plasma crystal. Models with feedback due to the overall magnetic field of the sun do not ensure stability of the cyclic activity either in amplitude or in phase.

10. One of the features of this model which distinguishes it from all other systematically examined models of cyclic activity is the spatial periodicity of the creation and amplification sources for the magnetic field due to the proposed stable multilayer convection.

We shall now discuss the agreement of the model with observations. The model was developed after an analysis of the mean-annual Wolf numbers by the inverse-problem method of the theory of oscillations led to the conclusion that the record of these numbers could be regarded as the signal of a relaxation oscillator similar to a neon lamp oscillator. This both confirmed and defined Waldmeier's eruptive hypothesis. The relaxation-activity oscillator could most easily be understood by stating that the cyclic variation in the activity is determined by the buildup of some sort of "seeds," by their rapid conversion into active regions, and then by the almost complete loss of the corresponding energy from the system which creates the activity. The model of the cellular belt also began by taking this property into account. Observations show that the mean-annual Wolf numbers are proportional to the number of groups of spots created per year (hence the importance of analyzing the variations in the Wolf numbers in constructing physical models of the cyclic activity), and in the cellular-belt model the number of cells which leave the belt is roughly equal to the number of groups of spots formed. Therefore, in the disintegrating belt model the number of cells which leave the belt per unit time is linearly related to an observed quantity (the mean-annual Wolf numbers) and the coefficient of proportionality is known to within a factor of 1-3 (which takes into account the fact that not all the cells which leave the belt reach the photosphere and are seen in the form of groups of sunspots). The model involves fully specific, observed indices. The plasma crystal whose upper layer decays at the edges into isolated emerging magnetized cells makes it possible to describe the behavior of the Wolf number with sufficient accuracy. As already reported, by choosing the function $\lambda(\theta)$ [where $\lambda(\theta)dt$ is the probability of a perturbing element's (cell's) breaking off at activity phase θ over a time interval dt] in a suitable fashion it is possible to match the observed and model characteristics, including the shape of the limiting cycle, the rigidity curve $N(\theta)$, and the variation of the average reduced (in fractions of the latitude of the activity zone) latitude of sunspot formation. Therefore, the model agrees with the observed variation of the Wolf numbers and Spörer's law. The Maunder "butterflies" also follow naturally from the model. A more subtle property of the cyclic activity, phase self-stabilization, which we have recognized over sufficiently long series of dates of solar activity maxima, also finds a natural explanation in this model.

One of the verifiable consequences of the model is the latent latitudinal periodicity of the centers of the groups of spots as they first appear. The observed period is about 1.4°, and data on the width of the activity zone allow us to estimate the number of cells in the belt, which in turn is also known (to within a factor of 1-3) from the number of spots generated during the cycle. The agreement between these two estimates is an argument in favor of the cell model.

The plasma crystal offers a natural explanation of Hale's law. The surprising ordering in the signs of the fields in the sunspots and their alternation follow from the geometry of the model and the successive disintegration of layer after layer. The rapid emersion of the magnetized toroids causes them to expand and cool adiabatically and causes a fairly strong motion of matter, all of which are possibly observed as the Evershead effect. Also, this model does not contradict the Wilson effect. Finally, slow fluctuations in the depth of the belt and in the convection parameters with longitude could lead to the appearance of "active longitudes."

With this model it has also been possible to derive a period of the cyclic activity in agreement with the observed $T \approx 1$ yr.

Therefore, at this stage the model explains almost all the principal features of cyclic solar activity. Of course, it is still far from perfect. There remain many questions of theory, interpretations of observations, and proposals for new observations in order to obtain the missing empirical characteristics. Perhaps the most important problems from the standpoint of developing the theory are to calculate the multilayer (stratified) convection and in this connection to investigate new properties of the convective zone of the sun, in particular the structure of the solar atmosphere. Among the most interesting problems in the interpretation of observations which are hardly explained as yet we must include the stability of the leading spot of a group, the orientation of the group's symmetry axis with respect to the equator, and the question of the virtual absence of spots near the equator. (We can only assume that the loss of the bulk of the cells in this region is due to the solar wind.)

It would have been desirable to discuss the proposals for new observations or the analysis of already existing data in more detail. We believe that the verification of predictions of spatial periodicity in the sources of activity is the most important point for analyses of the available information on the cyclic activity. It would first be necessary to refine the results on the latitudinal quasiperiodicity obtained from an analysis of data from four 11-year cycles. Such data are available for five more cycles, and if they were analyzed the reliability of the determination of the mean latitudinal period would be greatly increased. The search for a longitudinal periodicity would be still more cumbersome and complicated because the longitude of a center is determined less accurately than its latitude, differential rotation causes smearing of the longitudinal period, and the period itself depends on the longitude. A correlation analysis (temporal and spatial) of the first appearances of sunspot groups could aid in choosing between the hypotheses of emerging parts of long bundles of lines of force or of emerging discrete magnetized regions.

An analysis of the motions of sunspots in a group and of the motion of the groups as a whole could permit reconstruction of the picture of a magnetized torus as it floats to the top.

For a better understanding of the physics of the processes involved in the cyclic activity there is some interest in determining the statistical characteristics of the active longitudes and in comparing the indices of activity with the characteristics of the solar wind and with possible variations in the solar constant.

Finally, it is of interest to reproduce (and study) the behavior of the cyclic activity over long times on the basis of indirect data.

It can be seen from this very incomplete listing that there is still very much unanalyzed information about the cyclic solar activity which has direct relevance to the construction of a physical model. Besides analyzing the available data it would be useful to make special observations to study the fine structure of the magnetic fields of a large number of sunspots and to explain the significant differences in the sunspot fields in even and odd cycles and in the northern and southern hemispheres of the sun (besides those differences which follow Hale's laws).

Literature Cited

1. L. I. Gudzenko, Radiofizika, 5:572 (1962).
2. L. I. Gudzenko and V. E. Chertoprud, Astron. Zh., 39:758 (1962).
3. L. I. Gudzenko and V. E. Chertoprud, Astron. Zh., 41:697 (1964).
4. L. I. Gudzenko and V. E. Chertoprud, Radiofizika, 8:1213 (1965).
5. L. I. Gudzenko, Radiofizika, 4:267 (1961).
6. Solar-Terrestrial Physics and Meteorology: a Working Document. Issued by SCOSTEP Secretariat, Washington (July, 1975).
7. Yu. I. Vitinskii, Cyclicity and the Prediction of Solar Activity [in Russian], Nauka, Leningrad (1973).
8. V. E. Chertoprud and V. A. Kotov, Astron. Tsirk., No. 318 (1965).
9. V. E. Chertoprud, Candidate's Dissertation, Moscow State University (1966).
10. L. I. Gudzenko and V. E. Chertoprud, Radiofizika, 10:353, 363 (1967).
11. M. Waldmeier, Results and Problems in Solar Research [Russian translation], IL, Moscow (1950).
12. V. A. Vlasov, L. I. Gudzenko, and V. E. Chertoprud, Kratk. Soobshch Fiz., No. 12 (1974).
13. H. W. Babcock, Astrophys. J., 133:572 (1961).
14. J. H. Piddington, Solar Phys., 22:1 (1972).
15. B. M. Rubashev, Problems of Solar Activity [in Russian], Nauka, Moscow–Leningrad (1964).
16. G. E. Hale and S. B. Nicholson, Carnegie Institute of Washington Publication No. 498 (1938).
17. H. Alfvén, Tellus, 8:274 (1956).
18. C. De Jager, The Structure and Dynamics of the Solar Atmosphere [Russian translation], IL, Moscow (1962).
19. C. W. Allen, Astrophysical Quantities [Russian translation], IL, Moscow (1960).
20. C. Oesterwinter and J. C. Cohen, Celestial Mechanics, Vol. 5, No. 3 (1972), p. 136.
21. E. E. Slutskii, Dokl. Akad. Nauk SSSR, 4(1–2) (1935).
22. D. J. Schove, J. Geophys. Res., 60:127 (1955).
23. L. I. Gudzenko and V. E. Chertoprud, Astron. Zh., 43:113 (1966).
24. Yu. I. Vitinskii, The Prediction of Solar Activity [in Russian], Izd. Akad. Nauk SSSR, Moscow (1963).
25. L. I. Gudzenko, Radiofizika, 12:12 (1969).
26. L. I. Gudzenko and V. E. Chertoprud, Kratk. Soobshch. Fiz., No. 1, p. 7 (1975).
27. L. I. Gudzenko et al., Trudy FIAN, 90:198 (1976) [this issue].
28. M. Waldmeier, Astron. Mitt. Zürich, No. 133 (1935).
29. L. I. Gudzenko and V. E. Chertoprud, Kratk. Soobshch. Fiz., No. 9, p. 27 (1970).
30. L. I. Gudzenko and V. E. Chertoprud, Astron. Zh., 42; 267 (1965).
31. G. G. Mursalimova, Trudy Tashkent. Astron. Obs. Ser II, Vtoraya, 5:145 (1975).
32. Greenwich Photo-heliographic Results.
33. V. D. Granova, L. I. Gudzenko, and V. E. Chertoprud, Astron. Tsirk., No. 703 (1972).
34. L. I. Gudzenko and V. E. Chertoprud, Kratk. Soobshch. Fiz., No. 1, p. 3 (1975).
35. S. B. Pikel'ner, The Elements of Cosmic Electrodynamics [in Russian], Fizmatgiz, Moscow (1966).
36. R. J. Bray and R. E. Loughhead, Sunspots, Chapman and Hall (1964).
37. L. I. Gudzenko and V. E. Chertoprud, Kratk. Soobshch. Fiz., No. 8 (1975).
38. V. E. Chertoprud, Astron. Zh., 43:390 (1966).
39. K. Ya. Kondrat'ev and G. A. Nikol'skii, in: Sun–Atmosphere Coupling in the Theory of Climate and Weather Prediction [in Russian], Gidrometeoizdat, Leningrad (1974), p. 128.
40. S. A. Kaplan and K. P. Stanyukovich, Dokl. Akad Nauk SSSR, 95:769 (1954).
41. L. Bierman, Vierteljahrchr. Astron. Ges., 76:194 (1941).

42. C. D. De Jager, Bull. Astron. Inst. Netherl., 17:253 (1964).
43. S. I. Syrovatskii and Yu. D. Zhugzhda, Astron. Zh., 44:1180 (1967).
44. T. D. Cowling, Solar Electrodynamics, in: The Sun [Russian translation], IIL, Moscow (1957).
45. N. H. Baker and S. Temesvar, Tables of Convective Stellar Envelope Models, Goddard Space Flight Center (1966).
46. E. N. Parker, Astrophys. J., 162:665 (1970); 161:491 (1971).
47. H. Zirin, The Solar Atmosphere, Blaisdell, Waltham (1966), p. 388.

CHAPTER IX

PHASE SELF-STABILIZATION

The Phase of Motions

If $\mathbf{x} = \mathbf{x}_0(t)$ is the solution of the system of equations

$$dx/dt = \mathrm{X}(\mathrm{x}(t)), \quad \mathrm{x} \equiv (x^{(1)}, \ldots, x^{(n)}), \quad \mathrm{X} \equiv (X^{(1)}, \ldots, X^{(n)})$$

with the initial condition $\mathbf{x}_0(t_0) = \mathbf{a}$, then the condition $\mathbf{x}_0(t + \gamma) = \mathbf{a}$ evidently corresponds to the solution $\mathbf{x} = \mathbf{x}_0(t + \gamma)$. The set of motions of an autonomous dynamic system permits arbitrary shifts in time. Solutions of the form $x_0(t + \gamma)$ for arbitrary fixed values of γ are combined by the curve $[\mathbf{x}_0(\theta)]$ in $\{\mathbf{x}\}$ space by saying that the motions with a given trajectory $[\mathbf{x}_0(\theta)]$ differ only by a constant phase difference.† This concept is generalized to motion along different (nearby) trajectories. Denoting the image point of the motion $\mathbf{x}(t)$ on the curve $[\mathbf{x}_0(\theta)]$ by $\mathbf{x}_0(t + \gamma(t))$, $\gamma(t)$ is called the phase difference between the motions $\mathbf{x}(t)$ and $\mathbf{x}_0(t)$ [1, 2]. Phase considerations make it possible to explain properties of the motion associated with shifts in time, thereby easing the analysis of complicated systems.

According to ideas generally accepted until recently, in an autonomous oscillatory system phase shifts in the oscillations do not cause a restoring force (as opposed to shifts in the amplitude). The resulting nonselective (in frequency) fluctuating effect on the oscillator must always lead to an unbounded (practically proportional to t) growth in the dispersion of the phase with time. This idea contradicts the results [3] of an analysis of long-term observations of the cyclic activity of the sun, which demonstrated that the growth in the phase dispersion is limited while there is no reason to regard these oscillations as nonautonomous. The ensuing discussion [4-6] partially explained the significance of this unexpected statistical effect by proceeding from the previously formulated analogy [7] (also, later in more detail in [8]) between the oscillations in the solar activity and those of a neon lamp relaxation oscillator. This chapter is devoted to a discussion of the dynamic reaction which restores the displacement in phase of the oscillations in this simplest of autonomous relaxation oscillators.

In the following we shall need one further definition. Let a system be determined by the parameter vector $\mathbf{b}(b_1, \ldots, b_N)$. We shall refer to a property as coarse (stable) at $b = \mathbf{b}^0$ if it is conserved for sufficiently small deviations of the parameters from $\mathbf{b}^0$.

† One often goes from the phase $\theta \equiv t + \gamma$ to its dimensionless value $\psi = \theta/\theta_0$, where θ_0 is some characteristic time of the system. With periodicity, when $x_0(t + T) \equiv x_0(t)$, the phase is usually "reduced to unity over one period" and $\psi \equiv \theta/T$. Otherwise, in the case of sinusoidal $x_0(t)$ the phase is more often reduced to 2π or 360° and measured in angular units along the cycle.

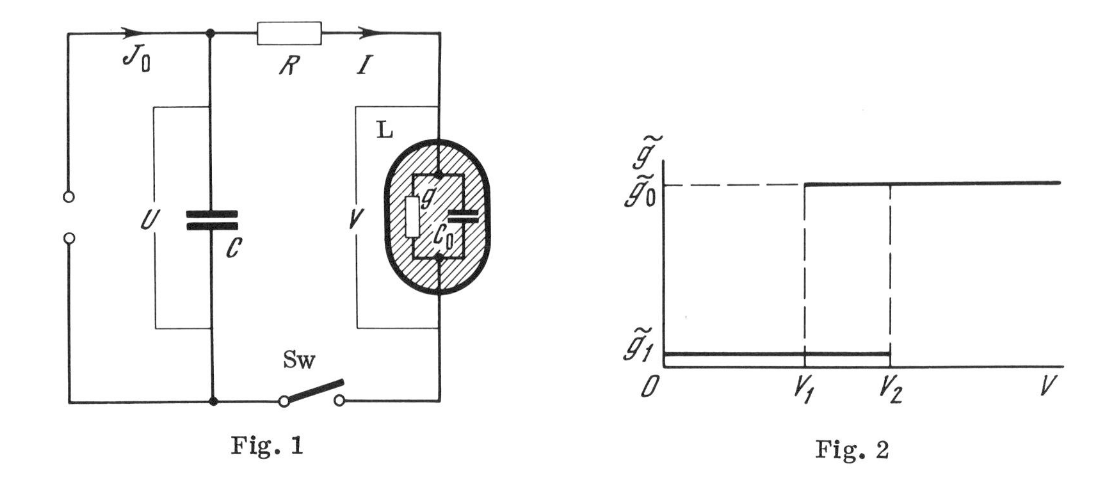

Fig. 1

Fig. 2

It will be shown below that dynamic compensation (partial or full) of large phase shifts is a characteristic of an ideal relaxation oscillator as well as of one with a slight nonidealness. Therefore, the phase self-stabilization property is coarse.

A Model of a Neon Lamp Self-oscillator

Beginning with the diagram of a neon lamp relaxation oscillator often cited in courses on the theory of oscillations, we shall discuss the dynamic model shown in Fig. 1 in comparative detail. Here J_0 is the constant current from a source shunted by a condenser, where C and U are its capacitance and the voltage across it; R is a resistance in series with the neon lamp L; I(t) is the current flowing through this resistance; and V(t) is the voltage across the lamp. We shall describe the lamp itself by a parallel capacitance C_0 and nonlinear resistance $\tilde{R} \equiv 1/g$. In the hysteresis dependence $\tilde{g}(V)$ of the lamp's conductivity on the voltage across it shown in Fig. 2 the values $V = V_1$ and $V = V_2$ are the extinction and ignition voltages for the discharge. Here $V_1 < V_2$.

In order to isolate the properties of interest to us we shall limit ourselves to analyzing the oscillations at limiting values of several parameters of the oscillator. We shall assume first of all that the conductivity of the extinguished (off) lamp is negligible compared to $1/R$ and $J_0/(V_2 - V_1)$ (i.e., $\tilde{g}_1 = 0$ if $dV/dt \geq 0$ while $V < V_2$) and that for a burning (on) lamp the value of g is very large, that is,

$$\tilde{g}_0 \equiv 1/R_0, \quad R_0 \ll R, \quad \text{if} \quad dV/dt < 0 \ \text{while}\ V > V_1.$$

With the dimensionless parameters

$$a \equiv \frac{C_0}{C + C_0}, \quad \alpha_0 \equiv \frac{R_0}{R(1-a)}, \quad v^{(1)} \equiv \frac{V_1}{aI_0R}, \quad v^{(2)} \equiv \frac{V_2}{aI_0R}$$

and variables

$$u \equiv \frac{U}{aI_0R}, \quad v \equiv \frac{V}{aI_0R}, \quad i = \frac{I}{aI_0}, \quad \tau = \frac{t}{aRC}$$

the equations of our model may be then be written in the form

$$\frac{du}{d\tau} = 1 - ai, \quad \frac{di}{d\tau} + i\left(1 + \frac{n}{\alpha_0}\right) = 1 + \frac{n}{\alpha_0}u, \quad v = u - i, \tag{1}$$

where

$$p = \begin{cases} 0, & \text{if} \quad \frac{dv}{d\tau} \geqslant 0 \quad \text{while} \quad v < v^{(2)}, \\ 1, & \text{if} \quad \frac{dv}{d\tau} < 0 \quad \text{while} \quad v > v^{(1)}. \end{cases}$$

We thus have for the voltage on the condenser C

$$u(\tau) = u(\tau_0) + \tau - \tau_0 - a \int_{\tau_0}^{\tau} i(\vartheta)\, d\vartheta. \tag{2}$$

The current through the resistance R during the time interval that the lamp is off (p = 0) is

$$i(\tau) = 1 + [i(\tau_0) - 1] \exp(\tau_0 - \tau), \tag{3}$$

and when the lamp is on (p = 1) it is

$$\begin{gathered} i(\tau) = \frac{1}{a} + \frac{1}{M}\left\{\left[m_1 i(\tau_0) - \frac{1}{\alpha_0} u(\tau_0) - 1 + \frac{m_2}{a}\right] \exp[m_1(\tau_0 - \tau)] - \right. \\ \left. - \left[m_2 \cdot i(\tau_0) - \frac{1}{\alpha_0} u(\tau_0) - 1 + \frac{m_1}{a}\right] \exp[m_2(\tau_0 - \tau)]\right\}, \end{gathered} \tag{4}$$

$$m_1 \equiv \frac{1}{2}\left(\frac{1}{\alpha_0} + M + 1\right), \quad m_2 \equiv \frac{1}{2}\left(\frac{1}{\alpha_0} - M + 1\right),$$

$$M \equiv \sqrt{\left(\frac{1}{\alpha_0} + 1\right)^2 - 4\frac{a}{\alpha_0}}.$$

Therefore, for an extinguished lamp (p = 0) we obtain variable dimensionless voltages of the form

$$u(\tau) = u(\tau_0) - (1 - a)(\tau - \tau_0) - a[i(\tau_0) - 1][1 - \exp(\tau_0 - \tau)], \tag{5}$$

$$v(\tau) = v(\tau_0) + (1 - a)\{\tau - \tau_0 + [i(\tau_0) - 1][1 - \exp(\tau_0 - \tau)]\}. \tag{6}$$

In view of the smallness of the resistance of the burning lamp we write

$$m_1 = 1/\alpha_0[1 + \alpha_0(1 - a)]. \quad m_2 = a(1 - \alpha_0),$$
$$M = 1/\alpha_0 [1 + \alpha_0 (1 - 2a)], \quad \alpha_0 \ll 1.$$

Then during the times the lamp is on (p = 1) Eqs. (1), (2), and (4) imply that the voltages vary with time as

$$\begin{gathered} u(\tau) = \frac{1}{a} + \alpha_0\left[\frac{1}{a} - au(\tau_0)\right] - \alpha_0 av(\tau_0) \exp[m_1(\tau_0 - \tau)] + \\ + \left\{u(\tau_0) - \frac{1}{a} + \alpha_0\left[2au(\tau_0) - ai(\tau_0) - \frac{1}{a}\right]\right\} \exp[m_2(\tau_0 - \tau)], \end{gathered} \tag{7}$$

$$\begin{gathered} v(\tau) = \alpha_0\left[\frac{1}{a} - au(\tau_0)\right] + [v(\tau_0) + \alpha_0(a - 1)u(\tau_0)] \exp[m(\tau_0 - \tau)] + \\ + \alpha_0\left[u(\tau_0) - \frac{1}{a}\right] \exp[m_2(\tau_0 - \tau)]. \end{gathered} \tag{8}$$

The motions of our oscillator are determined by the two initial conditions $u(\tau_0)$ and $v(\tau_0)$ (a point in the (u, v) plane). Here the constant $i(\tau_0)$ depends on $u(\tau_0)$ and $v(\tau_0)$. According to Eq. (1) $i(\tau_0) = u(\tau_0) - v(\tau_0)$. If the model included an inductance L in series with the lamp to account for parasitic losses in the circuits and the inertia of the discharge in the lamp, it would be necessary to analyze a third-order system of equations in (u, v, i) space instead of Eq. (1). For small values of L the projections of the trajectories of such a system on the (u, v) plane are formed by "slow" motions (steady-state burning or a completely extinguished discharge) joined by "rapid turning"arcs. In the limiting case of L = 0 the trajectory on the (u, v) plane consists of parts with p = 0 and p = 1 between which the image point instantly changes its direction of motion at $v = v^{(1)}$ and $v = v^{(2)}$.

The second limiting case of interest to us involves low resistance of the burning lamp ($R_0/R \ll 1$). We now consider this in more detail. Let τ_n be the successive times the lamp goes out, τ_n^* be the times the lamp is ignited, and $\tau_n < \tau_n^* < \tau_{n+1}$, $n = 0, \pm 1, \pm 2, \ldots$ The oscillations of a perturbed self-oscillator are often referred to as cyclic oscillations and the duration of the cycles (the time between two equivalent characteristic points of the signal) are called the instantaneous periods d_n. In our case $d_n = \Delta_n + \Delta_n^* = \tau_{n+1} - \tau_n$, where $\Delta_n = \tau_n^* - \tau_n$ and $\Delta_n^* = \tau_{n+1} - \tau_n^*$ are the durations of the off and on states of the lamp. We shall relate these times to the voltages at the moments the states of the lamp change. Using the notation $u(\tau_n) = u_n$ and $u(\tau_n^*) = u_n^*$, we write

$$u_n^* = u_n + \Delta_n (1 - a) - a (u_n - 1 - v^{(1)}) \cdot [1 - \exp(-\Delta_n)], \tag{9}$$

$$v^{(2)} = v^{(1)} + (1 - a)\{\Delta_n + (u_n - 1 - v^{(1)})[1 - \exp(-\Delta_n)]\}, \tag{10}$$

$$v^{(1)} = v^{(2)} \exp\left(-\frac{\Delta_n^*}{\alpha_0}\right) + O(\alpha_0^2), \tag{11}$$

$$u_{n+1} = u_n^* \exp(-a\Delta_n^*) + 1/a\,[1 - \exp(-a\Delta_n^*)] + O(\alpha_0^2). \tag{12}$$

It follows from Eq. (11) that

$$\Delta_n^* = \alpha_0 \ln\left(\frac{v^{(2)}}{v^{(1)}}\right) + O(\alpha_0^2), \tag{13}$$

that is, the intervals Δ_n^* during which the lamp burns are small and are practically independent of the initial conditions. The changes in the voltage $u(\tau)$ over these intervals are also negligibly small. According to Eqs. (12) and (13)

$$u_{n+1} = u_n^* + O(\alpha_0). \tag{14}$$

Equation (10) relates the cutoff intervals Δ_n to the voltages u_n at their initial times. The quantity α_0 does not enter here. As $\alpha_0 \to 0$ the values of Δ_n (as opposed to Δ_n^*) do not vanish and the signal $v(\tau)$ from the oscillator takes on a sawtooth shape. The motion over the (u, v) plane during the on times becomes more rapid as α_0 is reduced, and for $\alpha_0 = 0$ the flash of the lamp is described by an instantaneous jump from the point $(u_n^*, v^{(2)})$ in the slow motion part (p = 0) to the point $(u_{n+1} = u_n^*, v^{(1)})$. The oscillations of our ideal oscillator ($R_0 = 0$) are discontinuous (Fig. 3).

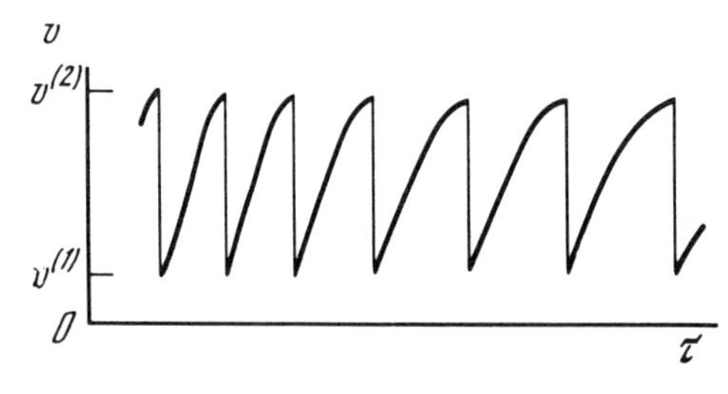

Fig. 3

Equation (10) makes it possible in principle to find the dependence $\Delta_n = D(u_n)$, with which it is possible [taking Eqs. (9) and (14) into account] to represent the transformation in the values of the voltage on the condenser C at the moment of firing by $u_{n+1} = F(u_n)$, where $F(x) \equiv x + (1-a)D(x) - a(x - 1 - v^{(1)})\{1 - \exp[-D(x)]\}$.

These established periodic, but not cyclic, oscillations correspond to a fixed point of this transformation $u_n = \bar{u}$: $\bar{u} = F(\bar{u})$. Denoting the steady-state interval between flashes by $\bar{\Delta} = D(\bar{u})$, we find

$$\bar{\Delta} = \frac{ab}{1-a}, \quad \bar{u} = \frac{b}{1-s} + 1 + v^{(1)}, \quad b \equiv v^{(2)} - v^{(1)}, \quad s \equiv \exp(-\bar{\Delta}). \tag{15}$$

When $u(\tau_0) = \bar{u}$, $v(\tau_0) = v^{(1)}$, and $i(\tau_0) = u(\tau_0) - v(\tau_0)$, (5) and (6) acquire the significance of parametric equations for the steady-state motion of the oscillator. Inverting Eq. (6) under these initial conditions yields

$$\tau = \bar{\tau}(v) + \tau_0 \tag{16}$$

for the change in phase of the oscillations along the limiting cycle.

Determining the Phase of an Ideal Oscillator

The voltage $v(\tau)$ on the lamp increases throughout the interval Δ_n and this means that the inverse function $\tau = \tau(v)$ is always unique between flashes. We shall determine the phase of the transient oscillations by projecting the straight lines v = const of the image point of the oscillator as it moves over the (u, v) plane onto the limiting curve which corresponds to $u(\tau_0) = \bar{u}$ and $v(\tau_0) = v^{(1)}$. Then Eq. (16) which gives the phase as a function of the voltage v on the lamp is conserved over each interval Δ_n. When during the oscillations there are N flashes after a given value the phase increases by $N\bar{\Delta}$.

The deviations of the initial voltages u_n and intervals Δ_n from their steady-state values, given by $\omega_n \equiv u_n - \bar{u}$, $\delta_n \equiv \Delta_n - \bar{\Delta}$, have different signs. In fact, it follows from Eq. (10) that

$$\begin{aligned} &\delta_n + Q_n\omega_n = -P_n, \qquad Q_n \equiv 1 - \exp(-\Delta_n), \\ &P_n \equiv \bar{\Delta}(a^{-1} - 1)\frac{s}{1-s}[1 - \exp(-\delta_n)]. \end{aligned} \tag{17}$$

Since $\Delta_n > 0$ we always have $Q_n > 0$. Let $\omega_n > 0$; then the assumption that $\delta_n \geq 0$ contradicts Eq. (17) since $P_n \geq 0$. If $\delta_n > 0$, then since $P_n > 0$ it follows from Eq. (17) that $\omega_n < 0$. The significance of this is obvious: An increase in the voltage u_n causes a reduction in the interval Δ_n between flashes.

According to Eqs. (9), (10), (14), and (15) the changes in the successive values of u_n for an ideal oscillator ($\alpha_0 = 0$) are equal to the deviations δ_n of the corresponding interval between breakdowns from $\bar{\Delta}$; i.e.,

$$\omega_{n+1} - \omega_n = \delta_n. \tag{18}$$

According to Eqs. (10) and (15) we obtain

$$\omega_n = \frac{b - \delta_n}{1 - s\exp(-\delta_n)} - \frac{b}{1-s}. \tag{19}$$

Substituting these values for ω_n and ω_{n+1} into Eq. (18) we find

$$\frac{b - \delta_{n+1}}{1 - s\exp(-\delta_{n+1})} - \frac{b - \delta_n}{1 - s\exp(-\delta_n)} = \delta_n,$$

which determines the relaxation intervals between flashes. With this formula it is possible to follow the changes in phase during the steady-state oscillations. However, to avoid cumbersome mathematical calculations we shall make this analysis more indirectly.

We shall use the notation $\xi_n \equiv -\delta_n/\omega_n$. From Eq. (19) it follows that

$$\xi_n = \frac{\delta_n[1 - s\exp(-\delta_n)]}{\delta_n + \frac{sb}{1-s}[1 - \exp(-\delta_n)]}. \tag{20}$$

From this when $|\delta_n| \ll 1$ and $\exp(-\delta_n) \simeq 1 - \delta_n$, we obtain

$$\xi_n = \xi + o(\delta_n), \quad \xi \equiv \frac{(1-s)^2}{1 - s(1-b)}. \tag{21}$$

Thus, for small $|\delta_n|$ we have $\xi_n \simeq \xi < 1 - s < 1$. Can ξ_n exceed unity as $|\delta_n|$ increases? In view of the continuity of the function $\xi_n = \Xi(\delta_n)$ given by Eq. (20), in this case there would exist a value $\delta^{(0)}$ at which $\Xi(\delta^{(0)}) = 1$. From Eq. (20) it would then follow that

$$-\delta^{(0)}\exp(-\delta^{(0)}) = \frac{b}{1-s}[1 - \exp(-\delta^{(0)})].$$

Since for real $\delta^{(0)} \neq 0$ the left- and right-hand sides of this expression have different signs, this equation means that always $0 < \xi_n < 1$. Thus, according to Eq. (18) we conclude that for an arbitrary initial deviation ω_1 the sequence $\{\omega_n\}$ (n = 1, 2, ...) corresponding to steady-state free relaxation consists of monotonically decreasing (in absolute value) quantities of a single sign. If we assume that $\lim_{n\to\infty} \omega_n \neq 0$, then $\lim_{n\to\infty} \omega_{n+1}/\omega_n = 1$, $\lim_{n\to\infty} \delta_n = 0$, and $\lim_{n\to\infty} \xi_n = 0$. However, this contradicts Eq. (21). Therefore, $\lim_{n\to\infty} \omega_n = 0$ and $\lim_{n\to\infty} \delta_n = 0$. Thus, the trajectory of the steady-state motion which passes through the point $(\bar{u}, v^{(0)})$ is the only limiting cycle. This is illustrated by Fig. 4.

We shall now show that not only the frequency but also the phase is established in an ideal oscillator. For clarity we shall consider two identical ideal oscillators (1) which perform synchronous steady-state oscillations with equal phases until $\tau = \tau_0$. At time $\tau = \tau_0$ the lamp circuit (L) of the first oscillator is instantaneously broken by a switch Sw while the second continues its stable oscillations. After a time Θ the switch Sw is instantaneously closed so a transient regime is established in the first oscillator with an initial phase lag relative to the first of $\varphi(\tau_0 + \Theta) = -\Theta$. From Eq. (18) it follows that $\sum_{n=1}^{N} \delta_n = \omega_{N+1} - \omega_1$. Since $\lim_{N\to\infty} \omega_N = 0$, we find that $\sum_{n=1}^{\infty} \delta_n = -\omega_1$. When the switch Sw is open the voltage across the condenser C increases linearly. According to Eq. (2), for i = 0 we have $u(\tau_0 + \Theta) = \bar{u} + \Theta$. We thus obtain $\omega_1 = \omega(\tau_0 + \Theta) = \Theta$ and, finally,

$$\sum_{n=1}^{\infty} \delta_n = -\Theta. \tag{22}$$

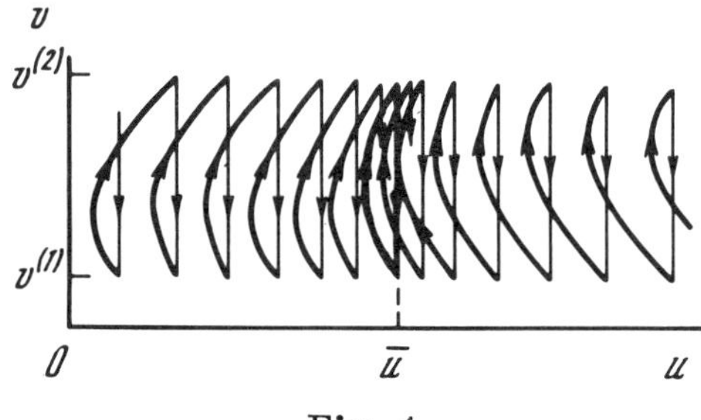

Fig. 4

Therefore, the intervals between flashes are shortened just enough that the phase difference between the oscillations of the first and second oscillators approach zero as time passes. An ideal relaxation self-oscillator completely eliminates the phase shift imposed on it. This property, which defines the "phase self-stabilization" effect, is not trivial. If we turn off a circuit in an ordinary "Thomson" lamp oscillator with an oscillator circuit or stop a clock pendulum for a long time Θ then the frequency of the succeeding oscillations does not speed up in proportion to Θ and the phase lost over this time is not recovered. In ordinary self-oscillators there is no phase feedback.

The physical significance of phase feedback, which produces phase self-stabilization in an ideal relaxation oscillator, is fairly clear. In each flash of the lamp there passes the same amount of charge q_n. According to Eqs. (3) and (10) we have

$$q_n = \int_{\tau_n}^{\tau_{n+1}} i(\vartheta)\, d\vartheta \simeq \Delta_n - [1 - i(\tau_n)] \int_0^{\Delta_n} \exp(\vartheta)\, d\vartheta = \Delta_n + (u_n - 1 - v^{(1)})[1 - \exp(-\Delta_n)] = q, \qquad q \equiv \frac{b}{1-a}.$$

From dimensional arguments we obtain an obvious expression for this charge:

$$Q = aJ_0\, aRC\, q = C_0 (V_2 - V_1).$$

Until the circuit is broken the "excess" (above the steady-state amount) charge accumulates in the condenser C. From $\lim_{n\to\infty} \omega_n = 0$ it follows that this excess is lost over a finite effective number of cycles. Since each flash carries off the same portion of charge, the liquidation of the excess is equivalent to restoration of the oscillator to the lost (because the circuit is off for time Θ) phase.

The time dependence of the phase difference between the oscillations of our two oscillators, $\varphi_0(\tau)$, is shown in Fig. 5. Up to time $\tau = \tau_0$ the oscillators are in phase, i.e., $\varphi_0(\tau) = 0$. Over the time interval $\tau_0 < \tau < \tau_0 + \Theta$ the phase of the oscillations of the perturbed oscillator is constant $(\vartheta_{(1)} = \gamma + \tau_0)$ while the phase of the second oscillator continues to increase linearly with τ: $(\vartheta_{(2)} = \gamma + \tau)$; thus, $\varphi_0(\tau) \equiv \vartheta_{(1)} - \vartheta_{(2)} = \tau_0 - \tau$. We now go to $\tau > \tau_0 + \Theta$. The

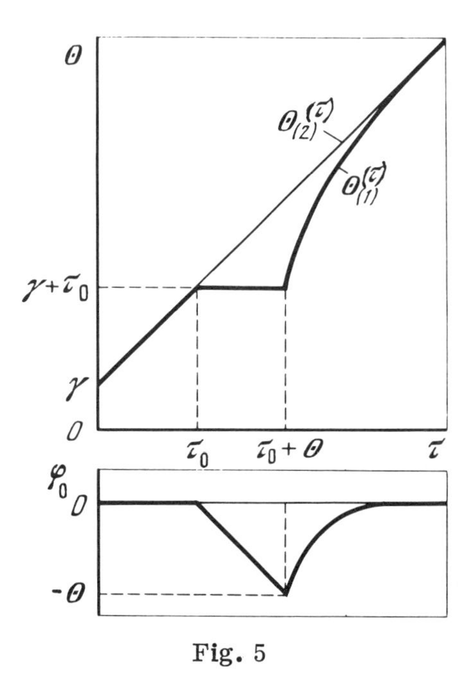

Fig. 5

phase is the "oscillator's own time" as measured by its oscillations. Let us suppose that at a time $\tau > \tau_0 + \Theta$ the lamp of the first oscillator flashes after turning on for the N-th time, that is, $\tau = \tau_0 + \Theta + \sum_{n=1}^{N} \Delta_n$. Then for the phases of the first and second oscillators we write $\vartheta_{(1)} = \gamma + \tau_0 + N\bar{\Delta}$ and $\vartheta_{(2)} = \gamma + \tau = \gamma + \tau_0 + \Theta + \sum_{n=1}^{N} \Delta_n$. Thus, for $\tau > \tau_0 + \Theta$ the phase difference is

$$\varphi_0(\tau) \equiv \vartheta_{(1)} - \vartheta_{(2)} = -\Theta - \sum_{n=1}^{N} \delta_n = -\Theta - (\omega_{N+1} - \omega_1) = -\omega_{N+1}.$$

For $|\delta_n| \gtrsim 1$ the function $|\varphi_0(\tau)|$ falls monotonically as τ increases in a complicated way described by elementary functions. As $|\delta_n|$ is reduced the fall in $|\varphi_0(\tau)|$ is slowed down somewhat. Limiting ourselves to the case in which $s|\delta_1| \ll 1 - s$, we write† $\omega_{n+1} \simeq (1 - \xi)\omega_n$ and, further (Fig. 5),

$$\varphi_0(\tau) = -\Theta \exp\left(-\frac{\tau - \tau_0}{\tau_\varphi}\right), \qquad \tau_\varphi \equiv \frac{\bar{\Delta}}{|\ln(1 - \xi)|}.$$

Therefore, for small $|\delta_n|$ the quantity τ_φ is the characteristic phase relaxation time. In particular, if $|\delta_1| \ll \bar{\Delta} \ll 1$, we find $\tau_\varphi = 1/a$. Going to dimensional quantities we obtain the natural result $t_\varphi = RC$ for the phase relaxation time.

Phase Self-stabilization in a Nonideal Oscillator

The idealness of the model for a relaxation oscillator given here is violated if we include the fact that the conductivity of the lamp is not infinite when it is on (i.e., $R_0 \neq 0$). In real neon lamp oscillators there may be other more significant reasons for nonidealness due to various deviations from this model: nonzero conductivity of the lamp when it is off, parasitic inductance in the oscillator circuits (and inertia of the gaseous discharge), finite internal resistance of the energy source (when it is necessary to speak of the emf of a battery instead of the source current J_0), noise interactions which appear primarily as chaotic changes in the firing and quenching potentials of the lamp, and so on. A nonideal relaxation oscillator no longer eliminates an imposed phase difference completely. In discussing the nature of its phase changes, it is important to first clarify whether phase self-stabilization is a coarse characteristic of the model or, in other words, whether this effect can be of real physical significance.

We shall call the absolute value of the steady-state (after cutoff of the perturbation) phase difference between the oscillations of the perturbed and unperturbed oscillators, $|\varphi(\infty)|$, the phase loss and the ratio $k = |\varphi(\infty)|/\Theta$, obtained by producing a phase loss by shutting off one of the oscillators for a time Θ, the loss coefficient. The coarseness of this effect is confirmed if when the ratio R_0/R (and therefore α_0) is small enough, it results in the two quantities $|\varphi(\infty)|$ and k being of specified smallness. We now rewrite Eqs. (11) and (12) expanding the terms linear in α_0:

$$v^{(1)} = v^{(2)} \exp\left\{-\frac{\Delta_n^*}{\alpha_0}[1 + \alpha_0(1 - a)]\right\} + \alpha_0\left\{\frac{1}{a} - u_n^*\left[a\left(1 - \frac{v^{(1)}}{v^{(2)}}\right) + 1\right]\right\} + o(\alpha_0^2), \tag{11'}$$

† The condition $s|\delta_n| \ll 1 - s$ is an indication of the relative smallness of Θ. However, for $\bar{\Delta} \ll 1$ we obtain $|\delta_1| \simeq \xi|\omega_1| \simeq a\bar{\Delta}\Theta \ll 1 - s \simeq \bar{\Delta}$, that is, Θ has only to satisfy the limitation $\Theta \ll a^{-1} = (C/C_0) + 1$.

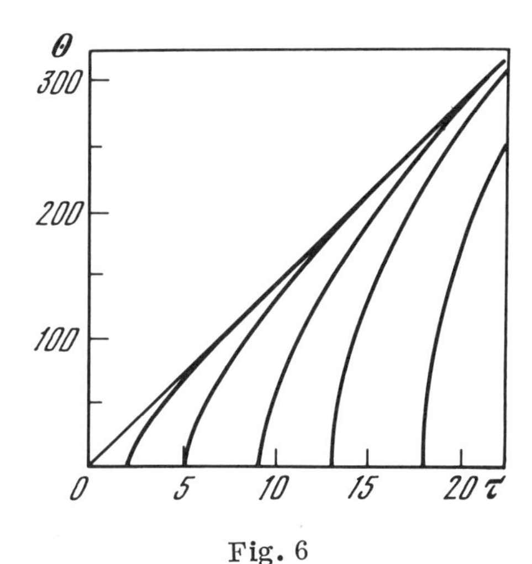

Fig. 6

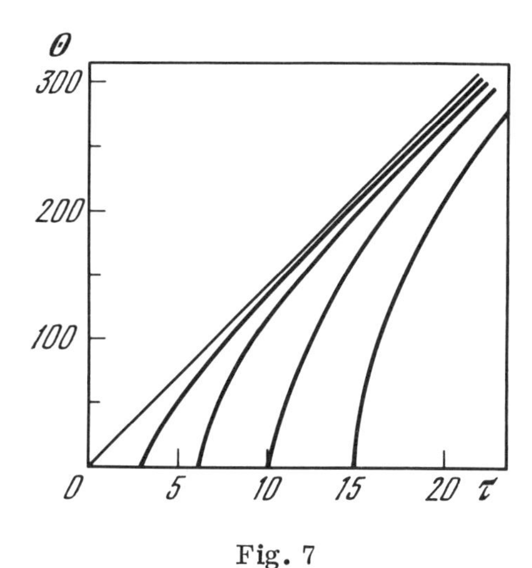

Fig. 7

$$u_{n+1} = u_n^* + \Delta_n^* a \left[-u_n^* + \frac{1}{a} + \frac{b}{\ln\left(\frac{v^{(2)}}{v^{(1)}}\right)} \right] + o(\alpha_0^2). \tag{12'}$$

From Eq. (11') we find

$$\Delta_n^* = \Delta^* + O(\alpha_0^3), \qquad \Delta^* \equiv \alpha_0 [1 - \alpha_0 (1 - a)] \ln (v^{(2)}/v^{(1)}),$$

that is, the burning intervals are independent of the voltages u_n in the second approximation in α_0 ($\alpha_0 \ll 1$) as well. Furthermore, from Eqs. (9), (10), and (12') we obtain a relationship between the deviations in the voltages u_n and the intervals Δ_n for the lamp when it is off to their steady-state values:

$$\omega_{n+1}(1 + a\Delta^*) - \omega_n = \delta_n + o(\alpha_0^2). \tag{18'}$$

Since for $\alpha_0 \ll 1$ only the intervals Δ_n in the instantaneous periods $d_n \equiv \Delta_n + \Delta_n^*$ depend on the voltages u_n, the considerations used to estimate the steady-state phase of an ideal oscillator remain valid. From Eq. (18') it follows that

$$\sum_{n=1}^{\infty} \delta_n = \sum_{n=1}^{\infty} (\omega_{n+1} - \omega_n + a\Delta^* \omega_{n+1}) + O(\alpha_0^2) \simeq -\Theta + \varphi_0(\infty), \; \varphi_0(\infty) \equiv a\Delta^* \sum_{n=2}^{\infty} \omega_n.$$

Since $\Delta^* \simeq \alpha_0 \ln\left(\frac{v^{(2)}}{v^{(1)}}\right)$, we then have $\lim_{\alpha_0 \to 0} \varphi_0(\infty) = 0$, that is, phase self-stabilization is a coarse property of the nonideal relaxation oscillators (1).

In evaluating $\sum_{n=2}^{\infty} \omega_n$ it is possible to use the equations for an ideal oscillator in the first approximation in α_0. It follows from them that $\varphi_0(\infty) < 0$; that is, the increase in the frequency of the oscillations of a nonideal oscillator after it has been shut off a time Θ does not completely compensate the phase lag produced by the shutting off. The physical significance of this is easily explained. For $\alpha_0 \neq 0$ the charge q_n which passes through the lamp in each flash is no longer constant but increases with the voltage u_n. Thus, the charge surplus built up over a time Θ is expended over a smaller number of pulses than when $\alpha_0 = 0$. We shall estimate the phase loss for the simplest case discussed above in which $|\delta_1| \ll \bar{\Delta} \ll 1$. Then

$$\varphi_0(\infty) = a\Delta^* \Theta \frac{1 - a\bar{\Delta}}{a\bar{\Delta}} \simeq \frac{\Delta^*}{\bar{\Delta}} \Theta, \qquad k \simeq \frac{\Delta^*}{\bar{\Delta}}. \tag{23}$$

The nature of phase-stabilization in a nonideal oscillator is illustrated by Figs. 6 and 7 in each of which the abscissa is the observation time and the ordinate is the instantaneous phase of the oscillations. The straight line which passes through the coordinate origin ($\tau_0 = 0$, $\gamma = 0$) corresponds to an oscillator whose steady-state oscillations are not cut off, while the curves which leave the abscissa from various points Θ_n correspond to oscillators which are turned off over the time interval $(0, \Theta_n)$. In Fig. 6 these curves correspond to an almost ideal generator with $\alpha_0 = 0.001$. In Fig. 7 the curves have been plotted for an oscillator with a higher nonidealness parameter $\alpha_0 = 0.01$.

From the figure it is clear that the phase loss increases with the duration Θ_n.

We shall not analyze the properties of relaxation oscillators which do not follow the chosen model (1). We limit ourselves only to recalling the simplest generalization of this scheme which takes into account the nonzero conductivity of the lamp when it is off. Then the small nonidealness will be characterized by yet another small parameter besides α_0, the parameter $\alpha_1 \equiv a(R/R_1)$, where R_1 is the resistance of the lamp when it is off. Using the above method we find in the first approximation in α_0 and α_1 when $|\delta_1| \ll \bar{\Delta} \ll 1$ that the loss coefficient is $k = |\alpha_1 v^{(1)} - \Delta^*/\bar{\Delta}|$. From this it is clear that the sign of the steady-state phase imbalance $\varphi_1(\infty)$ due to capacitive leakage in the lamp L is opposite in sign to the term $\varphi_0(\infty)$ due to the finite resistance R_0 of the burning lamp. In the case of nonidealness due to the finite resistance R_1 of the lamp when it is off, there is "overcompensation" of the phase shift induced by turning off the oscillator. This is because the total charge flowing through the turned-off lamp is proportional to the time. After a lamp circuit which has been turned off for a time Θ is turned on the intervals between its flashes are shortened and a smaller charge flows through the lamp during the same number of flashes. It is clear that (from the standpoint of dynamic phase self-stabilization) the nonidealnesses due to nonzero resistance of the lamp when it is burning and to nonzero conductivity of the lamp when it is off may cancel one another.

Literature Cited

1. L. I. Gudzenko, Dokl. Akad. Nauk SSSR, 125:62 (1959).
2. L. I. Gudzenko, Izv. Vyssh. Uchebn. Zaved., Radiofizika, No. 4, p. 5 (1962).
3. L. I. Gudzenko and V. E. Chertoprud, Astron. Zh., 43:113 (1966).
4. L. I. Gudzenko, A Mechanism for the Cyclic Activity of the Sun, FIAN Preprint No. 24 (1967).
5. L. I. Gudzenko, Izv. Vyssh. Uchebn. Zaved., Radiofizika, No. 12 (1969).
6. L. I. Guzenko and V. E. Chertoprud, Kratk. Soobshch. Fiz., No. 9 (1970).
7. L. I. Gudzenko and V. E. Chertoprud, Astron. Zh., 41:697 (1964).
8. L. I. Gudzenko and V. E. Chertoprud, Astron. Zh., 42:267 (1965).